Terms of
Service

Social Media and the Price of Constant Connection

Terms of Service

Jacob Silverman

NEW YORK • LONDON • TORONTO • SYDNEY • NEW DELHI • AUCKLAND

HARPER PERENNIAL

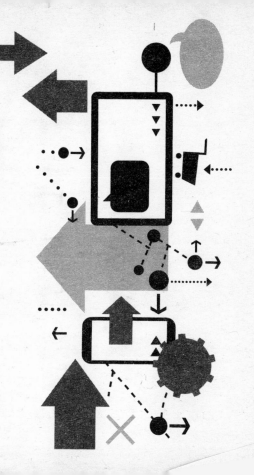

HARPER ● PERENNIAL

FIRST HARPER PERENNIAL EDITION PUBLISHED 2016.

Designed by William Ruoto

The Library of Congress has catalogued the hardcover edition as follows:

Silverman, Jacob.
    Terms of service : social media and the price of constant connection / Jacob Silverman.—First edition.
        pages cm
    ISBN 978-0-06-228246-0 (hardback)
    1. Information technology—Social aspects. 2. Information technology—Moral and ethical aspects. 3. Right of privacy. 4. Electronic surveillance—Social aspects. I. Title.
    HM851.S554   2015
        303.48'33—dc23   2014035599

ISBN 978-0-06-228248-4 (pbk.)

16 17 18 19 20   OV/RRD   10 9 8 7 6 5 4 3 2 1

Contents

Introduction

Instant messages between Mark Zuckerberg and a friend after Facebook launched:

Zuck: Yeah so if you ever need info about anyone at Harvard

Zuck: Just ask.

Zuck: I have over 4,000 emails, pictures, addresses, SNS

[Redacted friend's name]: What? How'd you manage that one?

Zuck: People just submitted it.

Zuck: I don't know why.

Zuck: They "trust me"

Zuck: Dumb fucks.

A quarter-century after the advent of the World Wide Web, communication has become synonymous with surveillance. The only unrecorded speech is the chatter of two friends spending a moment together, and one day soon that will change. Sensors and cameras proliferate through our homes and cities like spores, appearing in eyeglasses, phones, cameras, streetlights, cars, game systems, shoes, jewelry, and wherever else a signal may be found. Eventually, if the technology industry's most fervent boosters are to be believed, our whole world, and all of our sensations and thoughts within it, will be transcribed. Not because it is right or good, but because we can, and because this information, they promise, will be useful. In this temple, anything is worth sacrificing on the altars of efficiency and productivity.

In recent years, surveillance has become an inextricable part of our culture, from tech companies' lofty pronouncements that, by observing their users so closely, they can understand them and anticipate their needs; to the hundreds of millions of us engaging in mutual surveillance, crafting permanent online identities that allow us to see and be seen. This is the culture of social media, which has become a catchall term, describing more than just Facebook, Twitter, and the other big social networks in which millions of people broadcast to one another simultaneously and trade in the currency of attention. Social media now refers to a set of technical features and social practices—real names, sharing widgets, fixed identities and profiles, behavioral tracking, data mining, the spreading of media and personalities through viral channels, a process of relentless personalization—that have come to define the Web and our place in digital culture. The Internet is being thoroughly socialized, which is to say thoroughly monitored, whether by another human being or one of the ubiquitous tracking systems that supplies data to social networks, advertisers, and market researchers. Social media is the number one activity on the Web. There's nothing we spend more time doing online. One poll found that most British babies appear on social media within an hour of being born. Many human beings acquire a data trail almost as soon as they appear in the world—and sometimes before. Once created, it can never be destroyed, only modified, added to, parts of it made more visible while others are suppressed. To become part of the social web, then, is to join the networks of surveillance, tracking, and data circulation that now support a vast informational economy and increasingly shape our social and cultural lives.

Few aspects of contemporary life have gone unaffected by this shift, by the ability to publish immediately, freely, and to a massive audience. Shareability, and the drive to rack up likes and other metrics, guides the agendas of magazine editors and the budgets of marketers. Sentiment analysis—the mining of social-network data to determine the attitudes of individuals or whole populations—helps intelligence analysts learn where potential extremists are becoming radicalized. Advertisers collect social-media data and form consumer profiles with tens of thousands of pieces of information. Large corporations use social media to befriend customers, offer personalized customer service, and churn out friendly propaganda. Reporters publish breaking information on Twitter before they do in their own papers. Far-flung friends and families stay in touch, share news, fall in love, argue about politics, and ponder the trivial items of the day—all in what is essentially public view. Indeed, we are all public figures now, though few of us reap the benefits. We write and take more photographs than ever before, with these actions becoming less about capturing events than sharing them. The fullness of our lives is confirmed by our place in these networks and by the size of our audiences. I share, therefore I am—more interesting, more sociable, more desirable, more *myself.*

Social media is also part of a utopian vision that a number of major tech executives share. It is the means by which we will create better, more equitable societies, where problems will be solved by the harnessing of ever more personal data. Today's big technology firms stand ready to lead us there. As Eric Schmidt, Google's executive chairman, describes it: "It's a future where you don't forget anything . . . In this new future you're never lost . . . We will know your position down to the foot and down to the inch over time . . . Your car will drive itself, it's a bug that cars were invented before computers . . . you're never lonely . . . you're never bored . . . you're never out of ideas."

Cell phone sensors, search tools, GPS, and the like will provide some of the data that will inform this future. But social media will lead the vanguard, trailed by the personal information, thoughts, feelings, reflections, and raw data that users provide, whether we like it or not. It will be made up of the informationalized versions of ourselves, as every aspect of human affairs becomes digitized, tracked, circulated, mined for patterns. The paradigm of social media is one that Silicon Valley would like to extend to society at large: a technocracy of benevolent, but total, surveillance. In this kind of society, profits flow to platform owners, not those writing tweets and sharing YouTube videos.

This is a book about the technological moment and the one to come—about surveillance and celebrity; reputation and influence; the identity-confirming pleasures of visibility and the loneliness of overexposure; and the way we view our world through the Facebook Eye, looking for shareable moments and performing a kind of informational triage, apportioning our lives through digital networks. It's also a response to cyber-libertarianism, the reigning industry philosophy, which holds that large corporations, freed from the shackles of government regulation, claim to know what's best for us and that digital life is inherently emancipatory. (Mostly, we're only surrendering ourselves, in the form of data and personal autonomy, to oligarchic platform owners, who sell us to advertisers, data brokers, and intelligence agencies.)

Popular tech writing tends to fluctuate between the two poles of Luddite rejection and unvarnished techno-utopianism. The former is generally considered far less respectable and easily stigmatized. As Richard Byrne wrote in *The Baffler*, "In the straitened and highly ritualized discourse of tech boosterism, 'Luddite' has become a catchall dirty word for anything that stands in its way." Byrne's smart essay about the history of Luddism offered a necessary revisionist take, showing how it really began as a labor movement concerned with workers' rights and how automation would deprive skilled workers of their livelihoods. In other words, it's a surprisingly contemporary, flexible belief system, not the rigid extremism of the Unabomber.

But even if this book is fated to be categorized as the work of a digital skeptic or neo-Luddite, I would argue in return that such perspectives are needed. Techno-utopians have plenty of allies—in business, in government, in media, in every celebrity with a million hard-earned followers or the fanboys who wait in line to buy a new Apple product on the day it drops. If there is actually some recurrent dialectic between technological skepticism and utopianism that we are locked into as a culture, then all the better that the people belonging to the former should be able to launch volleys such as this one. Because digital skepticism is not about a reflexive criticism of the latest consumer gadgets and Internet technologies. It's about political economy, labor rights, how digital technologies change our culture and us as social beings, how they create new economic and social divides even as they claim to demolish others. This much-needed skepticism is against the parlous influence of those with money. It offers a doubting perspective on the grandiose claims of Silicon Valley titans and a sympathetic treatment of those who find the new digital culture stultifying or overwhelming. That's where the allegiances of this book lie—with those who don't have power, who don't have anything to sell.

Talking about these issues, it's easy to fall into what I call the "tech-only" trap, an excessive attention toward the transformative power of "technology"—a broad, vague term that could really refer to anything, but in these contexts usually describes social media, smartphones, the Internet, and other recent digital inventions. According to this line of thinking, the Internet is a cabinet of wonders, a place of magical baubles and trinkets, where life can only improve and where dissatisfaction is met with a dismissive wave of the hand. Those not having fun just don't *get* it. The other essential features of life—politics, power, culture, race, gender—are largely elsewhere, beyond the province of concern. When politics do enter the picture, it is only to proclaim the liberating potential of digital technology. Social media and smartphones will allow minorities and the marginalized to finally speak for themselves. They'll bring greater possibilities for democracy and personal freedom, as if governments from the United States to Turkey, Russia to Egypt, China to Azerbaijan aren't investing heavily in the latest surveillance and monitoring technologies and, in some instances, hiring fleets of paid Internet propagandists and censors. In reality, the old powers still have standing in this world; they have the money and the guns. Their influence remains formidable and won't be pushed aside by a popular hashtag.

As the scholar Trebor Scholz notes, "The essence of technology is not solely technological." Technology cannot be looked at outside of its relationship to politics, sociology, economics, or culture. Nor is technology something neutral, just a tool that can be put toward good or bad uses, as so many techno-utopians are fond of claiming. Digital technologies have certain capacities built into them, though some of them, as the U.S. military might say, are dual use. The GPS chip in a smartphone can help you find a local restaurant; it can also be used to track all of your movements. Other digital technologies are more obviously beneficent or pernicious. E-mail is mostly a useful, private communication tool—or it was private, until Gmail and the National Security Agency (NSA) got their hands on it. Facial recognition offers few obvious benefits and is, by design, inclined to serve the needs of advertisers, intelligence agencies, security contractors, and other potentially untrustworthy actors.

It would also serve us well to put some human agency back into the narrative. Too often defenders of the technological status quo—which is the same as the consumerist status quo—deflect criticism with a couple of banal, if tactically useful, responses: "You'll get used to it" or "This is just the way things are headed." This isn't really an argument, but it's effective. So too is the frequent claim that the introduction of new consumer technologies has created discord and confusion in the past, but we got over it then. We dealt with it and moved on; society adapted (for better or worse, one would rather not speculate). It's true that radio and the telephone inspired utopian and apocalyptic pronouncements in equal measure and that neither came to pass. But this is also an argument for passivity, for assuming that technologies have some inherent path and will find some natural accommodation with society. It is also, if one takes a more jaundiced view, an argument for deferring to powerful corporations to determine the proper role of their inventions. And it assumes that every application of new technology is coincident with progress.

Maybe some of this sounds familiar. But it is this very back and forth that is necessary, even if it recalls the debates of the past. We must shore up the levees against those forces—consumerism, irrational exuberance, the corruptions of power—that erode them. There is nothing predetermined about Facebook's role in public life or in compromising our privacy. It doesn't *have* to happen. It can be pushed back against by informed criticism, government regulation, and our own practices as consumers. We can help determine the course of these technologies and how they affect us as human beings. There's a reason that Facebook's facial recognition system isn't enabled for European customers but is for Americans. The EU and European nations, by and large, have stronger consumer privacy protections, owing to the Continent's history of authoritarian governments, invasive surveillance and, more recently, social-democratic governance. That doesn't mean that Facebook won't find opportunities to take advantage of European consumers in other ways, or to eventually make facial recognition a key part of its product there. But European government policies have worked, at least better than our own. There are things we can do to push back, and skepticism and criticism are a necessary part of this process. There is nothing assured about the march of Facebook; the company could collapse in a decade, like so many tech giants before it. What will we do then? Will we sit back and marvel at the next charismatic mogul pushing a magical product upon us?

The burden of proof lies on these companies to prove the ethics and usefulness of their products and methods. Their promises of "self-regulation" and "more relevant advertisements" satisfy neither category, not while a fat trade in personal data goes on in secret. Why should we rely on them to make innovative use of our data, at profit to themselves and debit—monetarily, socially, culturally—to us? Their job is to extract value from the intimate details of our lives. And you can be assured that shareholder value is more important to them than some sense of altruism or consumer rights.

Internet giants don't deserve our deference. As with government, our relationships with them should be adversarial, full of skepticism and critique. They show little loyalty to us, collecting, mining, and selling ever more of our data, and so they should receive little in return. As the writer David Golumbia said, cyber-libertarians "refuse to construe corporate power on the same order as governmental power." The task of this book is to surface that and other types of power. To show how social media has affected the culture and our lives. And to show the kind of world that is being made as the bulk of us enters a period in which we might never be disconnected, nor alone.

"There is not a lot of internal searching among engineers," the writer and computer programmer Ellen Ullman has said. "They are not encouraged to say, 'What does that mean for society?' That job is left for others."

This book attempts to answer that question, along with a few others: Where do we go from here? How does ubiquitous surveillance change our attitudes toward one another, and toward our very sense of self? How much should we worry about the lures of constant connection, or should we stop worrying and learn to love life in the data stream? To find some answers, we have to start by examining the rhetoric and beliefs of the tech companies providing us these services. The way that this class of innovators views the world matters for all of us, for tech culture has become our culture. As one critic remarked, "They are building their values into the infrastructure of your life." We should be wary, lest these values not match our own.

Terms of
Service

The Ideology
of Social

**And who would have suspected that as technology
and freedom were worshipped more and more,
it would become less and less possible to say
anything sensible about the society in which they
were applied?**

**—Richard Barbrook and Andy Cameron, "The
Californian Ideology"**

In 1995, two British media theorists, Richard Barbrook and Andy
Cameron, launched an extraordinary broadside against Silicon Valley.
In a work they titled "The Californian Ideology," Barbrook and Cameron described a "new faith" emerging "from a bizarre fusion of the
cultural bohemianism of San Francisco with the hi-tech industries of
Silicon Valley." Mixing "the freewheeling spirit of the hippies and the
entrepreneurial zeal of the yuppies," the Californian Ideology drew on
the state's history of countercultural rebellion, its role as a crucible of
the New Left, the global village prophecies of media theorist Marshall
McLuhan, and "a profound faith in the emancipatory potential of the
new information technologies." Survivors of the "Me" decade, weaned
on utopian sci-fi novels, self-help, and new-age spiritualism, adherents
of this faith forsook the street-side rebellion and civil actions of an earlier generation in favor of "a contradictory mix of technological deter-
minism and libertarian individualism." Freedom would not be found

in the streets but in an "electronic agora"—an open digital market-place where individuality would be allowed its fullest expression, away from the encumbrances of government and even of the physical world.

As Barbrook and Cameron recognized, the Californian Ideology was contradictory, but it also "[derived] its popularity from the very ambiguity of its precepts." Part of this belief system's appeal was its ability to combine a host of sometimes incompatible materials—radical individualism and digital community; neoliberal, free-market capitalism and an Internet industry pioneered by government-funded programs; countercultural rebellion and corporate conformity. For hackers-turned-systems-engineers or DIY artists-turned-graphic de-signers, this kind of belief held great appeal. It promised not only that their lives were authentic, but also that they had value and might make the world a better place. Joining Microsoft or AOL didn't mean selling out; it only meant recalibrating one's sense of how utopia might be achieved.

The Californian Ideology was based on a feeling—widely shared in the nascent years of the World Wide Web—that a revolution was un-der way, not only of communications and commerce, but also of social relations and culture. In fact, while few recognized it, there was little new about this. As Tim Wu and other media historians have chroni-cled, from the telegraph to the telephone to television, "the advent of every new technology of communication always brings with it a hope for ameliorating all the ills of society." In 1912, the radio pioneer Gug-lielmo Marconi announced, "The coming of the wireless era will make war impossible, because it will make war ridiculous." The failures of that and similar prophecies didn't stop techno-utopians from refresh-ing their revolutionary predictions for each new technology. The In-ternet, then, would be no exception. As Barbrook and Cameron wrote, Silicon Valley idealists believed that "existing social, political and legal power structures will wither away to be replaced by unfettered inter-actions between autonomous individuals and their software. These re-styled McLuhanites vigorously argue that big government should stay off the backs of resourceful entrepreneurs who are the only people cool and courageous enough to take risks." Another blast of Marconi bombast comes to mind: "The inventor is the greatest revolutionist in the world." Leave us alone and we'll change the world for the better, the industry promised.

Though he goes unmentioned in "The Californian Ideology," Steve Jobs was the archetypal Californian ideologue.* Steeped in the counterculture, Jobs had dated Joan Baez, regarded his experience with LSD as among the most important events of his life, hacked phone networks, and tinkered with computers as part of the Homebrew Computer Club. But he grew into a rigid capitalist, always wore the same austere uniform of black turtleneck and jeans, eliminated Apple's corporate philanthropy program, and presided over a technology firm renowned for its secrecy and strict controls over what users could do with its products. At the same time, Jobs somehow managed to make customers believe that Apple's mass-produced products were works of art and that by fueling the company's rise to becoming one of the most valuable corporations in the world, they were, to borrow Apple's catchphrase, thinking differently. Produced under grueling conditions in enormous Chinese factories, Apple's expensive, minimalist products came to be seen as tokens of individuality and cool, while the company's chief executive retained an associative halo of daring and rebellious élan. Jobs's notorious temper, for example, was often cast as a prototypical quality of the maverick chief executive.

Of course, while the Internet, digital communications, and Apple products are now ubiquitous, the revolution hasn't come.† Instantaneous global communication has failed to stop war, genocide, or

---

* The essay, while deeply perceptive, curiously doesn't mention many technology-industry figures by name. But in 1995, when "The Californian Ideology" was published, Jobs had been booted from Apple ten years earlier and wouldn't return until 1997, when Apple bought his company NeXT Computer.

† At the 2013 SXSW Interactive Festival, the science fiction writer Bruce Sterling launched his own sharp attack against techno-utopianism, asking the assembled crowd of true believers: "A billion apps have been sold. Where's the betterness?"

famine; women remain second-class citizens in large parts of the world; authoritarian propaganda travels as easily online as human rights reports (in some countries, more easily); smartphones have become the preeminent surveillance tool for corporations and governments alike. While many once foresaw digital capitalism as the harbinger of an era of widespread prosperity, legacy industries such as newspapers have crumbled, and income inequality is now higher than ever—particularly in San Francisco, home to many technology-industry employees who are shuttled daily on private buses to and from massive suburban campuses, where they're showered with amenities and services and never have to interact with residents of the surrounding communities. Barbrook and Cameron were prescient on this front, writing that "members of the virtual class and other professionals can play at being cyberpunks within hyper-reality without having to meet any of their impoverished neighbors." If you work for Google or Facebook, whose luxury buses now park at municipal stops, and thus crowd out the city's public transportation, you rarely have to set foot on anything not owned by the company.

One would think that the dot-com crash of 2000 and the failure of Silicon Valley's immense ambitions would have introduced some humility, along with some critical self-examination about the role of technology in human affairs. The opposite has been the case. The Californian Ideology chugs along, finding new forms for new times. Steve Jobs remains secure in his perch as an industry icon, his example now posthumous and unimpeachable. The industry's triumphant individualism has been augmented by the introduction of Ayn Rand as Silicon Valley's de facto house philosopher; her atavistic arguments for the virtues of selfishness and unfettered enterprise have found supporters from *Wikipedia*'s Jimmy Wales to Oracle's Larry Ellison. Unabashed cyber-libertarianism, combined with an avaricious and wholly unconflicted brand of consumerism, permeates America's digital elite. Evgeny Morozov, a fierce critic of the industry, highlighted two important strains of belief in his recent books: the congenital utopianism of

Silicon Valley moguls and their attendant faith in technological solutionism. Political strife, social injustice, economy inequality, the thorny challenges of human behavior and even the randomness of life—all might fall away when presented with a sophisticated technological fix.

Cyber-libertarians believe that digital communication inherently favors oppressed parties, particularly in authoritarian countries, despite widespread evidence that countries from Bahrain to China to the United States are quite adept at surveilling, policing, and propagandizing their populations using the latest digital technologies. They also contend that as our lives flower online, our communities and, indeed, our planet will eventually be brought together in harmony. Misunderstanding will disappear, difference will become superfluous, and peace and freedom will break out. If this sounds a few notches below messianism, that's because it is. Google, remember, is a company that plans to organize all of the world's information; its ambitions could not be greater. It also employs Ray Kurzweil, the chief promoter of the Singularity, a belief that machines will soon become self-aware, or that we will somehow combine with them, or upgrade our brains to the cloud—something amazing, we just don't know what yet. In short, this is an industry in which such far-fetched thinking is de rigueur. Some of these people actually believe they will live forever. Peter Thiel, a PayPal cofounder and major early investor in Facebook (and another Rand disciple), has derided the inevitability of death as an "ideology" while plowing millions into companies that might, as he said, "cure aging." Google's own forays into life-extension research, through a biotech subsidiary called Calico, reflects its belief that it can solve death—at least for a paying few.

So how does social media fit into these dreams of emancipatory digital technology? Social media is a means to the cyber-libertarian end. That it's only the latest hyped product to come down the pipe— that the inventors of the telegraph and the telephone and the Internet itself shared similar naïve fantasies—doesn't seem to matter. Facebook, Twitter, Google+, and Pinterest have the capacity—if we are to believe the companies marketing these products to us—to remake our world for the better. As Sheryl Sandberg, the COO of Facebook, promised, "People will share more and more of their lives online, transforming relationships on every level—personal, commercial and institutional."

It's not just that the executives of social-media firms—along with their enablers in the media, the consulting circuit, and certain sectors of government—believe in this deliriously optimistic, social media–enabled future. They believe that they are the people, and the companies, to get us there. And so they treat themselves with utter self-seriousness, while handling their customers with a heavy-handed paternalism. Twitter cofounder Biz Stone has called his firm "a triumph of humanity." (Apple apparently agrees, publishing a profile of the company with the title: "Twitter. Triumph of Humanity.") Mark Zuckerberg has claimed that connectivity is a human right and that Facebook could help stop terrorism. In her book *The Boy Kings*, Katherine Losse, an early Facebook employee who eventually became a speechwriter for Zuckerberg, recalls a day when Zuckerberg asked her to write an essay about "companies over countries." He explained: "It means that the best thing to do now, if you want to change the world, is to start a company. It's the best model for getting things done and bringing your vision to the world."

In this worldview, with its tendency toward self-aggrandizement, there is little room for humility, much less irony. It is a fantasy that has the audacity to call itself empirical. Through the mountains of data these companies are collecting, they have come to believe that they know us. (Google's Eric Schmidt: "We know where you are, we know what you like.") It's only a small step, then, for them to decide that they know what's best for us. And we've seen that play out already, both in the numerous daily reminders of people we might know or products we might like and, more important, in the larger strategic direction toward which these companies are moving. Google, for one, has been trumpeting Google Now, a personal assistant–like product that uses social media, search, calendar, GPS, and other data to provide you with information and suggestions throughout your day. You may not think you need a food recommendation while walking down a street of chic bistros, but Google Now—or the restaurant that paid it to push you a coupon—might decide as much. Google Now runs constantly and is always watching. As *Popular Science*, which named the product an "innovation of the year," declared: "Instead of telling your phone what you want, the phone tells you."

The example of Google Now demonstrates the extent to which the environments that social-media companies are building and herding us into are fundamentally manipulative. Armed with data that we provide (sometimes unwillingly or unknowingly), Google and its competitors are engaging in nothing less than social engineering on a broad scale, pushing us to share and share under the pretense of improving our lives and building global community when, in fact, they want nothing more than to target us with ads that they deem "relevant" and urge us to buy products from their partners. Whether they actually believe in their grand prophecies only matters insofar as it provides cover for their assaults on user privacy, identity, self-expression, and autonomy. Under the paternalistic hands of Google and Facebook, we have been building digital lives only to give them away wholesale, all because the services were convenient and free and we told ourselves that we didn't know better.

- - - - - - - - - - - - - - - - - -

HOW SHOULD A SOCIAL PERSON BE?

A key element of the new digital ideology is that everything can be personalized and made social. Nothing needs to be done alone or anonymously. Consequently, like the data-tracking firms they partner with (whose tracking systems you can withdraw from only by finding a carefully buried opt-out page, when one even exists), new features on social media tend to be opt-out rather than opt-in. Whenever a new feature will cause a user to share more data, it is likely to be opt-out.

From the point of view of these companies, forcing users to opt out of new features, or out of membership entirely, is merely a way of smoothing the path in front of them. In the case of Facebook, the company has long taken an attitude that every Internet user would eventually want to be a member. The company's Like buttons, for example, track browsing information even for people who haven't signed up for the site. And so, early in the company's history, they established profiles for people who hadn't joined yet. Katherine Losse, the early Facebook employee, explains:

> We were so convinced that Facebook was something everyone should have that when the product team created an experimental feature called 'dark profiles' in fall 2006, nobody even flinched. This product created hidden profiles for people who were not yet Facebook users but whose photographs had been tagged on the site. It reminds me now of the way members of the Mormon church convert dead people, following the logic that if they had known about Mormonism when they were alive, they would have been believers. Facebook was our religion and we believed everyone should be a member, even if they hadn't consented yet.

The invocation of religion is apt. For the true believers of Facebook, everything—dating, listening to music, browsing photos, playing games—is better with other people watching. Who wouldn't want to be a member? Who wouldn't want to be sharing as much of their lives as possible?

The accompanying belief is that before social media, life was lacking, and so Facebook is here to help us become more "authentic." As we continue to share our lives online, Sheryl Sandberg wrote in the *Economist*, we will produce "one voice," which is to say "the convergence of our real and virtual selves." This is our "authentic identity," one that has "a name and a face"—in other words, it's an identity that makes us no longer anonymous, an identity whose online habits and expressions can all be pegged to a single account, about which a profitable data profile can be developed. Seemingly conscious of the rhetorical bait-and-switch she's performing, Sandberg cautions her readers: "And, yes, this shift to authenticity will take getting used to and will elicit cries about lost privacy. But people will increasingly recognise the benefits of such expression." Converging your real and virtual selves naturally requires Facebook to know what you are doing and thinking at all times.

Authenticity is an unreachable goal. An important part of identity, as it's long been understood, is that we act differently in different situations. We put on different roles, we code-switch. We might speak differently with our parents than we would with our children or a coworker or someone we're flirting with at a bar. At the risk of slipping into something hazily postmodern, there is no single "self." But don't tell that to Peter Thiel. The Facebook investor has said that Myspace was "about being someone fake on the Internet; everyone could be a movie star." With Myspace's demise and Facebook's subsequent ascension, Thiel told a reporter, "The real people have won out over the fake people."

Social-media companies have worked hard to equate authenticity with being "real," and having a single identity, a single log-in associated with your real name and all of your social-media accounts, with being a safe and good citizen of the social web. The corollary is that, as Mark Zuckerberg has said, people with multiple online identities can't be trusted. Never mind that digital life should be much like analog life: a place for identity play, where we can find refuge in anonymity if need be, where we don't always have to expose ourselves as *this* person with *this* account living at *these* GPS coordinates. The push for authenticity is really a campaign to have every user be immediately identifiable, not only for who they are, but where they live, what they do, what they like, their whole data self immediately visible to the social network controlling it and selling it back to them in the form of advertising, coupons, and gifts. When social networks like Facebook and Twitter offer to "authenticate" users (such as by the blue check mark that appears next to verified Twitter accounts), they are not speaking in terms of users being able to fulfill their own sense of authenticity, their own ideas about what it means to be true to one's self. Rather, they are authenticating, or verifying, users for the network's purposes.

- - - - - - - - - - - - - - - - - -

THE SOCIAL GRAPH AND FRICTIONLESS SHARING

The "social graph" is a term popularized by Facebook, and it has spread throughout the industry, coming to stand in for the complex web of relationships we create and maintain, both online and off. You are probably connected to your parents online but also to friends, family, coworkers, strangers you talk to on Twitter, people you buy from on eBay, anonymous interlocutors on message boards, your college friends on Instagram, and so on. The social graph is who you talk to, how often, and what these relationships might say about you. And in the hands of Google, Facebook, and Twitter, it's a potential gold mine of data that advertisers love. It's human life as a clutch of data points, every feeling and expression and relationship recorded, mined, algorithmized.

The social graph, however, is only as useful as the data to which it's connected. And it's pretty limited if you don't share very often or if Facebook or Google or Twitter can't learn what you do when you're not on their sites. That's why these companies have led the way in socializing—or surveilling—the entire Web. Each social widget, each Like or +1 button on a Web page, of which there are millions now, acts as a tracking beacon, feeding information back to the company that owns it. The practice has its roots in the advertising and tracking industry, in which third-party companies have long used online advertisements to plant cookies—small files containing information about the user—on user's machines. (Google might have been late to the game with +1 buttons, but as the king of Internet advertising, it already had similar tracking mechanisms in place.)

In 2010, Facebook took the concept one step further by introducing Open Graph. Open Graph allows third-party apps and sites to use information from Facebook to personalize material for individual users. Yelp can recommend Indian restaurants in Boston because you've indicated on Facebook that you like Indian food and live by the Charles River. A news site can recommend stories about your favorite band because you posted about going to their show. Actions that you perform on Open Graph–enabled sites can also be syndicated back to the Facebook News Feed, meaning that a wide range of behaviors, from listening to music to online shopping, can become socialized. In other words, online experiences that would have occurred, if not privately, then more discreetly, are now exposed for viewing and commentary by one's network of friends (and by Facebook and its advertising partners). It can also reconfigure one's sense of what the Web, or certain sites, are supposed to do. For example, you might read a newspaper because you like certain writers, you think the paper shows good news judgment or offers a wider picture of the world than you receive elsewhere, or you're curious about news you might've missed. But if that newspaper decides to lean on information provided by Open Graph, you may see only what an algorithm decides is interesting to you. That may be the usual host of diverse, insightful articles, or it might be ten stories about golf (on Facebook you've been posting photos of your new clubs and recently shared an article about Phil Mickelson) and a couple of recipes (after all, you do like vegan food and it is almost dinnertime). Human agency, judgment, discernment, and curation is given over to an algorithm, and both reader and publication lose.

With Open Graph, Facebook has the capacity to know everything that you're doing online. The writer Adrian Short summed it up well: what Facebook calls "frictionless sharing"—Mark Zuckerberg's term for sharing that is swift, simple, and automatic—is better understood as a "euphemism for *silent total surveillance*." The idea is that, as one Facebook marketing slogan has it, "the Web is more fun with friends" and that sharing should be made as easy as possible, even if that means that much sharing becomes involuntary. It fits in with Zuckerberg's dream for Facebook, which as he told it to Katherine Losse, was "to make us all cells in a single organism, communicating automatically in spite of ourselves, perhaps without the need for intention or speech." We're certainly not there yet, but these efforts have been extraordinarily successful for the company and its growing hoard of user data. In March 2013, Facebook announced that users had shared 400 billion actions on Facebook through the site's Open Graph.

The overarching goal of frictionless sharing—and the direct tie it has to Facebook's ad-driven bottom line—means that the social network will do anything to get its users to share more. Facebook even goes so far as to track when users start typing status updates and delete them. The company call this kind of behavior "self-censoring." The choice of terminology is astonishing. People who choose not to post information aren't engaging in self-editing or thoughtful consideration about what they might want to post and how they want to say it. No, they are censoring themselves, denying themselves the pleasure and privilege of sharing. The logical conclusion is that this must be overcome, and it must be done for the user's own benefit. Upon learning this news, Katherine Losse followed up on Twitter, citing her own remarks about Zuckerberg wanting to turn us all into cells into a single organism, one that Facebook naturally would control. For Facebook, she wrote, "the privacy of thought is a problem for tech to overcome." That's why Facebook's status bar doesn't ask you what's new; it asks, "What's on your mind?" That is both a prompt and an explicit statement of intent: to know, whenever possible, what we are thinking and doing.

- - - - - - - - - - - - - - - - - -

THE SPREAD OF GOOGLE+

Facebook's promotion of the social graph—and its attendant features, such as Like buttons, third-party apps, and universal log-ins that allow you to use your social-media identity across services—has some competition in Google and its Google+ social network. The two companies are also ideological fellow travelers, with Google+ representing the search company's own effort to apply a social layer over the Internet and to capture and filter all user behavior through its own social network. Google+ received some criticism upon its launch— mostly from a tech commentariat skeptical that Google could possibly catch up with the more entrenched Facebook, which at the time had hundreds of millions of users. No one really wanted a Google social network. There was hardly a clamor for yet another place to share personal information and news. But in terms of what it knew about its customers, Google already was essentially a social network. It knew what people searched for, what was in their e-mails, their schedules, where they went, what they did. It had their photos, their messages, their social graphs. It had the most popular video site, YouTube, with its commenting system and wealth of sharing features. It made sense to erect the scaffolding of a traditional, profile-based social network around that. Google+ offered a familiar shape and central input for the kind of scattered but all-encompassing data collection that Google had been doing for years.

Both aesthetically and in terms of privacy features (its "Circles," which allowed users to tailor their updates to narrow audiences, was soon copied by Facebook), Google+ seemed a worthy social network on its own. But Facebook had network effects on its side—who would join Google+ if it seemed like an also-ran where nobody wanted to hang out? Google then had a challenge: how to get the hundreds of millions of people using its other products, like Search, Gmail, Docs, and YouTube, to embrace Google+. And it was a problem they desperately needed to solve, since Facebook's huge, active membership and its use of Like buttons and other tracking tools meant that it had a glut of information that advertisers wanted. No matter its various ambitious projects (self-driving cars, mapping the world) or its widely used e-mail service, Google is, at heart, an advertising company—about 91 percent of its 2013 revenue came from its advertising arm, down from 95 percent the previous year. So how could Google both rope more users into Google+ and maintain its informational supremacy in the advertising arms race?

The answer, it turned out, was to remove the element of choice. Adopting a key method of social ideology—opt out rather than opt in—Google began signing up its users for Google+, whether they liked it or not. By late November 2011, a person signing up for a Google account also had to create a Gmail account, a Google profile, and a Google+ account (unless the user was a minor). He or she was also prompted to input more personal information, some of which, such as gender, was required. Various Google products began to be "Plus-ified," with information from Google+ being integrated, for instance, into search results. Users searching for something would now see more personalized results, including things their Google+ contacts had shared, while Google+ profiles also began appearing in search results. This might be simply annoying for someone searching for "Cancun" and finding photos from a friend's vacation at the top of the results. It may also raise privacy concerns about two formerly separate functions—search and social—becoming intertwined. Others may worry about Google showing preference for results from its own network, potentially robbing users of useful search results from the wider Web (or from rival social networks). Moreover, with this product, called "Search, plus Your World," Google was guilty of a category error, presuming that content from social media—frequently ephemeral, of the moment, and with little lasting import—may be just as useful as other content. The push for personalization mistakes the familiar for what might be useful or accurate.

These and other criticisms didn't stop Plus-ification. The effort appeared to ramp up throughout 2012 and beyond. People who wanted to leave restaurant reviews on Google-owned Zagat or review an app in the Play store now found themselves required to do it through Plus. YouTube began to be integrated with Google+, with users pushed to adopt one login name across both platforms. (When I go to YouTube.com, I'm often greeted by a message at the top of the page encouraging me to shed my old, pseudonymous user name for my Google+ user name, which is, of course, also my real name.) Anyone wanting to leave a comment on the video site had to have a Plus account—a fact which angered Jawed Karim, one of YouTube's founders, who deleted most of his videos, writing, "Why the fuck do i need a google+ account to comment on a video?"

Google also started offering advice to corporations on how to improve their Plus profiles, information from which would appear in search results (the mixing of Google Search and Plus led to a Federal Trade Commission antitrust investigation). Google+ notifications soon were appearing in Gmail. Gmail's immensely popular chat feature found itself merged with Hangouts, Google+'s chat environment. That Google+'s Hangouts was not an improvement of Gmail's chat—for one thing, the new interface made it more difficult to see when contacts were online—didn't seem to matter. Instead, integration mattered, as the Google+ layer moved inexorably forward, swallowing all in its path.

While unpopular in some quarters, these policies worked. Or at least they seemed to, as Google proved itself adept at fudging its numbers. Within a few months of its launch, Google claimed to have 90 million users for Google+, with 60 percent of them "engaged daily," even though "engaged" only meant that they had logged into one of Google's many services. By December 2012, Google said that 235 million people had "used Google+ features" across its various products. While Google's definition of a Google+ user is vague, it's clear that pushing Google users onto its social network has worked. People might share less and spend less time on Google+ than on Facebook, but if the network itself becomes inescapable, then those metrics have far less meaning. More important, advertisers, Google's true constituency, were pleased: data from Google+ reportedly increased ad click-through rates dramatically.

While Google's story may be peculiar, it's not unique. The major social networks tend toward a totalizing view, trying to keep as much of their users' Web activities within their walled gardens or under their surveillance, while simultaneously preaching openness, transparency, and interoperability. They are not sites we occasionally visit but platforms designed to be the locus of as much activity and data as possible. For example, part of Twitter's early success owed itself to its willingness to accept suggestions from users and to allow outside apps to integrate with its network. Drawing on traditions long found in chat rooms and Internet forums, Twitter users, and not company programmers, invented hashtags, retweets, and @-replies—all now considered essential features of the service. Later, Twitter changed its API—the tools that outside developers use to build Twitter-compatible apps—to limit the scale of third-party programs. Some app developers fled in response. But the move was calculated to allow Twitter more control over the apps and media appearing in users' timelines, which would help in everything from collecting more data to making sure that news organizations, instead of creating their own Twitter apps, would feel compelled to form corporate partnerships with Twitter itself. (Sometimes rival networks fight back, as when Facebook bought Instagram—which was immensely popular with Twitter users—and later disabled the ability for Instagram photos to appear in the Twitter timeline and some Twitter apps.)

Facebook has achieved a similar effect by encouraging third-party sites, from online magazines to Spotify, to use its Facebook Connect tool for user log-ins and, in the case of crowdsourced service sites such as Airbnb, as a kind of ersatz background check. Your Facebook identity becomes the proof of who you are and your reliability. This may make it more difficult for non-Facebook users to access these services, but it also means that instead of having to sign up for an account to read *Foreign Policy* or use the Tinder dating app, a Facebook member can just join through his or her Facebook account. The relationship provides the sites themselves with a powerful partner: while Facebook gets some valuable data, the third-party site gets valuable data too, as joining a service through Facebook Connect often means authorizing it to read information from your Facebook profile and even post on your behalf. (In April 2013, Google released its own Connect competitor, called the Google+ Sign-In platform. Twitter and LinkedIn already had similar products.) The adoption of Facebook's commenting system by news sites and some blogs means that these sites don't have to worry about building their own commenting system or working with a small comment provider. Instead, they connect themselves to a larger community, while Facebook gets even more information about what its users do on outside sites. And with this comment activity often appearing on Facebook's activity feed, Facebook ensures more possibilities for interaction, as the user's friends may reply on Facebook to comments that originated elsewhere.

It begins to seem like a digital land rush, one which privileges the most aggressively expansionist companies. News and media companies establish partnerships, hoping to reach a mutually profitable détente with the Internet's empire builders. Smaller social sites are bought and integrated into the larger mother ship, or shut down entirely in a process that's been labeled "acquihiring"—buying a small start-up in order to gain access to its engineering talent. The large platform owners create their own smartphones and operating system, as Google did by purchasing Motorola and Android. Or they try to run the phone itself, as in the case of Facebook's failed Home initiative, which essentially overlaid Android with a Facebook layer but found little commercial interest.

As time goes on, the space for independent expression, for un-networked sharing and browsing, begins to feel squeezed. After the slow but inexorable rollout of Google+, Google began integrating more of its systems, including adopting one privacy policy across all of its products. A privacy policy, in this case, meant that Google would pool user data produced across its many products. It would track all of your Google-related activities and, through its vast advertising network and +1 buttons, practically all of your Internet activity.

It shouldn't be long before anyone using any of Google's services finds themselves, however unwillingly, to be a Google+ member. But if you are a Google employee, if you feel that your company is con-stitutionally opposed to being "evil," and that it is your mission to organize all of the world's information and make it useful, then there is no question that these changes are positive ones. By Google's defini-tion, the world's information also includes our relationships, our likes and dislikes, our feelings, what we share with friends, the e-mails we send, the videos we upload, the sentiments implicitly expressed by our browsing habits. Why wouldn't they do everything they can to try to organize that, too?

This isn't the digital freedom we were once promised. It's rather an environment of mass surveillance in which every action and incidental association is treated as consequential, all grist to be fed through the data-mining mill. Both Google search results and Facebook's News Feed algorithm are based on showing us what they think we want. The corollary of this arrangement is that no action you perform is neutral. Everything is permanently associated with you in the platform owners' database, and in turn affects the types of stories, suggestions, media, advertisements, and so forth that you're served up. This type of connection means that you can never have a truly private moment. You can't look, say, at *The Anarchist Cookbook* without stirring the at-tention of an intelligence agency or an advertiser of survival gear. Sure, you can search for and read Goebbels's speeches, but it's the platform owner's systems that decide whether you did this out of academic interest or admiration. Either way, this information is recorded in your personal graph; in an important sense, it becomes part of who you are in the eyes of the network.

And yet if you are a believer in connectivity for its own sake, if you have the odd fusion of techno-utopianism and libertarian resentment toward government and established industries that now defines Silicon Valley, then this is undoubtedly a good state of affairs. Everyone can be part of the great authentic sharing collective, but more important, they *should* be part of it. It is, in its way, a type of freedom—the freedom of an elite class of innovators to use our personal information however they choose and to push us toward a set of standardized behaviors and values. By critically exploring these values and behaviors, we can learn a lot about the digital world we live in, and the one that Silicon Valley moguls are, whether we like it or not, creating around us.

Engineered to Like

All this, of course, will be mere electronic wallpaper, the background to the main program in which each of us will be both star and supporting player. Every one of our actions during the day, across the entire spectrum of domestic life, will be instantly recorded on videotape. In the evening, we will sit back to scan the rushes, selected by a computer trained to pick out only our best profiles, our wittiest dialogue, our most affecting expressions filmed through the kindest filters, and then stitch these together into a heightened reenactment of the day. Regardless of our place in the family pecking order, each of us within the privacy of our own rooms will be the star in a continually unfolding domestic saga, with parents, husbands, wives, and children demoted to an appropriate supporting role. Free now to experiment with the dramatic possibilities of our lives, we will naturally conduct our relationships and modify our behavior toward each other with more than half an eye toward their place in the evening's program. When we visit our friends we will be immediately co-opted into a half-familiar play whose plotlines may well elude us.

—J. G. Ballard, "The Future of the Future," *Vogue*, 1977

What kind of people do we become when we're in the social web, where we're always tracked, prompted for more information, urged to share everything we're reading and thinking and feeling? When our timelines and news feeds are filled with advertising that increasingly looks indistinguishable from genuine content? When we can never be alone or anonymous or even sure if what we're reading was placed there for us by a human being or by an algorithm? When we are surrounded by the incessant chorus of likes, favorites, and a thousand bits of banal-but-cheerfully-good news?

Speed, radical transparency, confessionalism, exhibitionism, prideful consumerism, and, above all, a relentless positivity—these are the values and practices of today's social media. They are enforced by tribalist pressures (the need to fit in, the example set by friends and the famous) as much as by the programmers and moderators who manage these networks. These qualities also represent the tenor of advertising. In a medium dominated first-to-last by advertising, every expression tends to sell *something*—a product, a personality, a lifestyle, a relationship, a personal achievement.

We often use confessionalism, for example, not to admit something personal or analyze its importance but to curry attention by being willing to make the confession itself. The shock of disclosure—so familiar in a society in which memoir has become the genre of first resort for many writers, in which the therapist's couch has given way to the talk-show host's—becomes a reason for us to be watched. It allows for us to gain some of what is most scarce in this economy: time and attention. What's more is that in the struggle to appear authentic, confessionalism or over-sharing can be a way of mitigating the anxieties over seeming insincere or as if one isn't projecting one's true self. It might lead to further disclosures and revelations (the proverbial digging yourself a deeper hole) in order to try to regain control over how others see us. A general online tendency toward disinhibition also can cause us to reveal information that we wouldn't share face-to-face, where there might be more possibilities for clarification, reading social cues, and the useful intimacy of a shared, private moment.* When we share too much, we often think of ourselves as politicians or celebrities who have made some public faux pas. We might delete the post, retreat into digital silence, issue a remark with the expected mix of nonchalance and wry self-flagellation. A new post becomes necessary—a funny video, a learned article—something to show that you have moved on and that will make your past indiscretions disappear from the feed and your audience's memory.

In the performative context of social media, the language and techniques of branding, PR, and advertising take hold as we compete for the social spotlight with one another and with the corporations who pay to appear in our feeds. Calvin Coolidge called advertising "the method by which the desire is created for better things." Advertising is rooted in coercion, in civilizing the masses as good consumers. The rise of consumerism and advertising was closely tied to the institution of the eight-hour workday, the advent of leisure time, and the development of the working class as potential purchasers of mass-produced goods. Retailers saw advertising as a method by which to teach the newly empowered masses how to be good consumers and upholders of the capitalist ideal. In his book *The Culture of Narcissism*, Christopher Lasch commented that "advertising institutionalizes envy and its attendant anxieties." Lasch was writing in the late seventies, but his book—written in the aftermath of a disastrous war, during an economic malaise when people had little faith in government and when the counterculture had given way to a culture of ironic display and self-celebration—remains remarkably current. One could say that social

---

* While over-sharing is problematic, it's too often treated with derision and condescension. A better discourse around over-sharing would look at why it happens, its varied manifestations, and what it means for digital culture. Sites such as *STFU, Parents* or *Lamebook* have made small successes out of chronicling over-sharing, but they also treat their subjects with the kind of name-and-shame heavy-handedness often accorded to people (especially women) who trespass against social-media customs.

media, with its own focus on advertising (advertising is the principal revenue stream for practically every social-media company, including Google), institutionalizes a peer envy. As John Berger wrote in *Ways of Seeing*, "The purpose of publicity is to make the spectator marginally dissatisfied with his present way of life." We publicize our lives through social media to create an aspirational ideal for others and an idealized, possibly unfulfillable, version of ourselves. No one's life is as good and eventful as seen through Facebook, no one's life as shimmering and beautiful as viewed through the filters of Instagram.

The social web is suffused with an incessant enthusiasm, constant liking, and a culture of mutual admiration in part because those are the possibilities offered to us. As Robert Gehl writes, "Dissent, dissensus, refusal are not easily afforded in Facebook. Dissenters have to work for it: they have to write out comments, start up a blog, seek out other dislikers." Positive sentiments are considered conducive to sharing and, therefore, easily monetizable. Positivity is the mood of most effective advertising. Companies want to know what we like and what we might like enough to pass on to friends. Therefore, social-media sites are designed to make sure that users can express approval whenever possible. This is accomplished both by the ability to reshare or retweet most updates with a click or two and with the Like, heart, and favorite buttons. Social video sites like Hulu take it a step further by asking some users to choose and rate the ads they see—which in turn links liking to the field of "liking studies," an academic subfield of marketing in which researchers examine consumers' responses to advertisements in order to find out which ones they, yes, like. (Indeed, social media has given corporations a range of new tools to gauge consumers' real-time emotional responses to ad campaigns.)

Liking also allows users to stay positively engaged with the social network. It doesn't leave a bad taste in the user's mind. Liking is supportive, affirmative. There's no reason to leave the site and browse elsewhere after indicating you approve of something. The system is solicitous of you, promising to take your feelings into account. But the gesture is also disposable, practically compulsive. Writing for the *Atlantic*, Alexis Madrigal links liking to being in a "machine zone" analogous to the bovine stupor that people fall prey to at casinos, where they sit in front of slot machines, mindlessly pulling levers and pushing buttons for hours. "Because designers and developers interpreted maximizing 'time on site,' 'stickiness,' 'engagement,' as giving people what they wanted," Madrigal writes, "they built a system that elicits compulsive responses from people that they later regret." On Facebook, the machine zone is what causes us to lose an hour—only it seems like far less time—clicking through hundreds of an acquaintance's photos or poring over the News Feed. In the logic of Facebook, a like means "more please," and so the system becomes centrifugal. Don't like this? Okay, okay, we'll show you something else. Just please don't leave!

By pushing users to like, and to show that sentiment whenever possible, social-media sites have become awash in praise, so much so that to declare that one "likes" something now seems meaningless. It has become both the default gesture and an empty one, all the more so because it is the only one we are allowed. You can choose to like or to refrain from liking—a functionality so limited that it belies how successful it has become—but you can't do much else. In retrospect, Like, heart, and favorite icons seem rather indiscriminate, particularly when one considers that Facebook originally considered titling its now-ubiquitous icon the Awesome button. ("Awesome," with all its sophomoric exuberance, reminds us of Facebook's origins as a college student's dorm-room project.) A five-star system was also considered, but there was "concern that it would translate to 'I give this 1 star' which is a bad review," according to Facebook director of engineering Andrew Bosworth. Anything less than full approval might harm Facebook's carefully calibrated sense of equilibrium.

In any event, "Like" appears to have been a brilliant choice—although others might have served just as well, including the Recommend button that Facebook makes available to third-party sites as a substitute for Like. And perhaps it's because the term is so bland, lukewarm, and nondeterminative. "Like" can mean anything at all, so why not click it? It's always there, and in the absence of any other choices or notifications, it's an easy way to show our friends that we've noticed what they've posted, and that we support it. Social networks appreciate this functionality because it allows an ordinary encounter between a reader and a piece of content to generate some useful data. As likes accumulate and are visible to other users, an incentive grows to participate, to click Like and be part of the bonhomie.

According to *Mashable*'s Pete Cashmore, the Like button "lowers the psychological barrier to connecting with commercial entities on the site—while previously users could 'Become a Fan' of a brand, they now simply 'Like' that brand's page, resulting in higher engagement." About the prospect of a "Dislike" button, Cashmore sees no chance of it happening, because of the simple reason that "Like buttons are about connection; Dislike buttons are about division," a positive sensibility in tune with that of advertisers. More important, a "Dislike" button would not be conducive to showing allegiance to corporations. It would allow users to show their distaste for certain brands—to protest them and shift traditional power dynamics—something that is considered antithetical to the harmonious and advertiser-friendly environment social networks strive to create. Facebook has even gone so far as to block the word "dislike" as a possible action available to programmers making new Facebook apps, although they can use similar words such as "abhor" and "hate."

Some corporations see a like as a kind of contract. In April 2014, General Mills quietly updated its Web site's terms of service agreement, making it so that anyone who had received something of value from the company would be barred from suing it. This didn't just mean if you received gifts or were an employee. The provision included purchasing General Mills products, downloading coupons, and even liking the food conglomerate on Facebook. If you were one day poisoned by a tainted General Mills cereal, you'd be unable to seek damages in court. After a stream of bad press, the company reverted to its old policy. But the incident offered a warning for how companies try to embed pernicious language in their terms of service agreements and how the Like button can be seen as a de facto legal agreement, one that leaves little room for protest or dissent.

Other social networks have instituted measures to encourage likeability. A social-media start-up called Happier encourages its users to post only positive things, based on "research that shows focusing on the positive and sharing good things with people you care about makes you happier, healthier and more productive." Tumblr's messaging system invites users to "Send Fan Mail." A company designer explained the phrasing to the *New York Times Magazine*: "We don't want to allow you to have your feelings hurt on Tumblr." The site's founder, David Karp, enthused about Tumblr's heart icon and its tendency to create links of cheery sentiment: "Everybody loves everybody through the chain." For a platform built on notions of authenticity (the *Times Magazine* article, nodding to the site's faux-DIY philosophy, was headlined: "Can Tumblr's David Karp Embrace Ads without Selling Out?") and creative freedom, this kind of preciousness clangs.

Of course, as long as they continue to offer their services gratis, networks such as Tumblr will have to depend on advertising, which may conflict with its stated plan to be a platform for "creators" (a digital-era term, often used by the advertising industry, for artists, writers, and designers). This was in evidence when, in April 2013, Tumblr said that it "couldn't be happier" with its *Storyboard* blog's team of in-house journalists, and then, in the same announcement, fired them. The reasoning behind the move was lost in a sea of corporate pabulum ("What we've accomplished with *Storyboard* has run its course," David Karp wrote). But one answer might be that journalists, even those charged with celebrating the platform funding them, tend to tell complicated, even painful stories. Plus, the kind of church-and-state separation expected between editorial and advertising departments doesn't work as well at a cash-starved social-networking company. The *Storyboard* issue was also a matter of identity. Given that it covered a mix of the serious (drug violence, cancer support groups) and the lightweight (celebrity Q&As, short essays on favorite books), it was never clear what, exactly, *Storyboard* was. Was it branded content, sponsored content, advertorial, or any of the other neologisms denoting paid-for journalism? (*Storyboard*'s articles were published in partnership with other media outlets and advertising sponsors.) The site's former editor in chief called it "marketing as journalism" but also "real journalism." This haziness was part of the problem, and though sponsored content has found a new foothold in digital journalism, it's a form that tends to work—if one thinks it can at all—at publications where there are clear distinctions between what's sponsored and what's not. Tumblr's *Storyboard* was, in the end, promoting Tumblr itself, but its main audience was Tumblr users. Why market a platform to people who already use it? For that same reason, Facebook Stories, the blue giant's own storytelling platform, has gone almost nowhere.

The other problem with Tumblr's plan to be a platform for creators is that many of those creators are pornographers. Or at least, they're sharers of pornography: about 11 percent of Tumblr's top 200,000 sites are porn-related; 22 percent of visitors to Tumblr come from porn sites. The site's focus on lush photography and easy reblogging make it naturally suited to the sharing of porn, which is a problem for potential advertisers and for Yahoo, which bought Tumblr for $1.1 billion in May 2013. If Tumblr is supposed to be a gentlehearted platform for creators where no one gets hurt, can being one of the most popular social networking sites for porn fit that image? Prior to being bought by Yahoo, the company had tried to be open to adult material, though it forbade pornographic videos (they require too much bandwidth) and asked that NSFW material be labeled as such. Following the Yahoo acquisition, Tumblr has responded by essentially walling off its vast stables of porn. First, Yahoo's CEO declared that they would employ ad retargeting tools so that advertisers don't find their messages popping up next to explicit photos. Two days after David Karp declared on the *Colbert Report* that he didn't want "to police" porn, the company made adult material unsearchable, meaning that you won't be able to find it without a link. They did this by eliminating search results for any tags considered "adult," but for their mobile app—which must pass muster with Apple's censorious App Store—this list included terms such as "suicide," "depression," "gay," "lesbian," and "bisexual." Tumblr soon reversed some of these measures, claiming that the inability to search for adult tags on the site was a mistake, with one result being that its policies (and the future of porn on the network) now seemed even more uncertain. Karp explained that "gay" and some similar tags would still be blocked in the iOS app because they were associated, at least in the eyes of Apple, with pornography. The whole episode, which surely has several more chapters yet unwritten, points to both how difficult it can be for social networks to create advertiser-friendly environments and how, when companies choose the puritanism of Madison Avenue, users inevitably lose out. Crude binaries are employed: NSFW or SFW, adult or non-adult, safe or unsafe. "Gay" becomes equated with porn, which in turn is somehow toxic. "Suicide" becomes a verboten subject, a move of particular consequence for a site whose demographic skews young. Where does all this leave, say, a professional photographer who takes nude photographs and sells prints through Tumblr? Or a lonely teenager searching her Tumblr iPhone app for people talking about depression? Companies that extol the virtues of creativity and free speech should do better. Tumblr could start by acknowledging that porn, for better and worse, is a huge part of digital life. (It's also a big money-maker for search engines that sell text ads alongside porn searches.) It should find more sophisticated and respectful ways to handle the site's porn-browsing users, who contributed mightily to the site's rapid growth.

We now know whom the once high-minded Karp does respect: advertisers. In 2010, Karp said that his company was "pretty opposed to advertising." By the summer of 2013, he was courting ad execs at a festival in France, telling them that they were "heroes" and "more talented than anyone in the Tumblr office or in Palo Alto or Sunnyvale."

Like Tumblr's competitors, Karp is in hock to the purveyors of sponsored content and advertorials. It's their money that keeps the lights on, so their sensibilities, and their desires, must be flattered.

- - - - - - - - - - - - - - - - - -
TAG YOUR FEELINGS

In April 2013, Facebook introduced the ability to tag a number of emotions, activities, and products in status updates, while also including some emoticons. The result is that I might be able to say that I'm "feeling sad watching *Dances with Wolves* with my friend @JoeBiden." The point is not, as Facebook claimed at the time, to allow for greater possibilities of expression—I could post this kind of status update already—but rather to chop up this status update into little bits of taggable data. If I posted this update as plain text, Facebook wouldn't have gotten much out of it, unless they have an algorithm that scans the text for certain key words (e.g., "watching," which might indicate an interest in film and TV, especially if it picked up on the movie title). But parsing that kind of natural language is still pretty hard for algorithms. Under this new system, they'd have a whole range of data: that I'm watching a movie, which movie I'm watching, the friend I'm watching with, how I feel, when this activity is taking place, etc. Each status update may now be loaded with actionable information; for Facebook, it's a potentially momentous, if little heralded, innovation.

But for users, it means something much different. It represents another step toward the data-ization of the digital self, of making every online action into a set of data points to be mined. Some users may take up these features unthinkingly or with little hesitation, amending their status updates with options from Facebook's drop-down menus. But for others, it might give them pause, heightening the already existent sense of surveillance on Facebook and highlighting the fact that much of what Facebook touts as opportunities for expression are instead opportunities for them to collect more user information, enriching their ever-expanding profiles of each individual on the site. In that way, Facebook acts not as an aid to users' possibilities for expression or connection but rather as an impediment—mediating, surveilling, watching, prompting the user for more precise information. Instead of simply typing a status update, we have to fight through prompts that pop up, asking us to choose which movie we're watching from a list in Facebook's database. One could see Facebook instituting an autocorrect-like feature, such as those on smartphones, in Google Search, or in word processors. *You typed danses with wolves. Did you mean "Dances with Wolves?"* Advertisers may also pay to have their products appear in these lists. Given Facebook's tendency toward introducing op-out rather than opt-in features, one might not even have the choice at all, as all status updates become automatically reformatted so as to wring maximum useful data out of them. What is clear, though, is that this is another instance of Facebook asking users to structure data for them—in other words, to do work that helps aid their targeted advertising efforts.

These emoticons also allow Facebook to monetize negative emotions, creating a kind of counterpart to the Like button, albeit filtered through natural language. We have already reached a point where nearly any Facebook status update, from a wedding announcement to a mournful post about the death of a loved one, may receive a like, if only to indicate support or acknowledgment. Now, negative feelings are part of the Facebook architecture—just choose the appropriate emotion from a list—meaning that they can not only be expressed but used for finely targeted advertising. Feeling sad? Perhaps this cute puppy video, brought to your timeline by a pharmaceutical company, will cheer you up.

The trend toward presenting feelings as structured data presents new possibilities for advertising. In the eyes of Facebook, clicking Like essentially serves as a commercial endorsement, an indication to the company that you don't mind being associated with a product or brand mentioned in a post. A man named Nick Bergus found this out the hard way. A couple of years ago, Bergus was on Stellar, a small social-media aggregator, where he favorited a tweet about a fifty-five-gallon tub of personal lubricant for sale on Amazon. He browsed through some of the reviews—there's a cottage industry of people who post entertaining paeans to some of the more exotic or ridiculous products on Amazon—and then posted it on Facebook, along with a cheeky comment. It wasn't long before he was hearing from Facebook friends who saw Bergus's name and profile photo as a "sponsored story"—an ad—next to the huge drum of lubricant and a link to Amazon. The ad even included the text from his original status update: "A 55-gallon drum of lube on Amazon. For Valentine's Day. And every day. For the rest of your life." Both tickled and bothered, Bergus wrote, "In the context of a sponsored story, some of the context in which it was a joke is lost, and I've started to wonder how many people now see me as the pitchman for a 55-gallon drum of lube."

There are ways to opt out of these ads, but many users don't know that, nor do they know that their updates and personal information may be used in ads pushed to their friends. And of course, Facebook and its advertising partners have created an environment that encourages these kinds of incidents. A popular type of contest prods users to like—or tweet, retweet, pin, and so on—to win prizes. Many do; who would turn down an easy chance to win something? And for brands, it's an easy way to boost their likes, follower count, impressions—all the various metrics that they value and can show to journalists, supervisors, and corporate partners alike to say, *See, our marketing campaigns are working. People like us.* As for users, they are unlikely to give the contest a second thought, until, perhaps, they learn that their likeness has been used to market that product, or another one from that company, to their friends and followers.

By way of justification, a marketing executive for 1-800-Flowers.com told the *New York Times*, "The person has given their consent because they're engaging with your brand page, and you're boosting that engagement." This kind of rhetoric is found throughout the social-media marketing sphere. It's difficult at times to tell whether it's genuine.

How does an ad boost my engagement, and what does that even mean? By engaging with a brand, why must I consent, by default, to become part of their promotional apparatus? But if it's a genuine opinion, it seems to overlook a pretty obvious fact: that most users look at ads as a nuisance, if not a necessary evil that they must put up with in exchange for free services. With the exception of coupon services such as Groupon and Google Offers, Internet users rarely invite more advertising into their lives, nor do they see these ads as stories or something with which to engage. They ignore advertising whenever possible.

Facebook was on the receiving end of a class-action lawsuit for its sponsored ads. Eventually, the company agreed to a $20 million settlement. It amended the name of Sponsored Stories to Social Ads, but little else changed. The company simply, as the settlement dictated, added clarifying language to its data use policy. Facebook continued to have complicated controls, hidden under several layers of menus, requiring users to opt out of these ads. These mostly cosmetic changes didn't do much to stem criticism of Facebook's ad practices. And the next month, Facebook once again was excoriated for its ad policies, this time for an advertisement touting a dating site—an ad that featured a photo of a since-deceased seventeen-year-old girl. The girl had been gang-raped at age fifteen, with photos of the crime shared online; two years later, she committed suicide. The ad appeared about five months after her death, and following an outcry, Facebook removed the ad, placing blame upon the advertiser.

Facebook is not alone in engaging in these kinds of practices, but it is certainly the industry leader in dismantling the boundaries between user expression, reviews, and advertising. Google has introduced its own form of sponsored stories called Shared Endorsements. So if you approvingly post about a restaurant you visited or +1 a bar on Google+, that might later show up in an advertisement. To its credit, Google has instituted one simple control to opt out of this process, and when I visited the page, the necessary checkbox was unmarked, meaning that my information wouldn't be included in any endorsements.

In perhaps the ultimate form of social advertising, attendees of New York Comic Con 2013—a festival celebrating comics, movies, and all things geeky—found that their Twitter accounts were sending out tweets praising the event, complete with #NYCC hashtag and links to Comic Con's Facebook page. The tweets were written by someone at Comic Con and were posted without the knowledge of the users, many of whom were influential members of the press and Comic Con community. The fault was Comic Con's, not Twitter's. Attendees hadn't read the fine print—who does?—and didn't realize that by accepting Comic Con's offer to connect their festival badges to their Twitter accounts, they were allowing the event full access to their Twitter profiles, including the ability to post. Many apps and companies offering giveaways ask for access to customers' Twitter accounts, and sometimes a customer is required to tweet or post (often using a required hashtag) in order to enter a contest, though rarely are these posts automated. Other companies have gotten into trouble for being too eager to get users to tweet for them. BookVibe, a book-discovery service, was publicly chastised by Nick Bilton, a *New York Times* tech reporter and author of a history of Twitter, after he signed up for the service, only to discover that they later tweeted from his account. Bilton hadn't noticed a small bit of text, in gray font on a gray background, asking him to share BookVibe on Twitter. Naturally, the necessary box was checked by default, meaning that most users would overlook it and neglect to opt out.

Through these and other measures, corporations ask us to join in the marketing process, to become advertisers alongside them and on their behalf, just as Facebook and Google do with their various sponsored stories strategies. By clicking like or favorite, or retweeting a promoted tweet, a user becomes a vector in the spreading of advertising. You are selling both a product and your networked persona; in many ways, you *are* the product, particularly when it's your data being sold to advertisers and other unknown partners.

This is what's so problematic about ad-supported social networks.

Like viral media, the line between advertising and honest expression is continually blurred, sometimes to the point where they become one and the same. This is endemic to like/+1 culture. When everyone is constantly asked to review, rate, favorite, like, and retweet, praise becomes a fungible commodity. The language and mind-set of advertising take over. We don't just endorse one another on LinkedIn; we endorse products for use in future advertisements. (And we make clear that our retweets aren't endorsements.) This is the sort of thing that celebrities get paid for, but on social media, we do it without charge. Individual expression, after all, is worthless to Google and Facebook if it can't be matched with an ad, so why not make it *become* the ad? These companies will tinker with policies, especially after every public outrage and class-action lawsuit, but the end point remains the same: to retain rights over your data and expressions, and to make the transition from a status update to a related, paid advertisement as smooth as possible.

CONVERTING EMOTIONS INTO PROFITABLE DATA

Like buttons and taggable emotions are just two features of what has become a like economy, which depends on the growth of sentiment analysis, the examination of huge data sets to find out how people are reacting to news, products, or the events of their own lives. Retailers and advertisers want to know what individual consumers are thinking and buying, but they, along with investors, banks, consultants, and others, also want to be able to take the pulse of public opinion. To do this, they try to tap into the welter of data we produce on social media and blogs and also in traditional news media, review sites, message boards, and interviews with corporate executives. In its most rarified form, this stuff is called "market intelligence" or "social listening," and companies are willing to pay quite a bit for it. In a sense, what they're doing is just a far more sophisticated version of what you do on Facebook when you tell them your mood: structuring data so that it's in an easy-to-digest format (for a human or a computer, depending on the project).

An analytics firm might plumb Reddit and pass along to its client, PepsiCo, that its new, celebrity-driven marketing campaign is a flop with people in Arizona. (Better to pay to get this kind of information early than to read about it later in a newspaper's business section.) Or Shell might discover that its actions in the Niger Delta are stirring up discontent among nationalist bloggers inside Nigeria but that criticism outside the country is sparse (*We can handle that*, its PR department might say). Wall Street investors could load some sentiment analysis into their high-frequency trading algorithms; if a tide of negative sentiment builds, say, around the prices of some basic household items, the firm's supercomputers might decide to pull back on certain commitments or to short them. The system could also spew out particular alerts, pluck out important articles, or issue scores about the level of influence that sources appear to have. For example, on October 3, 2013, tweets began appearing saying that a shooting had occurred on Capitol Hill. Dataminr, an analytics firm, quickly sent messages to its customers, allowing them to take action before the stock market took a dip a few minutes later. It's a cynical approach to the news, but it could make some investors millions.

Thomson Reuters, the parent company of the Reuters newswire, has gotten into the sentiment analysis business, examining more than four million blogs and social-media feeds. They claim to be able to offer forecasts of how individual stocks will do, in addition to rating sources and offering big-picture analysis about the state (and sentiment) of the market. Their rival Dow Jones, parent company of the *Wall Street Journal*, also offers a "machine-readable news" feed that can be plugged into automated trading platforms.

This kind of information is chum in the water for hedge funds, which will pay to get practically any information that their competitors don't have—or pay to get it just a few milliseconds earlier than their peers. Derwent Capital Markets examined 250 million tweets daily and reportedly beat the market in its first month, earning 1.85 percent against a 2.2 percent drop in the S&P 500. That's an auspicious start, except apparently it got no further: Derwent closed out the fund after a single month and announced it would develop a platform to sell sentiment analysis to investors. They weren't the only hedge fund to make a splash with promises of a social-media-driven fund only to soon backtrack. MarketPsy Capital also briefly ran a hedge fund relying on social-media data before liquidating the fund and deciding to focus on selling its social-media analysis directly to clients. (Investors, some shaken by the recession, were reportedly leery of putting their money in such novel investment funds.) Most hedge funds looking at social-media data seem to be taking this kind of approach, buying packages of analysis from third-party firms. As the proverb about the California gold rush goes, it often pays more to sell the shovels than to use them to dig. But at least twelve quantitative hedge funds pay a firm called Gnip to pipe all of the over 500 million or so tweets produced each day directly into their platforms.

Sentiment analysis is a perfect product for a tech industry awash in data and searching for ways to make money off it. It's but another way in which the behaviors, actions, identities, and feelings of Internet users are being bought and sold, often without their knowledge, and put toward uncertain ends. For although these tools can decode some larger trends—several studies have found correlations between positive Twitter sentiments and overall bullishness in the stock market—they are also easily fooled. Some databases of word associations are dated, while others have been created using dodgy methods, such as assigning the same sentiment to many synonyms of a single word. Syntax, slang, idioms, sarcasm, irony, and other matters of tone, intent, and cultural context can throw off an algorithm. I might tweet that LeBron James's slam dunk was "sick"—meaning awesome, in my sophisticated vernacular—and an algorithm might think that seeing LeBron James made me feel ill, lowering James's estimation in the report that a high-priced consultancy is producing for Nike. According to the simple binary the algorithm employs, "sick" is either good or bad, and it might get it wrong. More complicated opinions are further susceptible to misreadings. If Facebook could understand sarcasm or humor, would it have packaged Nick Bergus's post into an ad for Amazon?

A related problem is that many social-media postings aren't carried out by humans. A seminal *New Yorker* cartoon put forth the adage—since canonized into Internet lore—that on the Internet, nobody knows you're a dog. Well, on social media, no one knows you're a bot. An executive at one analytics firm found that two-thirds of tweets they examined were from "a bot—such as an automated account broadcasting news—or an organization." Companies automate tweets, syndicating information across platforms or spewing out data as it's processed. Geological agencies send out notifications about seismic activity. Others are spam bots or conceptual art projects tweeting every word from Shakespeare. In fall 2013, a group of Italian researchers published a study determining that about 9 percent of active Twitter users are bots, mostly created to be sold as fake followers; Twitter's own estimate is that 5 percent of accounts are fake. In addition to fake accounts, people also post things that are intentionally insincere and misleading, including in their profiles, which further complicates the effort to divide people into the kinds of highly specific categories (e.g., single dads from major cities who don't belong to gyms) that market researchers like.

Of course, these analytical tools are getting better, incorporating the latest discoveries in computational linguistics and deep learning, a form of artificial intelligence in which computers are taught to understand colloquial speech and recognize objects (such as people's faces). Some sentiment analysis software now applies several different filters to each piece of text in order to consider not only the tone and meaning of the utterance but also whether the source is reliable or somehow biased. IBM has claimed that its Social Sentiment Index "can distinguish between sarcasm and sincerity." SAS, a massive, privately owned software firm, touts its ability to analyze sentiment in multiple languages. BehaviorMatrix says that its program examined cancer blogs and discovered that cancer patients are most optimistic just after receiving their diagnosis. This insight might be useful for therapists, doctors, and public health professionals, but the company's CEO told the *Wall Street Journal* that he drew on this information to advise drug companies in their ad targeting.

The most likely application of sentiment analysis, then, is to give a slight edge to hedge funds and advertisers. At the very least, a gaggle of digital media consultants are pulling down hefty fees selling these services to deep-pocketed corporate clients. But what happens when sentiment analysis is not just spilling out reports for an executive's consumption but is actually linked to potentially vital systems? And what happens then if a network becomes seeded with misinformation? You might just crash the stock market.

On April 23, 2013, the Associated Press's official Twitter account sent out the following tweet: "Breaking: Two Explosions in the White House and Barack Obama is injured." Almost immediately, the Dow plunged 100 points while a few percentage points were lopped off the S&P 500. For a brief period, it seemed like the stock market was in free fall. And then almost as quickly as it began, order was restored. The AP tweet was revealed to be the result of a hack (a pro-Assad group called the Syrian Electronic Army claimed responsibility), and the news organization soon regained control over its account. The incident though showed how closely linked systems—Twitter and stock markets, real-time sentiment analysis and automated trading—can be easily gamed, especially when someone in control of a heavily followed Twitter account clicks on a suspicious link, giving control to an unscrupulous hacker. It wasn't the first flash crash linked to automatic trading—that honor goes to the May 2010 Flash Crash, in which the Dow lost 1,000 points and swung back to equilibrium a few minutes later—but it was the first in which social media has played such an obvious role. Both Twitter and the AP were criticized for their lax security, and a few months later, Twitter introduced two-factor authentication, a security measure that should make such incidents less likely in the future. The financial industry didn't escape scrutiny either, as some commentators, already chastened by the 2010 crash, began to consider the consequences of automated, high-frequency trading.

The next frontier in sentiment analysis may be not in what we write but in what we say. Some call centers and customer-service lines are investing in computational voice analysis, allowing them to detect the moods of callers. The pitfalls of this practice are obvious—the software may not be accurate, customers may be sorted into categories in which they don't want to be, people's moods can change from call to call or even within a single call—but it gives companies more control over how they manage incoming calls. Instead of hearing a series of prompts and punching numbers on your keypad to indicate your intent, you might soon simply voice your problem and be algorithmically sorted, assigned to the proper representative who will have a brief, automated report about your needs, disposition, and history with the company. (They could also draw on TeleSign's PhoneID Score, which provides reputation scores for phone numbers based on proprietary criteria. The higher your score, the higher your supposed risk level. If your phone number scores above a certain threshold, the software recommends that you be denied service and your number blocked.)

Voice analysis has emerged as a growing part of a $214 million industry known as speech analytics. One company, Beyond Verbal, invites potential customers to test its software online by analyzing the emotions of Mitt Romney and Barack Obama in a presidential debate. Beyond Verbal also offers an API so that customers can add voice analysis to their own apps. The uses are potentially wide-ranging. Vine, Instagram, YouTube, and other video services could automatically analyze the audio content of uploaded material and then compile real-time reports that they sell on to advertising partners. Governments could mine videos for political opinions and create voice samples of troublesome citizens. Security agencies such as the FBI, which has the technical capability to remotely and surreptitiously activate the microphones in many smartphones, as well as the webcams in computers, could see if a surveillance target is lying or anxious. Google Glass could become a kind of roving emotion-meter, providing you with voice analysis of everyone you meet.

On a more conceptual level, voice analysis and sentiment analysis are about finding out what you think and feel: your "mood graph." Social-media companies really would like to know what you are thinking at all times, but they need the data to be machine-readable, which is why we're prompted to structure our data by tagging emotions, companies, people, and places and why forms of computational analysis promise to automate this process. The data can then be mined and sold on to advertisers, market researchers, and other partners. At its most expansive, this process is an EKG for not only individual opinions but also those of whole demographics, cultures, and communities. In this way, Facebook becomes like an opinion dial held by someone during a political debate—an always-on, always-tracking focus group, where individual opinions are subsumed by the greater marketing machine.

Already, researchers have found that you can tell a lot about people just based on their likes, including gender, sexual orientation, and ethnicity, even if users don't actually like pages explicitly related to those subjects. "Digital records of behavior may provide a convenient and reliable way to measure psychological traits," one group of scholars determined. "It can easily be applied to large numbers of people without obtaining their individual consent and without them noticing."

This process will likely develop into a two-way system. As networks begin to understand how we think and feel, they will prompt us for more information or suggest emotional responses, all of which will be machine-readable. They may also allow companies such as Facebook to help us stop self-censoring by pushing us to reconsider deleted updates or to post something when they detect a change in our mood. The writer Nicholas Carr envisions a system that "automates the feels":

"Whenever you write a message or update, the camera in your smartphone or tablet will 'read' your eyes and your facial expression, precisely calculate your mood, and append the appropriate emoji. Not only does this speed up the process immensely, but it removes the requirement for subjective self-examination and possible obfuscation. Automatically feeding objective mood readings into the mood graph helps purify and enrich the data even as it enhances the efficiency of the realtime stream. For the three parties involved in online messaging—sender, receiver, and tracker—it's a win-win-win."

Recent innovations have shown that social-media companies see human beings as one element in a vast, semiautomated system of communication and data production. From customer service bots to personal assistants (Siri, Google Now, Cortana, and a similar Yahoo-owned app called Aviate), we are increasingly delegating responsibilities to autonomous systems. And we are trusting them to understand us. Describing the next generation of virtual assistants, an executive at a company that contributes technology to Siri said, "The more proactive, the more it knows about you, the more empathetic the interaction will be." Google owns a patent for software that learns how users behave on social media and then suggests responses for them. The program would analyze all of the user's social networking feeds, sorting and flagging new messages and notifications based on what it rates to be their relative importance. As the software gets to know the user better, the user's role would decrease to that of a rubber stamp, approving prefabricated responses. One could imagine a future version of this product taking people out of the loop entirely, offering to maintain their social-media presences in their absence. Such a tool would be useful to those who feel overwhelmed by the labor of social media, who go on vacation and don't want their online presence to diminish, or public figures who want bots to do their promotional work for them.

Besides digital assistants, there are precedents for this kind of program—out-of-office replies, canned/suggested responses to text messages, companies that promise to maintain your social-media pres-

ence after you die, remote personal assistants with whom our relationships are so mediated (by software and distance) that they essentially
serve as bots. On many customer service lines, we already use our voices
to navigate menus, and some telemarketing operations have advanced
this practice, using robots to give a sales pitch before transferring the
customer to a human sales associate. In recent years, apps that mimic
your Twitter or Facebook posts, often in vaguely accurate but also
amusingly bizarre ways, have become an Internet phenomenon. It's the
Turing test as entertainment. Soon, one might choose a Google bot
that promises verisimilitude or one of these more ham-fisted creations
that would entertain you and your friends with a funhouse-mirror version of your online persona. In the eyes of the platform owner, the
difference is likely to be immaterial: ads are still being shown, data
will be created. Given the currencies of digital life—data, attention, ad
impressions, likes—bots may prove the more reliable moneymakers.

Pics or It Didn't Happen

Modern life is so thoroughly mediated by electronic images that we cannot help responding to others as if their actions—and our own—were being recorded and simultaneously transmitted to an unseen audience or stored up for closer scrutiny at some later time. "Smile, you're on candid camera!" The intrusion into everyday life of this all-seeing eye no longer takes us by surprise or catches us with our defenses down. We need no reminder to smile. A smile is permanently graven on our features, and we already know from which of several angles it photographs to best advantage.

—Christopher Lasch, *The Culture of Narcissism*

The numbed enthusiasm of constant liking—and its frequent companion, the white noise of perpetual outrage—can be enervating. Nuance, difference, and complexity evaporate as one scrolls through an endless feed, vaguely hoping for something new or important but mostly consigning oneself to variations on familiar material. These networks, particularly Facebook, have a banality problem. The cultural premium now placed on recording and broadcasting one's life and accomplishments means that Facebook timelines are suffused with what seem to be insignificant, trite postings about meals, workouts, non-accomplishments, the weather, recent purchases, funny ads, the milestones of people three degrees removed from you. On Instagram, one encounters a parade of the same carefully distressed portraits, well-plated dishes, and sunsets gilded with smog.

The ones who are most blissfully, unself-consciously happy baffle me most. Where is their neuroticism, their self-consciousness about sharing and laying their lives bare? How easily they've assimilated themselves to this lifestyle, tending to their profiles, little gardens of personality in which only pleasantries bloom and life's setbacks, even a death in the family, are presented with such overwrought sentimentality that it's possible to think that such tragedies are welcomed, because they offer an opportunity to share and be embraced by the social-media cocoon.

Couples who engage in public displays of affection have now found the ultimate stage on which to demonstrate their love and become insufferable toward everyone they know. And once they get engaged, whole new avenues of annoyance open up. The wedding-industrial complex gets its hooks into them, as do the practically erotic enthusiasms of their friends, who begin cooing over every newly shared detail—the close-up of the ring, the overlit engagement photos, the publicly shared anxieties over wedding-planning minutiae, the countdown to when one is finally, *finally* able to marry his/her best friend. Once a baby is born, this process becomes institutionalized, a daily sentimental scrapbooking of the little one's growth. It also ensures that the child will have a data trail practically from birth, with some parents choosing to register e-mail, domain, and social-media accounts in the kid's name—the late capitalist equivalent of starting a college fund.

For some in your social graph, these updates may have value. But so often, they please or inform no one; they're the inner monologue, emptied for public consumption. They receive comments or likes out of kindness, out of a desire of future reciprocity, or simply out of Pavlovian reaction. The updates themselves exist for their own sake. In a digital landscape built on attention and visibility, what matters is not so much the content of your updates but their existing at all. They must be *there*. Social broadcasts are not communications; they are re-cords of existence and accumulating metadata: "The point of being on social media is to produce and amass evidence of being on social media. (*Look at me, I'm liking some stuff! And I got retweeted 14 times! Seven new followers!*)" This is further complicated by the fact that the feed is always refreshing. Someone is always updating more often or rising to the top by virtue of retweets, reshares, or some opaque al-gorithmic calculation. In the ever-cresting tsunami of data, you are always out to sea, looking at the waves washing ashore. As the artist Fatima Al Qadiri has said, "There's no such thing at the most recent update. It immediately becomes obsolete."

Why, then, do we do it? If it's so easy to become cynical about social media, to see amid the occasionally illuminating exchanges or the harvesting of interesting links (which themselves come in bunches, in great indigestible numbers of browser tabs) that we are part of an unconquerable system, why go on? One answer is that it's a by-product of the network effect: the more people who are part of a network, the more one's experience can seem impoverished by being left out. Every-one else is doing it. A billion people on Facebook, hundreds of mil-lions scattered between these other networks—who wants to be on the outside? Who wants to miss a birthday, a friend's big news, a chance to sign up for Spotify, or the latest bit of juicy social intelligence? And once you've joined, the updates begin to flow, the small endorphin boosts of likes and repins becoming the meager rewards for all that work. The feeling of disappointment embedded in each gesture, the sense of "Is this it?", only advances the process, compelling us to con-tinue sharing and participating.

The achievement of social-media evangelists is to make this urge—the urge to share simply so that others might know you are there, that you're doing this thing, that you're with this person, that you've had this thought, that you have some urgent opinion on what's trending—second nature. This is society's great phenomenological shift which, over the last decade, has occurred almost without notice. Now anyone who opts out, or who feels uncomfortable about their participation, begins to feel retrograde, Luddite, uncool. Interiority begins to feel like a prison. The very process of thinking takes on a kind of trajectory: how can this idea be projected outward, toward others? If I have a witty or profound thought and I don't tweet or Facebook it, have I somehow failed? Is that bon mot now diminished, not quite as good or meaningful as it would be if laid bare for the public? And if people don't respond—retweet, like, favorite, etc.—have I boomeranged back again, committing the greater failure of sharing something that wasn't worth sharing in the first place? After all, to be uninteresting is a cardinal sin in the social-media age. To say "he's bad at Twitter" is like saying that someone fails to entertain; he won't be invited back for dinner.

In this environment, interiority, privacy, reserve, introspection—all those inward-looking, quieter elements of consciousness—begin to seem insincere. Sharing is sincerity. Removing the mediating elements of thought becomes a mark of authenticity, because it allows you to be more uninhibited in your sharing. Don't think, just post it. "Pics or it didn't happen"—that's the populist mantra of the social networking age. Show us what you did, so that we may believe and validate it.

Social media depends on recognition—more specifically, on acts of recognition. Rob Horning writes: "Social media functions as a giant scoreboard to confer significance to events that are more or less meaningless in the moment. Getting likes on a photo of the meal you made yourself is more important and more significant than eating it." Katherine Losse offers a similar gloss on "the logic the social net depends on: That because something or someone you know was filmed, it becomes interesting, worthy of watching." The thing itself is less interesting than the fact that we know someone involved, and if it is interesting or important, we can claim some tenuous connection to it. We enact the maxim of the great street photographer Garry Winogrand: "I photograph to find out what something will look like photographed." We document and share to find out how it feels to do it, and because we can't resist the urge. Otherwise, the experience, the pithy quote, the beautiful sunset, the overheard conversation, the stray insight, is lost, or seems somehow less substantial.

Sharing itself becomes personhood, with activities taking on meaning not for their basic content but for the way they are turned into content, disseminated through the digital network, and responded to. The walk in the park becomes less meaningful than the Instagrammed and crowd-approved photo of a tree in that park (with, of course, a caption explaining what you're doing—unless you want to be mysterious and withholding). In this context, your everyday experiences are limited only by your ability to share them and by your ability to package them appropriately—a photograph with a beautiful filter and a witty caption, or a tweet containing an obscure movie reference that hints at some hidden, more interesting depths. For some users, this process is easy: snap a photo, write below what you're doing, send it out on one or several networks. For others, it can lead to paralyzing self-consciousness, a sense that no social broadcast is good enough, no tweet or Facebook status update reflects the mix of cool, wit, and élan that will generate feedback and earn the user more social capital. Along the way, we've developed an ad hoc tolerance for these gestures, as well as a shared familiarity and understanding. Who hasn't stopped an activity mid-stride so that a friend can send out some update about it? Who hasn't done it himself?

On the social web, the person who doesn't share is subscribing to an outmoded identity and can't be included in the new social space. If not off the grid, he or she simply isn't on the grid that matters—he may have e-mail, but he's not on Facebook, or he's present but not using it enough. (The prevailing term for this is "lurker," an old online message board term, slightly pejorative, describing someone who reads the board but doesn't post.) It's not uncommon to ask why a friend is on Twitter but rarely tweets, or why she often likes Facebook statuses but never posts her own. Why aren't they busy accumulating social capital? Still, being in the quiet minority is far better than not participating at all. Worst, perhaps, is the person whose frequent tweets and updates and posts earn no response at all. In the social-media age, to strive for visibility and not achieve it is a bitter defeat.

- - - - - - - - - - - - - - - - - -

CALL AND RESPONSE

We become attuned to the pace and rhythm of sharing and viewing, building an instinctual sense of the habits of our followers and those we follow, those we call friends and those we just stalk. We develop what some social scientists have termed "ambient awareness" of the lives of those in our social graphs and intuit, Jedi-like, when they've been absent from the network. Our vision becomes geared toward looking at how many likes or comments a post has received, and when we open the app or log onto the network's Web site, our eyes dart toward the spot (the upper righthand corner, in Facebook's case) where our notifications appear as a number, vermilion bright. We might also receive e-mail alerts and pop-ups on our phones—good news can arrive in a variety of ways, always urging you to return to the network to respond.

The alerts and notifications have the potential to be constant, unless we turn them off, which only serves to take us back to the network's privacy settings and which might cause us to log onto Facebook or Twitter more often, so that we can see who might be trying to interact with us. Dalton Caldwell, founder of the alternative social network App.net, calls this "data dread"—the constant, insistent influx of information through updates, push notifications, and alerts. And when we stray out of cell service or are forced to turn off our phones, the dread turns into the fear of missing out. It's this anxiety that causes us to grab for our phones as soon as the subway car comes above ground, or when the movie ends, or the plane lands and the flight attendant, sounding like a magnanimous teacher releasing students early from detention, assures us that it's okay to turn on our cell phones now. When connectivity becomes the default state, outages induce a minor panic.

The problem with alerts is that, like our updates, they never end.

They become a way to be permanently chained to the network. We are always waiting to hear good news, even as we ostensibly are engaged in something else. Just as urban spaces threaten to do away with silence or with stars—the city's sound and light, its primordial vibrancy, become pollutants—notifications crowd out contemplation. They condition us to always expect something else, some outside message that is more important than whatever we might be doing then. When the phone lights up, it must be dealt with immediately, if only to banish the alert from the screen. In these moments, we tend to think more about how the alert wasn't important, rather than what the alert interrupted.

The writer and former tech executive Linda Stone calls this phenomenon "continuous partial attention." She differentiates it from multitasking, though there is some similarity. Continuous partial attention, she says, "is motivated by a desire to be a *live* node on the network. Another way of saying this is that we want to connect and be connected. We want to effectively scan for opportunity and optimize for the best opportunities, activities, and contacts, in any given moment. To be busy, to be connected, is to be alive, to be recognized, and to matter."

Psychologists and brain researchers have begun studying these problems, with some dispiriting conclusions: Multitasking is largely a myth; we can't do multiple things at once, and when we try, we tend to do a poorer job at both. Frequent interruptions—such as your phone starting to vibrate while you're reading this paragraph—make it harder to return to the task at hand. In fact, office workers experience an interruption about every three minutes. It could be an e-mail popping up or your friend coming by your desk. But it can take more than twenty minutes to shake off the interruption and get back to the job at hand. That means that many of us are being interrupted too often to regain focus, with the result that our work and mental clarity suffer. On the other hand, some of these same studies have found that when we expect interruptions, we can perform better, as we train ourselves to become more single-minded and complete a task in a limited period.

While management consultants may see that as a shortcut to productivity, I think it reflects something else: how as we become increasingly harried by our own digital communications, we force ourselves into a state of anxious focus. We may get more done, or because we now expect some alerts to pop up on-screen, we may be better able to ignore them, but that doesn't mean that we're happier doing it.

You can turn off your Twitter's e-mail alerts or tell your smartphone to stop pushing you Facebook updates or the latest news from Tumblr. But alerts are the critical symbol of the call and response, the affirmation and approval, that tie a social network together. They let us know that we're being heard, and if we don't have them forced on us, we still have to reckon with them when we log into the app or onto the network's Web site. It's important not only to have them but also to have them in sufficient number—or at least some amount which, we tell ourselves, justifies the update. *Four people clicked Like; that's enough, I guess.* After posting an update, we might return to the network several times over the next hour, hoping for some validating reply.

The window for this kind of response is painfully brief. We know fairly quickly whether our beautifully filtered photo of a grilled cheese sandwich or our joke about a philandering politician was a dud. According to a 2010 study by Sysomos, only 29 percent of tweets receive a response—a reply, retweet, or favorite—while 6 percent are retweeted. Ninety-two percent of retweets happen within the tweet's first hour, meaning that if sixty minutes have passed and no one's picked up on your tweet, it's likely disappeared into the ether. Even when they do appear, likes and favorites have been mostly drained of meaning—a sign of approval and popularity, sure, but also now a rather conventional way of telling a friend that he was heard. The favorite has become a limp pat on the back.

This ephemerality contributes to social media's tendency toward self-consciousness and constant calibrating of one's public persona. We know that we don't have much time—or characters—and better make it count. "If I don't get more than ten faves in [the] first three minutes after tweeting something, I'll probably delete it," an amateur comedian told the *Wall Street Journal*. Otherwise, the tweet hangs there, a minor emblem of its author's unsatisfied ambition. That few will see it, or, if they do, make great judgments about the man himself, doesn't matter much. The possibility remains that they will judge him for it, and for as long as that feeling remains, it's intolerable. It reminds him that he wasn't able to distinguish himself from the other aspirants in his followers' timelines. What will they think of him? The joke's quiet landing is downright embarrassing. Perhaps he knows that the tweet may already have been scanned and collected by some social search engine, semantic algorithm, or the Library of Congress, that our data tends to find a life of its own, but it still must be expunged from his personal record. The mask must always be kept on, perfectly straight; you never know who's watching.

What that comedian really fears is the loss of followers and social capital. We take it for granted, perhaps, that social media comes with metrics. We are constantly told how many people are following us, how many approved of an update, how many people follow those people (naturally, favorites from people with more followers are more valuable, with retweets still another notch higher on the ladder). Metrics help create the hierarchies that are embedded in all social networks, and that often replicate offline hierarchies. If you don't know immediately how popular someone is on social media, then it's only a click away. The digital elite are affirmed in their place by way of a verified account, while the truly famous are granted the privilege of trending. These practices trace in part to message boards and forums, where users often have their post count appear next to their handle and power users receive special avatars, badges, or titles. In the process, the goal of the digital social space becomes not to enjoy yourself or aimlessly interact with others but to rise higher in the game, which can't really be won, unless you're hired by the Web site/social network/developer—an occasional reward for elite users in the Web's early social spaces and one that continues to be doled out in communities like Reddit or *Gawker*, which has experimented with various tiered commenting systems and hired popular commenters as writers.

This hovering awareness of rank and privilege helps drive the insecurity and self-consciousness that result from an environment suffused with the language of PR, branding, and advertising. Describing his experience on Twitter, the satirist Henry Alford writes that "every time someone retweets one of my jokes, it sets off a spate of fretting about reciprocity." He's unsure how to respond:

"If the person is a total stranger whose feed I do not follow, then I will look at this feed and consider climbing aboard. I'll look at the ratio of how many tweets to how many followers that person has: if it exceeds 10 to 1, then I may suddenly feel shy. Because this person is unknown to me, I will feel no compunction to retweet a post of hers, though I may be tempted to 'favorite' (the equivalent of Facebook's 'like' button) one."

Alford is demurring here. What he really means is that someone with a tweets-to-followers ratio of 10 to 1 is probably an unknown, one of the innumerable Twitter users whose many tweets go pretty much ignored, and not one of the journalists, comedians, or writers who likely belong to his intended audience. (He goes on to mention, happily, that one of his jokes was recognized by the comedians Merrill Markoe and Rob Delaney, the latter a Twitter superstar.) There's nothing wrong with that, of course, except that Alford's own admissions speak to the difficulty of negotiating the odd social pressures and anxieties that come with every utterance being public. Is he on Twitter to promote himself, to meet people, or to endear himself in the eyes of colleagues—or is there a conflicting mix of motivations? In Alford's case, his concerns about visibility and reciprocity intensify when he's responded to on Twitter by someone he knows: "Suddenly the pressure mounts. I'll proceed to follow her, of course, if I don't already. Then I'll start feeling very guilty if I don't retweet one of her posts." Each exchange requires a complex cost-benefit analysis, one which, for anyone who has experienced this for himself, may seem wildly disproportionate to the conversation at hand. Just as metadata (that is, the number of retweets or likes) can matter more than the message itself, this process of meta-analysis, of deciphering the uncertain power dynamic between two people, can seem more important than the conversation on which it's based.

- - - - - - - - - - - - - - - - - - -

PHOTOGRAPHY AS A MEANS OF CONSUMPTION

Maybe Alford would be more comfortable with photographs, which, in their vivid particularity, seem to demand less of a response. They can live on their own. We don't need to justify them. Photographs "furnish evidence," as Susan Sontag said. Or, in Paul Strand's words: "Your photography is a record of your living." You met a celebrity, cooked a great meal, or saw something extraordinary, and the photograph is what remains: the receipt of experience. Now that every smartphone comes complete with a digital camera as good as any point-and-shoot most of us had a few years ago, there is little reason not to photograph something. Into the camera roll it goes, so that later you can perform the ritual triage: Filter or no filter? Tumblr, Instagram, or Snapchat?

The ubiquity of digital photography, along with image-heavy (or image-only) social networks such as Instagram, Pinterest, Tumblr, Imgur, Snapchat, and Facebook, has changed what it means to take and collect photos. No longer do we shoot, develop, and then curate them in frames or albums in the privacy of our homes. If we organize them into albums at all, it's on Facebook or Flickr—that is, on someone else's platform—and leave them there to be commented upon and circulated through the network. Photos become less about memorializing a moment than communicating the reality of that moment to others. They are also a way to handle the newfound anxiety over living in the present moment, knowing that our friends and colleagues may be doing something more interesting at just that very moment, and that we will see those experiences documented later on social media. Do we come to feel an anticipatory regret, sensing that future social-media postings will make our own activities appear inadequate by comparison? Perhaps we try to stave off that regret, that fear of missing out, by launching a preemptive attack of photographic documentation. *Here we are, having fun! It looks good, right? Please validate it, and I'll validate yours, and we'll take turns saying how much we missed each other.*

Photography has always been "acquisitive," as Sontag called it, a way of appropriating the subject of the photograph. Online you can find a perfectly lit, professionally shot photo of nearly anything you want, but that doesn't work for most of us. We must do it ourselves.

Think about the pictures of a horde of tourists assembled in front of the *Mona Lisa*, their cameras clicking away. It's the most photographed work of art in human history. You can see it in full light, low light, close-up, far away, x-rayed; you can find parodies of parodies of parodies; and yet, seeing it in person and walking away doesn't suffice. The experience must be captured, the painting itself possessed, a poor facsimile of it acquired so that you can call it your own—a photograph which, in the end, says, *I was here. I went to Paris and saw the Mona Lisa.* The photo shows that you could afford the trip and that you're cultured, and offers an entrée to your story about the other tourists you had to elbow your way through, the security guard who tried to flirt with you, the incredible pastry you had afterwards, the realization that the painting really isn't much to look at and you've always preferred Rembrandt. The grainy, slightly askew photo of Ms. Lisa signifies all these things. Most important, it's yours. You took it. It got twelve likes.

This is also the unspoken thought process behind every reblog or retweet, every time you pin something that has already been pinned hundreds of times. You need it for yourself. Placing it on your blog or in your Twitter stream acts as a form of identification—a signal of your aesthetics, a reflection of your background, an avatar of your desires. It must be held, however provisionally and insubstantially, in your hand, and so, by reposting it, you claim some kind of possession of it.

Something similar can be said about people you see at concerts, recording or photographing a band. Unless you have a fancy digital SLR camera and are positioned close to the stage, the photographs will probably be terrible—mussed up by the combination of low house lights and the lens flare of stage lighting, the performers a blur, a smudge of some audience member's head jutting into the frame. Recording it is even more of a fool's errand, because it will show you only how poor your iPhone's microphone is and how this experience, ostensibly so precious at the time, can't be captured by the technology in your pocket. No, the video will be shaky, as you struggle to hold your phone above the heads of people in front of you, and the audio will be fuzzy, low-fi, like someone played the track at full volume inside a steel trash can. But of course, many of us do this, or we hold our phones aloft so friends on the other end of the line can hear—what exactly? Again, we find that it doesn't quite matter. We'll probably never watch that video later, nor will we make that photo our desktop wallpaper or print it out and frame it. The crummy photos, the crackly recording, the indecipherable blast of music a caller hears: these aren't personal remembrances or artistic artifacts. They're souvenirs, lifestyle totems meant to communicate status—to *be* your status update. They don't describe the band being captured; they describe us.

People often ask: Why don't concertgoers just live in the moment and enjoy the show? Don't they know that the photos won't turn out well and that focusing on their iPhone, staring at its small screen rather than the thing itself, takes them out of the experience? The answer is yes, but that's also beside the point. Taking photos of the band may show a kind of disregard for the music, an inability to enjoy it simply as it is, but it also reflects how the very act of photographing has become part of just about any event or evening out. In the same way that Sontag says that "travel becomes a strategy for accumulating photographs" (e.g., stand in a certain spot in front of the Leaning Tower of Pisa, where you can position your arms so it looks like you're holding up the building; or stop at a lookout point because your guidebook says it makes for a good picture), life itself becomes a way of accumulating and sharing photographs. Taking photos gives you something to do; it means that you no longer have to be idle and that, in the normally passive experience of concert-going, for example, you can become a participant. Living in the moment means trying to capture and possess it. We turn ourselves into tourists of our own lives and communities, our Instagram accounts our self-authored guidebooks, reflecting our good taste. At the same time, the use of filters and simple photo-editing software means that any scene can be made into an interesting photograph or at least one that has the appearance of being artistically interesting. The process of digitally weathering and distressing photographs is supposed to add a false vintage veneer, a shortcut to nostalgia, and it has the added benefit of making it look like the photo went through some process. You worked for it, at least in a fashion.

This kind of cultural practice is no more clearly on display than during a night out with twentysomethings. The evening becomes partitioned into opportunities for photo-taking: getting dressed, friends arriving, the pregame, a taxi ride, arriving at the bar, running into friends, encountering funny graffiti in the bathroom, drunk street food, the stranger vomiting on the street (or friend, if you're being uncharitable), the taxi home, maybe topping it off with a shot of the clock before bed. A story is told here, sure, but more precisely, life is documented, its reality confirmed by being spliced into shareable data. Now everyone knows how much fun you had and offers their approval, and you can return to it to see what you forgot in that boozy haze.

All of this also offers evidence of your consumption, of the great life you're living and the products contained therein. In a medium loaded with product placement and advertisements euphemistically called "stories," this is no small thing. To document your life on social media is also to become an advertisement for yourself and for all of your possessions that you put on display. You're not just sharing a photograph of your stuff or the bar you went to; you're saying, buy this stuff, go drink at this place. You are advertising a lifestyle—buttressed by the advertisements that draw from your updates and appear along-side them—the rightness of which is affirmed with every like and little expression of affirmation.

Photography's acquisitive aspect—the part of it that turns life into one long campaign of window-shopping—finds its fullest expression on Pinterest, Instagram, and other image-heavy social networks. There we become like the hero of Saul Bellow's *Henderson the Rain King*, a man who hears a voice within himself saying, "I want, I want, I want." Like Henderson, we don't know what we want exactly but have some sense that it's out there, in the endless feed of shimmering imagery, and that if we click through it long enough, maybe we'll find satisfaction. We'll find something we want to buy; or, poring over a hundred images of some acquaintance, we find someone we want to know. In this ruminative browsing, where time becomes something distended, passing without notice, our idle fantasies take flight.

Along the way, we post and share images of products and lifestyles because they reflect our tastes and pinning them is a desultory substitute for buying them. Brands understand this, and now their marketing departments invest in making sure that photos of their products are on prominent display on sites such as Tumblr, Pinterest, and Instagram. From household goods to fast food to designer fashion, these images circulate through the networks, often with no evidence that the photo of the Gucci dress or Mercedes SUV is a carefully crafted piece of advertising. The photo may not even feature the product, but it's representative of a certain lifestyle; it creates a mood, an image of the good life. They hope that you click on the link to the source and decide to follow it, so that they'll be able to shill to you directly. Or perhaps the photo is something funny or eye-catching "brought to you by," say, Oreo, which adds to the company's brand awareness, so that the next time you're in the convenience store, when you spot a package of Double Stuf Oreos, you remember the moment of pleasure you got from watching that cute animal video they shared. You might not buy it then, but at least they've nudged you a step closer.

Companies love this sharing model because it turns users into vectors for their own viral marketing. They enter your feed on the same terms as your friends and the celebrities you follow, appreciating the same kind of content. (This is why many memes die after they enter the visual field of advertising agencies, who, in an effort to appear hip, appropriate them for some corporate campaign.) They offer luxurious, beautiful, aspirational images, and users take them up and share them, making no distinction between what's original, what's a popularly shared meme, or what was created in an advertising department. And even if we know we're complicit in this process, maybe it doesn't matter. It's not as if we're buying the thing, and advertising is everywhere on the Internet, right? Many of us have our Amazon wishlists publicly searchable anyway. We write about what we want to buy and ask our followers to weigh in on a possible purchase. And so we submit. We share the image, which in turn sends a message that the network and the advertiser take to mean—even if we don't want them to—"more of this, please."

- - - - - - - - - - - - - - - - -

THE FACEBOOK EYE

The documentary lifestyle of social media raises concerns about how we commoditize ourselves and how we put ourselves up for public display and judgment. That doesn't mean that fun can't be had or that this kind of documentation can't coexist with an authentic life. It's just that the question of what's authentic shifts, sometimes rather uncomfortably, and not just in the Zuckerberg/Sandberg sense of frictionless sharing, of disclosing everything, always, completely. Instead, it's that our documentation and social broadcasts become the most important thing, an ulterior act that threatens to become the main event. We care less about having fun at a party than we do about checking in on Foursquare and appearing in others' photographs, our bodies and clothes arranged just so, all a means of telling people that we're having fun at a party. This is the real danger of not living in the moment, or of feeling like your attention is constantly elsewhere, your very sense of self divided. And when you don't find these things, when the party isn't good because you don't recognize anybody you know and they don't recognize you, or that nagging feeling takes over when you're at a baseball game, just trying to enjoy the scene in front of you, well, then you take out the smartphone. It's always there, always updating. The smartphone is the Swiss Army knife of social-media culture. It's also the ultimate site of social retreat. That's where the true audience is, waiting to be entertained, and when your surrounding environment fails to measure up, you can return to them, like a standup comedian taking the mic and putting in a few minutes of work. If you're lucky, the flattery of likes and favorites will flow your way.

Nathan Jurgenson describes social-media users as developing "a 'Facebook Eye': our brains always looking for moments where the ephemeral blur of lived experience might best be translated into a Facebook post; one that will draw the most comments and 'likes.'" We might feel this phenomenon in different ways, depending on which networks we use and which activities constitute our day. I feel it acutely when reading articles on my smartphone or my computer—this sense that I'm not just reading for my own enjoyment or edification but also so that I can pull out some pithy sentence (allowing enough space for the twenty-three or so characters needed for a link) to share on Twitter. I began to feel an odd kind of guilt, knowing that my attention isn't being brought to bear on what I'm doing. It feels dishonest, like I'm not reading for the right reasons, and self-loathing builds within me—the self-loathing of the amateur comedian who deletes his tweets if they don't quickly attract enough likes. I can offer myself justification by saying that I enjoy sharing or that I am trying to pass along interesting information to others—it's practically a public service!—but that strikes me as insincere. The truth is more depressing, as this impulse to share reflects, I think, the essential narcissism of the Facebook Eye (or the Instagram Lens, or whichever filter you prefer). Our experiences become not about our own fulfillment, the fulfillment of those we're with, or even about sharing; rather, they become about ego, demonstrating status, seeming cool or smart or well-informed. Perhaps if you're a young journalist, looking to increase your esteem in the eyes of peers and a couple thousand followers in the digital ether, the goal of reading is to be the first to share something newsworthy, sometimes before you've even finished reading the article. Like a village gossip, you want to build social capital by becoming a locus for news and information. But with this role, there's an inevitable hollowing out of interiority, of the quietness of your thoughts, as reading becomes directed outward, from a period of private contemplation to a strategic act meant to satisfy some nebulous public.

Our reading habits change, and so do the stories we tell, the way we share these things. We make our updates more machine-readable—adding tags; choosing brands and emotions supplied to us through Facebook's interface; shortening jokes to fit into Twitter's 140-character limit, itself a legacy of a time when some SMS were limited to that length by certain cell providers. We wait to post on Tumblr after 12:00 p.m., because we've read that usage rises during the lunch hour, or we schedule tweets to go up during some period of disconnection.

On social media, it can seem as if time and data are always slipping through our fingers. Particularly when we follow or befriend many people on a network, our updates seem illusory, as if they're never inscribed into anything but rather shouted into the void. (Did anyone really hear me? Why is no one replying?) In this way, social media can resemble traditional, preliterate societies, where communication is purely oral and everything—culture, news, gossip, history—is communicated through speech. Since nothing is written down, all memory is contained within individuals and communal memory must be passed down accordingly (think of Homer telling his tales, themselves passed orally through several generations before being written down). When we retweet someone, we're just speaking their words again, like a group of Occupy activists employing the "people's microphone." In repeating someone's words, we're ensuring that they're passed on and don't disperse in the flurry of communication. This act of amplification is a small one, but by contributing our voice—by causing that tweet or post to accumulate a little bit of metadata, in the form of a retweet or a like—we give it a small lift.

Media theorists refer to these eruptions of oral culture within literate culture as examples of "secondary orality." Social media's culture of sharing and storytelling, its lack of a long-term memory, and the use of news and information to build social capital are examples of this phenomenon. While records of our activities exist to varying extents, secondary orality shows us how social media exists largely in a kind of eternal present, upon which the past rarely intrudes. Twitter is a meaningful example. It's evanescent: posts are preserved, but in practice, they're lost in one's rapidly self-refreshing timeline; read it now or not at all. Twitter is also reminiscent of oral storytelling, in which one person is speaking to a larger assembled group and receiving feedback in return, which helps to shape the story. One of the digital twists here is that many storyteller-like personas are speaking simultaneously, jockeying for attention and for some form of recognition that gives their utterances social ballast. The point is not that social media is atavistically traditional but that it returns elements of oral societies to us. Our fancy new digital media is in fact not entirely new, but a hybrid of elements we've seen in past forms of communication. The outbursts of tribalism we sometimes see online—a group of anonymous trolls launching misogynist attacks on a female journalist; the ecstatic social-media groupies of Justin Bieber; the way one's Twitter timeline can, for a short while, become centered around parsing one major event, as if gathered in a village square—are evidence of a very old-fashioned, even preliterate communitarianism, reified for the digital world.

Think about the anxious comedian or your most over-sharing Facebook friend. Their frequent updates come in part from the (perhaps unacknowledged) feeling that in social media nothing is permanent. Of course, we know that these networks, and the varying sites and companies that piggyback off of them, preserve everything. But in practice, everything is fleeting. You must speak to be heard. And when everyone is speaking, you'd better do it often and do it well, or be drowned in the din, consigned to that house at the end of the village, where few people visit, and when they do, they don't expect to hear much of consequence.

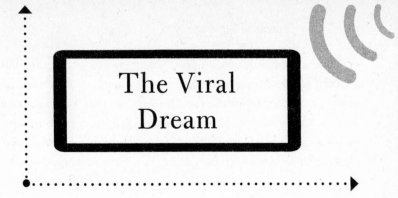

The Viral Dream

But telling the truth kills virality.

—Farhad Manjoo, *The Wall Street Journal*

On June 8, 2013, a twenty-seven-year-old woman confidently walked into a Dunkin' Donuts in Broward County, Florida. Her name was Taylor Chapman, and she had a problem that she wanted the world to know about. Brandishing her smartphone, Chapman approached a cashier and warned him that he was "under video surveillance." Soon Chapman began a racist, expletive-filled tirade—all of which she filmed—accusing the cashier's colleagues of treating her poorly the previous day, including not giving her a receipt. Acting erratically throughout the eight-minute video, she ordered from large parts of the menu, demanded that she not have to pay a thing, and continued to treat the polite cashier in an awful manner, calling him a "fucking sand nigger" and saying, "I'm about to nuke your whole fucking planet from Mars." Several times in the video, Chapman promised that she'd post it on Facebook and that it'd soon "go viral," which would some-how vindicate her original complaint.

Chapman's video did go viral, though not in the way she intended. The footage received millions of views and spread rapidly through social media. She was written up on *The Smoking Gun* ("boorish, ar-rogant"), *Gawker* ("worst person ever"), and a host of like-minded sites for whom the latest embarrassing viral phenomenon is often an opportunity to humiliate the subject even more and whip readers into a frenzy of schadenfreude and snark. A few hundred comments appeared on Reddit, before the popular social news site, which serves as ground zero for many a viral phenomenon, disabled "comments due to personal info/witch hunting." The Reddit thread was part of a familiar pattern: a video of a person acting in a surprising or awful manner quickly becomes popular through social and traditional media, as journalists and Internet users alike dredge up and spread the subject's personal information, including her social-media accounts. The Dunkin' Donuts video generated hundreds of thousands of page views for some lucky Web sites, while social-media users enjoyed piling on, playing armchair critic or laughing at the video before sharing it with friends. The Dunkin' Donuts employee who acted so charitably in the video was rewarded by the parent corporation. As Taylor Chapman's name and likeness spread everywhere, she decided to close her Facebook and Instagram accounts; she was also fired from her job. Soon she and her boyfriend were dodging reporters outside their apartment, her face concealed by large sunglasses, as if she were a celebrity ducking a paparazzo. Within a few weeks, she was totally forgotten—likely much sooner for many Internet users, including those who reshared her Facebook photos and added comments such as, "What a trashy, redneck bitch" or (below a photo that featured Chapman and her boyfriend) "DUMP THAT BITCH. Save yourself homie!" All that hilarity and vitriol melted away, as bored office workers waited for the next media sensation to take Chapman's place.

There's no doubt that Chapman acted reprehensibly and that she was complicit in creating such a level of infamy for herself. Chapman displayed a peculiar lack of self-awareness, thinking that she could treat someone so terribly and, just by virtue of filming it and posting it online, "go viral" and earn the appreciation of some imagined public. In that way, she's very much emblematic of the mentality behind virality, which its evangelists treat as a good in and of itself. According to the Internet guru Jeff Jarvis, "The new American dream is to go viral," to achieve celebrity by having your image explode across the digital firmament.

And for many, it doesn't matter whether that image is from a sex tape, a strange video, or an inspiring story. It's the repetition that matters. As Tim Wu writes in *The Master Switch*, "One of the stranger consequences of the electronic age is that almost any word or image, reiterated a million, or a billion, times, can become a valuable asset." For those savvy or lucky enough, the idea goes, virality is a shortcut to traditional fame and success. If you handle your newfound fame properly—hire the right agent and manager, who always seem to materialize in these situations; arrange for some paid appearances or endorsement opportunities; sell a book proposal as quickly as possible; maybe do porn—then you might be able turn your moment in the limelight into cash and even lasting notoriety. In the best of circumstances, you become Paris Hilton or Kim Kardashian, both of whom embody Daniel J. Boorstin's original definition of celebrity as "a person who is known for his well-knownness." Both Hilton and Kardashian turned meme fame (in each case, using a sex tape as a media launch pad) into something more lasting and extremely profitable.

In less fortuitous circumstances, however, you become Taylor Chapman: infamous, vilified, harassed, and then forgotten, except by Google Search, which never forgets. Viral fame quickly fades; that's in its nature and the nature of the systems and culture we've created to enjoy it. But the hangover it produces can be long. And given the speed with which stories such as Chapman's spread, a fuller picture rarely appears. But more than two weeks after Chapman's video erupted into view—two weeks being a generation in Web time—the *New Times*, an alt-weekly newspaper in Chapman's Broward–Palm Beach area, published a long story, based on interviews with Chapman and some of her friends and family members, which revealed that the young woman had a history of mental illness that included two hospitalizations, an arrest for erratic behavior, and signs of multiple personality disorder. To be sure, this wasn't a vindication of Chapman, but it did offer more information about who she was and how someone could act with such a mix of self-righteousness and utter depravity. It also cast some light on Chapman's certitude that filming the confrontation was the proper choice and on some of her more bizarre remarks.

For the *New Times*, the article seemed like a kind of penance. As the story's author, Terrence McCoy, noted, the paper had "published five articles calling Chapman a 'local racist.'" Several of those articles had been written by McCoy. Now he was writing sympathetically about Chapman being off her medicine, while noting that she had earned "a digital scarlet letter that won't soon disappear." A follow-up interview with Chapman, published in the same paper, offered more information, though among some apparently true statements (she thought about changing her name; she'd been arrested three times for exhibiting mental distress) appeared some comments that were hard to believe (an admissions officer at Harvard Law School said she could "definitely get in"). Perhaps because he was writing for a tabloid-style alt-weekly, McCoy also played up Chapman's claim that she had stripped at most of the strip clubs in the Fort Lauderdale area, along with one club's denial that they had ever employed her.

- - - - - - - - - - - - - - - - - -

THE PROBLEMS WITH VIRALITY

As the Taylor Chapman story illustrates, viral fame often descends upon those least equipped to handle it. Sometimes, like Chapman, these people sought out a certain degree of notoriety. But once these processes get under way, they're exceedingly difficult to control, especially for a mentally ill, intermittently employed woman from south Florida. They also tend to reveal some troubling issues about the wider culture. While virality is often treated as an inherent sign of quality or relevance, it's anything but. What goes viral tends to be something that will satisfy the id of readers—and of the overworked, under-resourced journalists and bloggers tasked with ginning up page views—or that is part of a covert marketing campaign. There's little distinction or nuance with a viral commodity. A person is either a hero or a villain; a photograph inspiring or horrific.

All viral media requires a certain amount of credulousness and self-seriousness. We have to be willing to believe, and by sharing, to nudge others to believe as well. The viral milieu doesn't accept skeptics—they spoil the fun. But this is a disingenuous posture, too, since viral media tends to be discussed in pretty market-based terms—number of shares, amount of money it generated, ranking compared with past viral phenoms. Longer examinations look under the hood at the mechanisms and personnel that helped engineer the content's sensational accumulation of views and attention. For a consumer, it's all drained of meaning, isn't it? It's pure metadata, running through the circulatory system of networked capitalism. The underlying material is interchangeable. We might as well call viral videos widgets or credit default swaps. Maybe that's a reason why we feel so detached from them as people, even as we consume them so readily. It may also be why, even though a disproportionate number of viral human-interest stories seem to be about people suffering from mental illness, we rarely consider the humanity, and the personal challenges, of those whose stories we laugh at and share.

Local newscasts provide a bountiful source of ready-made memes. With their tendency toward low production values, on-camera snafus, and interviews with odd characters, these broadcasts seem designed to produce odd, deliciously shareable moments. Whether it's the popular video of a reporter who starts stomping on grapes at a vineyard, only to fall over and start crying, whimpering pathetically; or an addled newscaster slipping into a stream of expletives, these videos entertain because they allow us to see beyond the absurd trappings of the self-important newscaster-cum-entertainer. They show us something ridiculous behind the curtain. The pretense of authority and professionalism falls away, and suddenly they're just like us—or worse. More generally, these videos are part of an inexhaustible supply of "fail" memes—photos and videos of people screwing up, falling down, or acting like doofuses while trying to achieve some goal or pull off some performance. They satisfy our desire to see others brought low, to ritualize humiliation.

In recent years, some of the most popular local news stories have featured poor African Americans who witnessed, or were the victims of, violent crimes. Some of these interviews appear newsworthy; after all, they present witnesses who can relate the story in specific detail and are often lively personalities. Without more information, it's difficult to accuse producers of setting up interviewees for ridicule. But the manner in which these videos circulate—the comments they attract, the ways in which they are remixed and presented by news aggregators—is nothing short of racist.

Antoine Dodson was the first in this trend. In July 2010, the then twenty-five-year-old resident of an Alabama housing project was interviewed on TV news after someone broke into his house and tried to rape his sister. Dodson intervened, the intruder fled, and he later gave an animated account to a reporter, saying, "Hide your kids, hide your wife, and hide your husband 'cause they rapin' everybody out here." Before long, the Dodson clip bloomed into a full-fledged meme, plastered all over news and viral video sites, remixed, Auto-Tuned, turned into a song sold on iTunes, his likeness and catchphrase printed on T-shirts. Since then, we've seen a woman named Sweet Brown, who survived an apartment fire in Oklahoma City, and Charles Ramsey, whose 911 call helped rescue several women his neighbor was holding in gruesome captivity, rise to similar levels of fame and ironic celebration. Less heralded encounters have produced Kai (full name: Caleb McGillvary), a white hitchhiker who helped a woman being attacked in Fresno but was later arrested in a separate murder case; and Michelle Clark, a black woman who gave a dramatic description to a local TV station of a hailstorm in Houston, Texas. YouTube commenters and journalists alike labeled Clark "the next Sweet Brown," though YouTube comments tended to be more explicitly racist or condescending.

"This is what is in our white house right now," one viewer remarked; wrote another: "She was actually trying to sound articulate and intelligent. EPIC FAIL!!!"

The page-view- and advertising-driven digital economy means that these stories inevitably become commoditized, with YouTube, iTunes, and the Web sites hosting these media pocketing most of the cash, and the blogs and other publications who post them (even in the spirit of honest commentary) profiting from the flood of traffic. In some cases, money trickles down to the original subjects; the Gregory Brothers, who produced the popular Auto-Tune the News series of videos, split profits from "Bed Intruder Song," their iTunes track, with Antoine Dodson. Lawyers, managers, and agents appear with regularity, promising appearance and interview fees to their new clients. Sometimes a friend or family member anoints himself as a spokesman and promises to wring some profit from the exposure. These viral stars might make a little money for their troubles, but they seem to receive all the notoriety of celebrity with almost none of the benefits. Charles Ramsey, for example, lost his job because of all the people coming in wanting to shake his hand. So many people came by his house that he decided to move out and stay with friends, leading to reports that he was homeless. While he did reportedly make $18,000 from appearance fees and had plans to set up an online store, speaking with a reporter a few months after his initial burst of fame, he said, "What I've made is a whole bunch of friends, and what I've been doing is borrowing money from a whole bunch of friends."

At the same time, the trials of poverty and urban crime are turned into yet another spectacle, in line with a popular culture that has made sport of satirizing the folkways of the poor, minorities, rednecks, and assorted eccentrics on reality TV and in tabloid media. Commenting on the "Bed Intruder Song," an NYU music professor told NPR: "There's a way in which the aesthetics of black poverty—the way they talk and they speak and they look—sort of becomes this fodder for humor without any interest in the context of the conditions in which people actually live." But at least Honey Boo Boo and her family get paid, are shown in something approaching the fullness of their lives, and are willing participants in their own degradation. Dodson and company seem to have few options but to hold their noses and try to make the most of their moment. (Sweet Brown, like some other viral stars of yore, has appeared in a couple of local commercials.) They become priests without pulpits, politicians without constituencies. They have followings, sure—30,000+ on Twitter for verified user Kevin Antoine Dodson; 670,000+ likes on his Facebook page. But they don't have dignity, having become a burlesque in the public eye. Perhaps, like Jack Rebney, a pre-Internet viral video star (in the early nineties, outtakes of his swearing and harrumphing through an industrial film shoot for Winnebago became a cult commodity on VHS), they'll retreat to a mountaintop—Shasta, in Rebney's case. And then, when they show up in the public eye, we'll laugh and throw them a few bucks. Rebney, whose story was beautifully depicted in the documentary *Winnebago Man*, was greeted at San Francisco's Found Footage Festival by excited fans, one of whom called him "a modern-day freak show." By the end of the festival, Rebney had won over the crowd, having shown off his wit and, yes, the propensity to anger that earned him a crude kind of fame in the first place. But an aura of indignity remained.

The problem with viral fame is that there's no real way to be rehabilitated from it—unless you achieve the kind of mainstream, corporate-branded respectability of Kim Kardashian or Paris Hilton, or if you change your name and cultivate a different appearance. Not all viral fame is humiliating—the anesthesia-addled child depicted in the "David after Dentist" video took it well, and his family profited off of T-shirt sales. But like Internet-famous cats, young children don't have much choice over how their guardians choose to depict them online, nor whether their sudden popularity is then run through the entrepreneurial apparatus of merchandise, follow-up videos, interviews, paid mall appearances, and crappy songs bought by most purchasers in a moment of carelessness or stoned amusement.

Some end up like Ghyslain Raza, the so-called Star Wars kid who, as a high school student in Quebec in 2002, filmed himself playing with a golf ball retriever as if it were a light saber. A fellow student found the video, shared it with a friend, and eventually it appeared online, where it began racking up millions of downloads through P2P sharing platforms. Raza was bullied and had to change schools. "No matter how hard I tried to ignore people telling me to commit suicide, I couldn't help but feel worthless, like my life wasn't worth living," he told a journalist a decade later, after putting his life back together and graduating from law school.

Aleksey Vayner was less fortunate. In 2006, Vayner, a Yale graduate who had emigrated as a child with his mother from Uzbekistan, sent a video résumé as part of an application to the Swiss bank UBS. The video, which Vayner later explained he had made in tribute to his deceased martial arts master, depicted the aspiring banker doing all manner of feats: lifting huge barbells, breaking bricks with his hands, hitting a tennis ball at 140 mph, ballroom dancing with a beautiful woman. Echoing the Adidas marketing slogan, Vayner called the video "Impossible Is Nothing." The young man had a history of embellishing his life story—he told college friends that he had worked for the CIA and was certified to handle nuclear waste in Connecticut—which had earned him some derision from *Rumpus*, a Yale campus humor magazine. The tolerable teasing of college turned into something much worse when someone posted Vayner's video résumé on YouTube. The video soon swept from *IvyGate*, an Ivy League gossip site, on to hundreds of news sites, including others besides the traditional purveyors of the viral flavor of the moment. Vayner appeared in a *New Yorker* "Talk of the Town" story, and in the *New York Times* under the headline, "The Résumé Mocked 'Round the World."

You can imagine what followed: a thousand jokes, no job, and permanent Internet infamy associated with his legal name. Over the next few years, he reportedly bounced between different ventures, trying to start investment businesses and a charity or two. He got married and moved to Queens. According to his LinkedIn profile, where he was using the name Alex Stone, he did some postgraduate studies at the London School of Economics. (On Facebook, he began using the name Alex S. Vayner.) In 2010, he appeared at ROFLCon, a biennial convention for the Internet famous and those who love and mock them. In an interview with *Vice* magazine, he recounted his difficulties: "Getting so much negative feedback you start to question yourself: well, maybe there is some truth to what all these people are saying. Maybe you can't do this. Maybe you aren't that good. You start questioning your core beliefs and convictions." After the "Impossible Is Nothing" video appeared, he said, he received death threats and messages from people clamoring for him to be deported. People "don't understand how much pressure and stress each drop of negativity places on that person," he explained. "They're laughing at a few seconds of video. They don't know the person. They don't know the story."

On January 18, 2013, a friend posted on Alex S. Vayner's Facebook wall: "Do not, anyone, sell this idiot ANY pills!" the note read, before exhorting Vayner, in Russian, to pick up his phone. Vayner responded later that night, telling his friend to go to hell. The next day, Vayner apparently took an overdose of pills, suffered a heart attack, and was pronounced dead at a hospital. He was twenty-nine.

Vayner was a serial fabricator and exaggerator, although he was accomplished in many of the disciplines portrayed in his video. Ironically, many Internet communities, even those such as 4chan, a raucous, often crude message board whose defining value is that every user must remain anonymous, have become intolerant of any sense of fakery, identity play, or possible deceit and obfuscation. In a digital culture in which everyone is increasingly pressed—by both social forces and the fiat of Facebook, Google, and LinkedIn—to present themselves *as* themselves, the appearance of dishonesty has become a capital offense. And should your crimes be made public, the mechanisms of virality and the economics of Web publishing mean that there's little room for modulation. Once a video finds its way to a popular aggregator—*Gawker, Business Insider*, the *Huffington Post*, Reddit,

BuzzFeed, et al.—it's practically assured that it will catch fire, satiating readers' hunger for a fool of the day.

- - - - - - - - - - - - - - - - - -

SUFFERING FROM INTERNET CELEBRITY—
AND ITS AFTERLIFE

Lena Chen went through a similar gauntlet, but made it out the other side. Chen began writing about her sex life after arriving as a freshman at Harvard in 2005. Young and happy to be away from her conservative upbringing, she didn't think that anyone there would judge her too harshly for writing honestly about herself. But her writing was posted publicly, both in Harvard publications and on her own site, *Sex and the Ivy*, and it eventually spread beyond the circle of friends and peers who she thought were her main readers. "My intended audience was very different from my actual audience," she told me.

For a time, it worked well, as she earned some freelance work writing about sex, gender, and feminism, and spoke at some campus events. But within about a year, things started to go awry. After becoming a celebrity at Harvard and, thanks to *IvyGate*, throughout the Ivy League, she began to feel the pressure of being observed and judged wherever she went. Critics appeared all around her, from anonymous members of the men's rights community harassing her online, calling her every derogatory term one might think of, to fellow students and members of the national press. *Gawker* listed her in a guide to "compulsive over-sharers of the Internet"; a few months later, another *Gawker* writer told her to "get off the Internet." She appeared in a *New York Times Magazine* article, in which she was depicted as a "small Asian woman," sultry and self-confident in her miniskirt, the foil in a debate against a thoughtful, chaste member of a campus virginity society.

Most of the things that Chen wrote about weren't so unusual— light drug use, vivid descriptions of erotic adventures and sexual experimentation, dating Republicans, the daily ups and downs of college life. It's just that she did it at times with a candidness, playfulness, or naïve pride that led to people being angry and suspicious. It's the kind of behavior that excites the sensibilities of bullies and misogynists who think, *Here's a target.* Even fellow students were rarely sympathetic: "The whole system is under the impression that if something happens to you, you 'asked for it,'" she said.

As much as she tried to resist her critics, harassers, and those who, under the guise of disseminating news or gossip, examined her actions with excruciating scrutiny, she still broke down. The worst came after a vindictive ex-boyfriend uploaded nude photos that he had taken of her. He tried to use SEO tactics to boost the search profile of the photos and posted links to them in *Gawker*'s comments section. Thanks to these efforts, the photos spread widely and were dissected and mocked on message boards and gossip blogs. Stress began to overtake her, although she said that she didn't understand the extent of it at the time. "It was kind of a mystery to me why was I tired all the time, why did I get anxious in social situations," she said. For the first time in her life, she began having panic attacks. Although she could close her computer and escape at least some of the vitriol she was receiving online, she couldn't ignore the fact that she was confined to a campus where everyone knew her and, as she was too well aware, were often discussing her behind her back. Eventually, she had to take a break from school.

"There were plenty who reacted with unmitigated glee that the campus jezebel had received her proper comeuppance," Chen would later write.

After this, while still at Harvard, Chen began dating someone new. His identity was sussed out by some of her harassers and posted on Juicy Campus, a college gossip site. And when it was revealed that her boyfriend was a graduate student, a campaign emerged to have him fired from his position. Members of other online communities spread rumors that her boyfriend was beating and raping her. His name also started attracting an ominous, and wholly libelous, Google trail.

The reaction toward Chen mostly was about shaming and humiliating her—for taking sexual photos with her boyfriend, for disclosing how many men she had slept with, for being comfortable with her sexuality, for discussing issues of sexuality in public and being seen as cultivating undeserved attention (a judgment that attends many women in the spotlight). Despite her efforts to harden herself, it took a toll. "When slut-shaming works—and even when it doesn't work—you end up losing a considerable amount of trust in people," she said in 2010.

It's worth stepping back for a moment and looking at what the digital/cultural landscape was like in the mid–2000s. Beginning in the early aughts, a generation of students was matriculating just as some of the traditional barriers between the cocoon of college and the outside world were breaking down. A media theorist would call this a form of context collapse—thanks to new technologies, different worlds were suddenly mixing, reading about one another. Stories about college life filtered through blogs and social media to the wider world. These stories weren't new, of course, but firsthand accounts became easily disseminated to audiences well beyond campus. Some gossip blogs, hungry for content, began reporting on the goings-on of college students—hazing scandals, wild parties, local celebrities—without the traditional circumspection of mainstream media. Larger publications started piggybacking on this work or running sober-minded trend stories on hookup culture. Meanwhile, sex, dating, and advice columnists, no longer confined to university publications, suddenly had the potential to reach national audiences (or beyond). A relatively novel form of celebrity appeared—Internet fame, a kind of micro-fame. Someone micro-famous at Harvard might become a media story mostly because she had a salacious or unusual backstory and attended an elite institution.

At the same time, changing social mores, new options for publishing, and a culture that had produced both Dr. Ruth and Julia Allison, one of the first college sex columnists to achieve something like mainstream fame (culminating in her appearing on the cover of *Wired* magazine in July 2008), meant that it was becoming fashionable for young women to write frankly about their sex lives. From *Sex and the City*, to LiveJournal and other blogging services, confessionalism was in—a way to express oneself, connect with an audience, and perhaps to launch a career as a writer. And yet, the reach of digital publishing brought in some uninvited visitors. Whether it was aggressive bloggers or the natural tendency of a sufficiently large audience to include some awful people, female sex columnists began being harassed all over the country. Speaking with Lena Chen and Zoe Yang, who wrote a sex column around this time at Pomona College, I heard about a half dozen or so young female writers whose college writing, ranging from sex diaries to op-eds on feminism and progressive politics, earned them continued harassment and abuse. (These are just the ones they became friends with, bonded by their difficult experiences; there are, unfortunately, others who faced similar treatment.) There was a clear disjunction between the environments in which these young writers lived— which were supposed to be liberal and welcoming, places of personal growth and experimentation—and the treatment they received from audiences both on campus and online.

More than five years later, both Chen and Yang still experience harassment. It's slowed in recent years, perhaps because both women have largely stopped writing for the public, but it continues. The two women believe that they have the same stalker (or stalkers) who has tried to ruin their reputations and get them and their boyfriends fired from jobs, bothered their families, and posted defamatory, threatening, and personal information on a number of blogs and message boards. The result has been both a campaign of recurrent harassment—from what source, they've never found out, despite enlisting lawyers and private investigators for help—and a permanent data trail of unearned infamy.

Both women are listed on the sex bloggers page on *Encyclopedia Dramatica*, a kind of depraved *Wikipedia* filled with bottom-of-the-barrel misogyny and homophobia. Chen believes that the women listed, all of whom wrote about sex during the period of about 2005 to

2009, have been targeted by the same person or group of people. Anonymous abusers have posted hundreds of messages, most of them cruel and grotesque, on a site called AutoAdmit, which deems itself "the most prestigious law school admissions discussion board in the world." Blogspot, Google's blogging service, played host to several more, including a site titled *Lena Chen Enablers*. There were others, including lenachenskank.blogspot.com and zoeyangskank.blogspot.com. Some of this material remains online. Writing on these blogs, their stalker often used tags and linked to past news articles and posts, which included photos on still other Web sites as well as contact information for professional and educational organizations. The authors seemed to be trying to create a linked record of these women's writings and sexual activities, and to use it to insult and silence them and ruin their reputations. Yang, a foodie who for a time after college kept a restaurant blog, found herself posting on her blog to refute rumors that she had been fired from her job at McKinsey, the consulting company. (Someone did contact McKinsey trying to defame Yang and get her fired, but the company's human-resources department supported her.)

The harassment has extended to their friends, family, and even their readers. When still writing *Sex and the Ivy*, Chen began warning her readers not to comment on her site or reblog her posts unless their identities weren't traceable. Miriam Lazewatsky didn't know Lena Chen, but she posted some supportive comments about her on some articles. Soon someone—likely the same person who was harassing Chen, Yang, and others, including some of their readers—found Lazewatsky's Tumblr, where her regular subjects included issues of body image and self-acceptance. That, she believes, spurred the troll into action. She eventually discovered that she had been written about on a rage of third-party sites, including AutoAdmit, Blogspot, Reddit, Fornits, Tumblr, and Twitter. "Most of the postings on these sites talked about my supposed sexual proclivities (like stating that I want to be raped) and my physical appearance (that I'm fat and ugly)," Lazewatsky told me. "Several also contained accusations that I had been fired from previous jobs for looking at pornography at work and that I'd been arrested for sexual misconduct, both of which are totally untrue." Lazewatsky contacted some sites and was able to get them to take down posts, but others were uncooperative or never responded. She received threatening e-mails: "The one I remember most clearly contained nothing but my parents' home address and phone number with the comment 'interesting,'" she said. She filed a police report and took on a pro-bono lawyer, who was able to offer only limited help.

All three women—Chen, Yang, and Lazewatsky—say their experiences changed how they behave online, though age has something to do with that. Lazewatsky reflected on how she approaches digital life now: "I certainly find that I share less information in general, especially personal information about where I live, who I'm dating, and where I work. I'm less likely to see or use the Internet as a place to vent; I don't think it used to even occur to me that there might be real people out there who could get enjoyment out of using another person's pain or frustration against them."

She'd love to defy her abusers and keep a more public profile online, but it's not that simple. "As a feminist, I find it deeply frustrating that I sometimes have to censor myself online," she said. "Unfortunately, at this point, my own safety and well-being have to come before my indignation."

Chen attributes some of her treatment, at least from the media, to the novelty of her experiences. At the time, sex-tape scandals were fairly new; they hadn't graduated to becoming a kind of rite-of-passage on the road to reality-TV-type fame. College columnists had only recently become objects of prurient interest from gossip blogs and national publications. Even so, the continued posture of many big blogs toward viral stars doesn't seem to have changed much. The general attitude appears to be that any public writing or appearance in a video makes one a public figure and that all attendant opprobrium and infamy is somehow deserved. Given how quickly gossip blogs cycle through their targets, a sense of detachment reigns (though, as in the case of Chen and many others, they occasionally return to old subjects with an acidic nostalgia).

Describing her stalker, Chen said, "This person has made me realize much more clearly that there is no divide between online and offline." Yang offered the same sentiment. For Chen, this perspective has been difficult to come by. "It was very hard for me before because I thought of it as something I was going through completely by myself," she said, "and it's not until recently that you hear about revenge porn and other people's sex scandals and there are far more psychos out there than previously expected."

Chen also attributes some of the blowback she received to being an Asian woman. She noted that she never wrote the kind of fratty, scatological, boorish accounts that earned Tucker Max such a lucrative audience. And she thinks that readers and writers have matured somewhat, becoming more aware of cyberbullying, media stereotypes of women, and the perniciousness of slut-shaming. She's still grateful for all she learned. While she felt alienated at Harvard, some of her readers became her friends and confidants. Today she lives in Berlin with her boyfriend—the same graduate student from Harvard—and his bulldog. "I just wanted to be away from America mentally," she said. For a time, she tried writing less about herself and focusing on issues of gender and feminist politics, but found that world to be rather ugly as well.

"I feel very jaded in a certain way," she said. She's withdrawn almost entirely from social media, keeping her accounts open but not posting anything viewable by a wider public. "It's weird, but at the same time very freeing," she said. "I don't represent anything anymore, I am not of the zeitgeist." She doesn't have to be "Lena Chen, Internet-famous sex blogger," though on her personal Web site, she still identifies herself, a bit self-deprecatingly, as "a walking case study on bad publicity."

Chen decided against changing her name—too difficult, she said. Her boyfriend paid a lot of money to ReputationDefender to clean up his data trail, but the results weren't positive. While some of her stalker's actions may qualify as criminal, police aren't usually very interested unless there are threats of violence, and lawyers are costly.

Zoe Yang doesn't do personal writing anymore. "He wanted me to shut up and I did," she said, noting that that hasn't been enough to get the stalker to go away. She does do some professional writing. She works as a researcher for a Harvard professor and plans to attend business school. She's not sure how her data trail will affect her in the future—or if she'll even learn if or when it does. "It is something that I worry about from time to time," she admitted. The damage has extended to her friends, some of whom have been targeted apparently just for being associated with Yang. One of these friends is worried because he's a doctor, and doctors frequently have professional reviews appear online—an easy target for sabotage. "People Google their doctors," Yang said.

"All of us went through periods of anxiety and depression and fear and hopelessness and feeling like we couldn't trust people," Yang said, referring to herself, Chen, and another writer. Yang doesn't use Twitter anymore and has tried to lock down the privacy settings on other accounts. Realizing her "name is going to be out there," she has tried to strategically improve her search profile, hoping that people find her LinkedIn page, for instance, instead of AutoAdmit.

"I think one thing we learned is that you can't make mistakes on the Internet," Yang said. "We've both moved away from who we were then—so many times over—but on the Internet, we're forever going to be eighteen-, nineteen-, twenty-year-olds writing about sex."

- - - - - - - - - - - - - - - -

TRENDING IS FOREVER

This is the dark side of viral fame, and certainly many of the issues depicted here were present in previous eras, when unwanted or distorted media attention could help ruin lives. Social media didn't originate this process—the tabloid style is associated with raising anonymous citizens to sudden fame—but it made it more common and easily transmittable. It also created a permanent afterlife in the form of a data trail. But to the partisans of social media, a sudden burst of viral fame is a path to success. The contours of this path may be elusive, but that is all the more reason to spend more time on social media, honing your brand, developing your voice, cultivating your audience, and all of the other vaguely buzzy practices around which a flourishing consulting industry has developed. But as with semantic analysis, the people achieving success here are not those trying to go viral at all costs, but rather those selling the shovels. It's the writers of how-to books, the leaders of viral-marketing seminars, and the highly paid advisors, experts, and keepers of metrics who have profited handsomely from the hunger to go from unknown to famous through some social-media alchemy. And most of all, it's the platform owners.

Some of us have something tangible to sell—the rise of self-publishing has spawned an ancillary industry telling newly minted authors how to get their books noticed through social media—but many of us are just selling ourselves. And so the murky quest to go viral has filtered down to the proles, who ask celebrities, for no reason but for the distinction of the thing itself, "Can I get a retweet?" A related custom: pushing to get someone trending on Twitter. At its most pathetic, this request is made on behalf of someone sick or dying, as if Make-a-Wish had established an arm to achieve viral fame for the terminally ill. Ephemeral renown for an ephemeral life. Let's get him trending, his supporters say. Let's get his name out there. But why? For what? So that it can pop up in the list of trending topics? People should be allowed their particular (or peculiar) forms of solace, but there's something very sad about this one: how it depends so much on the goodwill of a disinterested higher power, how meaningless and contingent the desired result is.

Like the concept of going viral, a "trending" topic (a term native to Twitter that has since percolated throughout social media) is treated as a good unto itself, a sign of public interest in a subject, without much deeper exploration of the technological and commercial forces that determine what trends. As Jared Keller, the director of programming at Mic, a media company catering to millennials, told me: "Just because it's trending doesn't mean it really matters. But we think of social media as a really great mood ring for popular culture." But there are some unexamined assumptions with this kind of thinking. Twitter is "self-selecting," Keller said. "People opt in to social, so they opt in to those kinds of conversations. Is this really what matters, or is it just what matters to the Twitter audience?"

There's nothing inherently good, worthwhile, or organic about a supposed trend. One study found that "most trends in Sina Weibo [a Chinese micro-blogging service] are due to the continuous retweets of a small percentage of fraudulent accounts." As Evgeny Morozov writes, "The hidden initial manipulations of the PR industry are only made worse by the business incentives of platforms such as YouTube and Facebook, which have their own reasons to promote memes: they create some shared culture and, more important, lead to more page views, more user interaction (i.e., users reveal more about their interests to the company), and, eventually, more and better advertising. Memes, then, are what happens when one greedy industry meets another."

In other words, social networks play favorites, while PR companies, producers, journalists, and others have the power to influence what rises to the top. A video featured on YouTube's home page will accrue tens or hundreds of thousands of views simply by virtue of being placed there. Kickstarter, an ostensibly meritocratic crowdfunding platform, regularly features selected projects in its newsletter and on its home page. Such a distinction can bring thousands of dollars into a project's coffers, which is often the difference between achieving a fund-raising goal—and being seen as a success—and walking away with nothing. In the same way, when Twitter shows "related headlines" below tweets about news events, the site is making deliberate choices—choices made in the design of an algorithm, rather than by an editor working on the fly—about which news outlets to privilege above others. Sponsored posts jostle alongside regular editorial mate-rial, the difference between the two barely marked, if noticed.

Views, likes, mentions, hashtags, followers, all the various indi-cators of virality—these can be bought. Seek out the right forums or tucked-away online shops, and for about the cost of dinner at a fancy restaurant, you can create the illusion that your previously unknown video or Twitter account has a huge fanbase. A site called YTView has a full menu of purchasable popularity; for YouTube alone, it offers views, likes, subscriptions, or comments. You can buy Facebook likes or raw Web traffic—indeed, the digital media industry is rife with ru-mors and accusations of various publishers using click farms to boost views and, by extension, advertising revenue. If you want to raise your status among your peers, you can send some cash through PayPal for Instagram or Twitter followers. It's a large, albeit shady industry, be-lieved to generate millions of dollars in revenue. Relying on bots, pur-veyors of views and likes—call them sellers of attention-on-demand—argue that theirs are *real* followers, while the other guy's are phony. Or, they promise, they can provide better engagement or views only from the United States, which are less likely to attract unwanted scrutiny. (Fake views often come from eastern Europe and east Asia, part of the broad archipelago of the Internet's black market.) That is, assum-ing that YouTube, Twitter, and other platform owners care about the fakery. While these sites are known to purge bots, fake followers, and fake sentiment—a practice that serves as a public affirmation of their networks' authenticity—they probably also don't mind the traffic.

Wanting to find out how this process worked, I went in search of followers of my own. A simple Google search for "buy Twitter fol-lowers" led me to dozens of sites. Some looked like hastily thrown together operations; others had more sophisticated graphics, bright logos, and fancy Wordpress templates. Despite the differing aesthetic presentations, it was hard to sift out which was more reliable, given that the entire industry operates in the shadows and is built on a cer-tain amount of deception. Even those that claim to provide more fully fledged marketing services don't seem like more than fly-by-night operations. I decided to let my wallet guide me. A site called Social Burst offered 2,000 Twitter followers for $15—a few dollars below the going rate on some other sites. I entered in the name of my Twitter handle, @silvermanjacob, and sent $15 to someone named Matthew Nuttall. Because I never had to enter my password or authorize the site to access my Twitter account, and because I paid the fee through PayPal, it all felt relatively secure. It was also astonishingly easy. Soon a confirmation e-mail arrived, promising that my followers would appear within one to two working days.

A few hours later I had forgotten that I had even made the transaction, until I opened my Twitter page and realized that my two thousand followers had already appeared (almost 2,200, in fact). Then I embarked on that peculiar, Internet-era game: Are they *real*? What does real even mean for these purposes? I scrolled through the list, clicking on some individuals, and gathering a sense of the rest. Most had full names, avatar photos, and had tweeted hundreds or thousands of times—signs of verisimilitude but not necessarily of life. A number of them appeared to be from overseas—Latin America, the Middle East—and followed far more people than followed them, though that may have been a sign only of their relative lack of popularity. What seemed like a disproportionate amount had locked their accounts, and I did find the occasional account that was tweeting spam or outright gibberish. Other messages were simply incomprehensible to me, shared as they were in languages I didn't understand.

Given the speed with which they appeared and this sketchy collection of characteristics, I'd have to say that most of the accounts were fake. But did it really matter? Is someone who tweets once a month in Tagalog more real than someone else who tweets in English about cricket ten times a day? On the other hand, maybe there is a meaningful distinction between a bot that tries to emulate its followers and one that spews out deals for porn sites. Still, there's a range of behaviors we expect from not only our followers, but also those we consider real and those we want to interact with; someone might be quite real but totally uninteresting to me, and I to them. Consider also that Twitter, in March 2013, reported that it had 200 million active users out of 500 million registered accounts—meaning that 60 percent of registered Twitter users no longer logged in (a sedentary posture that could also be confused for bot-like behavior). A year later, Twopcharts, an analytics firm, reported that 44 percent of 974 million registered Twitter accounts had never posted anything. What's more, past studies have found that many of the most popular Twitter accounts of politicians and celebrities are filled with followers that are actually fake or bots pushing spam. During the 2012 GOP presidential primary, Newt Gingrich trumpeted his 1.2 million or so Twitter followers as a sign of his popularity with the electorate, until several media outlets reported that they were fake and that his campaign used its war chest to pay companies to push followers his way. In short, the range of what's real/fake/active/inactive/a bot/not a bot becomes more complex the more you dive into the data and into varying definitions of what should be expected from a Twitter user.

For my purposes, questions of authenticity didn't really matter. I could imagine some other journalist scolding me for misrepresenting the size of my audience, but that also would strike me as picayune. I had no boss to answer to, and no one really cared about how many people follow me on Twitter. But I admit that I cared a little bit; no matter how much I tried to avoid the implicit measurement contest of comparing follower counts, it was hard not to think that this information did matter, even if for the wrong reasons. These are kinds of metrics that get dropped in hiring memos or that writers, with some shallow jealousy, marvel over when talking about competitors and colleagues. My inflated follower count gave me a small lift while also confirming my suspicion that these theatrics over metrics of popularity were both ridiculous and easily manipulated. Now that I had so easily, if perhaps irrelevantly, gamed the system, I was able to share in the sense that it was all a farce. But at the same time, I had this nice beefy follower count, cresting over 4,600. It felt like what I imagined Botox might feel like: an artificial-but-hard-to-detect boost to both my ego and the face that I showed the world. That is, until the face started to droop. Over the next several days, I lost about 400 followers, though it slowed in the following weeks. Some services claim that they allow you to "top up" followers in instances precisely like these. I decided to check on Social Burst and see what its policy was. Social-burst.com now redirected to instapromotion.co.uk, which in turn claimed a copyright in the name of Get Socials 2014. It was becoming clear what kind of world this was: one in which Web sites and followers alike flitted in and out of existence, changing their names and profiles as the opportunity presented itself. For a customer like me, it still didn't matter much. I had pretty much gotten what I paid for, and besides confirming for myself how easy it was to gin up attention on demand, I had the mild satisfaction of knowing that I was fooling those people who might care about things as arbitrary as follower counts.

For musicians, actors, and other artists, as well as their managers, the goal of botting, as this practice is known, is to quickly acquire enough attention to create the impression of legitimate interest. And it's important that that interest seem organic and rapid, as if the media object is already incipiently viral. It then only needs to catch the attention of an "amplifier"—a YouTube moderator, a popular Twitter account, a celebrity, some bloggers or journalists trawling for content—after which it can take off and be truly viral. The despicable "Innocence of Muslims" video, which eventually caused riots in several Muslim-majority countries, lurked unnoticed on YouTube for months until Arabic subtitles were added, and it then found amplifiers in an Egyptian newspaper and the Islamophobic pastor Terry Jones. The "Harlem Shake" only became a world-spanning meme after being promoted by the popular DJ Diplo and the site *CollegeHumor*.

It's a form of "fake it until you make it," although the fakery of buying attention is, in some regards, a more honest form of PR. Instead of flattering journalists or sending them swag, or cozying up to the owner of a popular YouTube channel, or exchanging favors with a celebrity with a big Twitter following, the attention is bought directly. No matter that the audience is likely composed of bots: the assumption is that the real views will come later. And if not, the agent or PR rep can still tell his or her boss that they got *x* number of views in such a short period of time; no need to say that these weren't real people, if it's all the same to the higher-ups. In the meantime, if the party buying YouTube views is a YouTube partner, they might even make back some of their investment based on the ad revenue that the fake views generate.

Botting and related practices are just the seamier side of how attention is traded, bought, sold, or pilfered on social media. You could pay a celebrity such as Paris Hilton $4,600—the reported going rate for that archetypal viral star—to tweet about your video or product. There are Internet-famous social-media users who charge far less. You can try to piggyback on an already popular hashtag, as some corporate marketing departments are wont to do, particularly with hashtags celebrating national holidays (Memorial Day is a great opportunity for a fast-food chain to tweet that it supports veterans). A range of Web sites selling piecemeal labor, such as Fiverr or Mechanical Turk, can also bring you "legitimate" traffic.*

Even Twitter engages in a version of this. Its algorithms can pick up which topics people are talking about and determine if they should be declared "trending." But there's a lot they can't do. Like semantic-analysis programs, they have trouble with tone or unorthodox meanings. And Twitter's algorithms can't always correlate breaking news events with the queries that people are searching for on the service. So the company built what they call a "human-computation engine"—a

---

* Or at least they promise it. As I write this, one of Fiverr's top users is a UK man calling himself Giblerto Samba, who for $5 will promote your music to his 100,000-odd Facebook fans. He also will include various hashtags to help get the post trending on Twitter and Facebook. His disclaimer cautions: "We do not guarantee activity on FB/YouTube/Twitter as we don't force anyone to react."

fancy term for a distributed network of low-paid contract workers, some of whom are guaranteed to be available at any time of day. When new search terms are spiking, the system sends questions about the terms to a team of "human evaluators" recruited from Mechanical Turk, an Amazon-owned company that allows people around the world to perform menial tasks, such as transcription or tagging images, from their computers for a tiny fee. The Turk workers then determine that, say, #bindersfullofwomen (the example provided by one Twitter engineer) is a reference to Mitt Romney's odd elocution in an ongoing primary debate being watched, and tweeted about, by millions of people. The workers classify the term appropriately and send it back to Twitter, which allows the company to determine whether the topic qualifies as trending.

Twitter offered some reasons why they work with Mechanical Turk: "In-house judges are unfortunately hard to scale as they require standardized hiring processes to be in place. They also tend to be relatively more expensive, it can be harder to communicate with them, and their schedules can be difficult to work with." In short, Mechanical Turk workers are cheap and always on call. They don't demand much, such as regular work hours or minimum wage. Cheap scalability is paramount, even for a corporation worth more than $20 billion. The company adds: "Our custom pool of judges work virtually all day. For many of them, this is a full-time job." But it's full-time work for Twitter Inc. without any benefits, institutional support, room for advancement, or pay that should come with a full-time job. The tasks available for Mechanical Turk do offer some people important flexibility, especially for stay-at-home parents, as well as a source of income for people who need a little more cash or are unable to find permanent work. But for a company as large as Twitter, hiring an in-house, full-time "human computation" staff shouldn't be prohibitively expensive. Twitter would presumably gain from bringing the staff closer to home; it'd be easier to respond quickly to breaking news with on-site employees and to learn from their experiences. These employees would in turn have opportunities to learn new skills and gain responsibilities, and presumably, some would be happy to work night shifts, as the company clearly requires. It would also help affirm Twitter's place as a media company, as many tech commentators have long considered it. Any legitimate media organization has staff working a night shift to respond to events around the world. Twitter has such a staff, but they've chosen the easier, cheaper option—one that ensures that one of Twitter's most distinctive features will always have a whiff of exploitation underlying it. At least until Twitter decides that its human computation engine can be replaced with automated algorithms.

- - - - - - - - - - - - - - - - -

VIRALITY IS FOR ADVERTISERS

One of the earliest Facebook groups to achieve some measure of viral fame was started by a college student named Brody Ruckus. Appearing on Facebook on September 5, 2006, the group had a long, attention-grabbing name: "If this group reaches 100,000 people my girlfriend will have a threesome." Supposedly started by a young man whose girlfriend, Holly, had promised him a sexual bounty if only he could achieve some dubious (and perhaps, she thought, impossible) Internet milestone, the group quickly drew interest. Within three days, its membership swelled to more than 100,000 people. It was then that Ruckus—who, in his initial post, said that a threesome was "perhaps one of the greatest things imaginable (right behind midget tossing)"—said that he would institute a mechanism to allow the group members to vote on who would complete the threesome. Ruckus threw in some sweeteners (today, on Kickstarter, this would be known as adding a "stretch goal"). If the group reached 300,000 members, his girlfriend would let him take and share pictures of the event. And if the group became the largest on Facebook—surpassing the 850,000-member group that at the time was the site's largest—then, there'd be a video in it for all of Ruckus's new fans. Ruckus also offered links to a site he had made selling T-shirts and other merchandise promoting himself and his sexual odyssey.

A problem soon emerged: Brody Ruckus didn't exist, nor did Holly. The whole campaign was a fiction, concocted by the marketing department of Ruckus Network, a digital music service. Through this spectacle, Ruckus Network had been able to obtain e-mail addresses for thousands of college students ("Brody Ruckus" had encouraged people, especially women, to e-mail him) and then to push them to sign up for their music service. Ruckus was sketchy in other ways, too. Soon after the group appeared, he began contacting students through Facebook, telling them that they were already eligible for free downloads from the service; they just had to activate their accounts.

In its use of an opt-out ploy (you're already a member!) and scammy viral marketing techniques, Ruckus Network, which eventually went under without making a splash in the digital music scene, was a forerunner of the kinds of techniques that permeate social media today. For years, Facebook has dealt with groups that pose as a legitimate interest group but are often just a way to bombard users with advertisements. This problem was made worse by Facebook allowing friends to add you to any "open" group to which they belonged—another use of opt-out to grow a user base at the expense of users' privacy rights. Spammers and marketers have adopted clever tactics, such as starting a group about a social or political issue and then, after its membership grows, changing the group's name to something else. Suddenly, that Free Tibet group you joined a few months back, or that other one you signed up for in order to be eligible for a contest that looked pretty above board, is posting advertisements in your News Feed for discounted iPads, and your friends are seeing sponsored stories saying that you're a fan of CheapIpads.biz. The original group owner either changed the group's name, intending to run this con all along, or sold the group to someone else—a social-media twist on the long-held practice of spammers trading massive e-mail lists among themselves.

These stories show social-media marketers at their worst. They build on a tradition of deception that lies at the heart of viral media and viral marketing, in which sponsored stories, videos, or even a sticker campaign are used to try to present a piece of advertising as something other than what it is. The goal isn't authenticity exactly. They don't necessarily want to convince you that that magazine article looks like any other—until you notice the different font or the tiny "sponsored story" label at the top of the page—or that the young woman in the YouTube video, talking about her lonely life, is actually down on her luck. Rather, they simply want you to believe it long enough, or to be amused at your own gullibility, to click. Just enough to register another page view, add another e-mail address to the mailing list, or create some brand awareness. Google, not immune to this mind-set, has taken this practice to its inevitable conclusion by planting fake e-mails in users' in-boxes. Google promises relevance and convenience in these ads, but they're mostly about tricking users, who, if they don't already feel like their privacy is being violated, might take a longer look at an ad that resembles an e-mail in their in-box than one that scrolls across the top of the screen.

Viral marketing has also benefited from being folded into ordinary consumer culture. Every day is the Super Bowl, with advertising parsed, shared, and tolerated as if it were, rather than an attempt at suasion, some essential part of how we live and see the world. And so it is, for example, that the Old Spice guy, a brawny avatar of masculinity who charms women with chatty monologues and a pleasant bouquet, becomes a pop culture meme. We no longer fast-forward through his commercials; we pass them around and ponder what they mean in chin-scratching blog posts. Then the question shifts to whether the company can do it again—i.e., achieve viral liftoff with a new video—or if the actor, Isaiah Mustafa, can graduate to TV or film work. In the process, the fact that this is advertising, an attempt at consumer manipulation rooted in fairly heteronormative assumptions about beauty and masculinity, becomes lost. For Old Spice, the campaign's a great success. Like a politician who posts an aggressive advertisement on YouTube, knowing that it will be covered as a controversy by the mainstream media, Old Spice has successfully made their advertising into something even more valuable: a commodity that we circulate voluntarily. It's the sort of treatment that they'd normally pay for, but in this case, journalists and consumers are doing it for free. These two groups may be asking different questions—"What does this say about our culture?" versus "Isn't this funny?"—but either way they are helping Old Spice and sending a message to every Fortune 500 firm that this is the way to reach consumers. (Plenty of other advertisers have followed suit, especially by investing in branded entertainment or, in the case of Chipotle, creating a TV show designed to highlight the company's moral business practices.) As with much else in an attention-based economy, recognition is key. Even criticizing a piece of advertising ensures its success for its stakeholders; the best way to handle offensive advertising is to ignore it entirely.

Hashtags, that small but vital connective tissue of social media, have become fully co-opted by advertisers. During the 2013 Super Bowl, 50 percent of ads included a hashtag. The symbol, adopted by more and more platforms, is useful for organizing conversations. It's also become a metric by which a marketing campaign's success is gauged. Companies often try to take advantage of hashtags that have nothing to do with them. On Twitter, at least, they have a built-in advantage, as their verified status and high follower counts would likely lead to them being listed among "top tweets" for the hashtag. During the series finale of *Breaking Bad*, the #BreakingBad hashtag was filled with messages from J.C. Penney, Clorox, a T-shirt company, Miller Lite, and several major political organizations (archconservative commentator and perennial provocateur Ann Coulter took to the hashtag to explain the need for gun ownership). This behavior may seem like someone awkwardly inserting himself into a stranger's conversation, but it's quickly become standard practice in an industry where companies try to place themselves in users' timelines and content streams in any way possible, including through paid native advertising that looks much like something produced by the news organization hosting it. On its advice page for brands, Tumblr recommends that corporate users reblog posts from other users—the message being, "We're just like you. Let's chat." These interactions create some peculiar dissonance. Do I respond to the cell phone manufacturer that has just replied to my tweet or reblogged my post? Why does it feel like I'm now chummy with the anonymous person (or persons) behind a shoe company's social-media feed? Or as App.net founder Dalton Caldwell put it, speaking about a bottle of water, "How are we supposed to talk about it now that all of these physical objects want to be your friend?" Corporations are turning everyday products into strange media personalities, asking that we recognize them. The vexing question is how to respond.

For marketing departments, these concerns don't mean much. Engagement is what matters. Media impressions. Conversation between customers and companies on social media, even if they're to lodge complaints, help drive up metrics. How many people visited our micro-site for the new tablet computer? How many people retweeted our funny hashtag? These are the sorts of numbers on which budgets are based and promotions handed out. It's the logic of a statistician.

If only the statistics themselves were more reliable. We've already looked at how followers and page views can be bought. An even bigger problem is that they're difficult to measure in the first place. Rarely are the numbers you hear cited about a Web site's audience actually true. Big Web analytics firms such as Quantcast, comScore, and Alexa, despite the useful services they can provide, often come to wildly different conclusions about the size and composition of a Web site's audience. Mobile versus desktop, unique visitors versus page views. The variety of new Internet-connected devices, from watches to video game systems to cars to tablets galore, make it so that "mobile" or "home" aren't always discreet categories. Savvy Web surfers also use anonymizing software or block cookies. And even if their visits are properly detected, they might employ ad-blocking software, obviating the main reason to measure page-view statistics in the first place. The high degree of variability undercuts the notion that digital media has led to the great revolution in audience measurement that many industry evangelists claim. For example, in August 2013, Quantcast announced that it had rejiggered its system to better track users on mobile devices. The result: some publishers discovered that their numbers dramatically shifted. The gossip and pop culture blog *Gawker* found that, during the previous June, 11.1 million unique users in the United States had visited their site. But the original statistic was pegged at 6.8 million unique visitors—63 percent less. So which was it? And what else might Quantcast be missing?

Around the same time, a spate of news stories appeared announcing that Yahoo had passed Google as the most visited Web site in the United States. For Yahoo, which hadn't been number one since 2011 and was attempting a turnaround under its new mediagenic CEO, Marissa Mayer, it was a big deal. But a little digging proved that Yahoo had leapt ahead of Google only if you considered unique visitors according to comScore. (ComScore also didn't include unique visitors to Tumblr, the mega-popular blogging platform/social network that Yahoo had recently purchased, which would've pushed Yahoo further ahead.) Other analytics companies had Google with more unique visitors or Yahoo with more unique visitors but Google with more total visits. Plus, all of these numbers were estimates of traffic from desktop computers in the United States, leaving the rest of the world, along with a lot of mobile traffic, uncounted. In short, page views—that essential metric for apportioning ad revenue, Internet fame, trending topics, lists of most e-mailed articles, and on and on—were exceedingly tough to pin down.

It's not just Web-site publishers who are in the dark. Even Twitter and Topsy, a social-media analytics firm with which the social network has partnered, have been caught revising statistics in rather odd ways. After a London Fashion Week campaign, Twitter, using metrics from Topsy, said that Burberry had received 245,762 "mentions"—one of the social-web metrics du jour. Later, that number changed to 10,000, as the Twitter company blog post about the campaign underwent several revisions. These mentions also don't get at some other important questions: How many people saw these tweets? How many people follow people who mentioned Burberry? How many of those people were logged on, and how many are fake or inactive accounts? How many people actually clicked on something or decided to make a purchase?

These statistics turn out to be profoundly important for how advertising dollars are apportioned out (as well as, in some cases, bonuses for writers and others in the industry). They also influence what you read online and how you read it—a tremendous amount of journalism on the social web is built on a kind of feedback loop, inspired by what is already trending and what might soon trend. And although some journalistic outlets have shifted toward single-page formats, for years the industry standard has been to chop up stories into as many pages as possible, which helps to boost the number of page views and ad impressions. Slideshows, aggregation, listicles, and short, provocatively headlined articles are other click-heavy features which, while they have roots in print media, have come to dominate online publishing. Online publishers have invested heavily in video not necessarily because readers want it but because the ad rates are far better than those for text articles.

At the same time, 50 percent of all video ads go unseen, some because they auto-play in tiny video players or far down a Web page, out of a reader's view. Other ads run simultaneously or appear on top of one another. Because they work through large ad networks, advertising agencies, or several layers of middlemen, many companies don't even know on which sites their ads appear or whether people see them at all. The same goes for text and display ads, which are also rife with fraud. Some industry studies have found that as much as a quarter of all online advertising money goes to botnet owners and other fraudsters manipulating the market by creating fake traffic or showing ads to bots. Because many ads are automatically sourced through ad exchanges or, in the case of big brands, spread across thousands of Web sites, it's often hard to track where ads appear and which ad impressions are fraudulent. Companies placing ads on behalf of brands have little interest in cracking down on bogus traffic, as it would eat into their bottom line. Some publishers are complicit, looking the other way or buying fake traffic on their own, knowing that verification remains difficult and editorial budgets depend on ad impressions. Attention remains the only commodity that counts and sometimes the simulacrum serves just as well.

Page views, like previous attempts to quantify attention, are problematic and easily gamed, and they don't really reflect article quality or reader interest. How many times have you clicked on a link only to be disappointed and close the tab immediately? Those fleeting visits count. An article that's promoted heavily is guaranteed an audience of at least some size. If the *New York Times* sends out a push alert to NYT app users about a certain article, that article will definitely get major traffic. But it doesn't mean that it will get more traffic, or is more important, or is more appreciated by readers, than articles that earn their traffic through other means. In truth, no article or piece of content earns traffic organically. Placement on the page, social-media promotion, inclusion in newsletters or push alerts, updating articles, refashioning headlines to be more eye-catching, enlisting PR reps—all of these and other methods can be used to drive more traffic, particularly for an article that seems to be lagging. News organizations also don't engage in publish-and-wait. Reliant on this feedback-driven system, they constantly examine what's being read and what's not and try to push important, or expensively produced, articles to the forefront. As the dominant metric, then, page views are both the goal and the guiding force for these kinds of behaviors. In the words of the data journalist Brian Abelson, who's worked at the *Times*, "when [page views are] widely adopted—when an industry seeks to optimize its activities for a given metric— it ceases to be a mere reflection of reality. Instead, the measure comes to actively shape the industry, oftentimes leading to unforeseen manipulations and externalities."

With digital publishing dependent on advertising, the search for accurate metrics remains of prime importance. Yet because the industry knows these metrics are unreliable and is ever-hungry for more data, there is an added interest to find more ways to track users, such as by utilizing the MAC (media access control) address, a unique identifier in networked devices, or Apple's proprietary Identifier for Advertisers. This growing surveillance is complemented by a crop of analytics firms and other consultancies that have appeared to vet Web-site traffic, pick out bots, and make sure companies don't get swindled or find that their ads for children's toys appeared on porn sites. At the same time, publishers are now interested in monitoring how long users stay on a page, where they click, and even the position of their eyes.

This hunger for metrics—for the brute certainty of quantification—helps to turn our devices around so that they watch us. E-book publishers, for example, know how far into books customers read, where they usually stop, which passages they highlight, which terms they search for, which phrases they share, and how long they read in a single sitting. (This information is now being offered to some authors to encourage them to change how they write—for instance, making chapters shorter to ensure that readers don't get tired and give up on the book.) Volkswagen and other companies have experimented with video ads that stop playing when you look away from the screen, forcing you to watch the ad. The technology is similar to that used in some smartphones, which allow you to unlock the phone with a glance or to scroll through an article with your eyes. In other words, the same technologies that provide us with a measure of convenience can easily be manipulated to exert control over users. And by making their businesses reliant on advertising dollars, digital publishers are underwriting these systems of control.

Virality, then, isn't just about apportioning fame or anointing that day's popular meme. It is intrinsically connected to the larger system of quantification, surveillance, and feedback loops that underlies social media. This same system is now roiling the journalistic world, transforming our relationship with media from passive readership to something far more complex and, perhaps, harmful to the culture.

Churnalism and the Problem of Social News

With such a tendency toward deception and chasing illusory metrics, you'd think that digital advertising and viral media would serve as a cautionary tale to news publishers. In fact, journalists, editors, and media executives have taken many of the worst aspects of virality and applied them to their craft. High among them is the premium journalists place on anointing traffic leaders and trending articles as those that win the day. If the articles are shameless clickbait, false, or not good at all, it doesn't much matter, because they are justified as subsidizing a publication's more in-depth work, which is seen as worthier and more prestigious, even if it gets less attention.

News organizations have put a lot of effort into quantifying their online audience (the old, back-of-the-envelope calculation had it that each magazine would probably be read by 2.5 people), only to encounter the subsequent difficulty of figuring out just how many people are reading an article and how to best sell ads against them. For years, online journalists worried about total number of visits, or hits, eventually discovering that that was foolish ("How Idiots Track Success" become one acronym for "hits"). Then unique visitors emerged as the proper metric, only to be supplanted by page views, the total number of pages across a site. Now publishers worry about engagement or engaged minutes, while a number of publishers and analytics firms have begun touting their own proprietary metrics. In the case of Upworthy, the site tracks how much time a user spends on a page, the movement of his or her cursor, and how far down a page he or she scrolls.

More toxic than this focus on quantification has been the way in which digital media has adopted the general tenor of the viral Web— its speed and wayward attention, its unrelenting profligacy, its treatment of every piece of content as another bit of ephemera to consume, without context or much explanation, before moving on to the next one. With rock-bottom advertising rates, journalists must produce immense amounts of content and flog it relentlessly on social media, where hyperbole is standard practice. Quick takes on the day's news are praised as "must-read." The word "breaking" is thrown around indiscriminately, usually in all caps, as if each micro-scoop is revelatory and must be read immediately. "Exclusive" is another widely used bit of inflationary rhetoric; what it usually means is that the reporter rewrote the press release before any other outlet or that a PR rep turned to him first, expecting favorable coverage.

It might be foolish to pine for some bygone golden age of journalism—the industry has always had its challenges and its discontents. But one can say that the advent of digital media has done little to improve journalists' lot. Fewer jobs, less pay and stability for those who remain (unless you're an executive), no separation between work and home life (a condition cemented by the expectation to be always on social media), declining rates for freelancers, and a general societal disregard for the value of journalism—these are the conditions of journalism in the social-media era. For every newly minted digital journalist establishing his personal brand with ten or twenty thousand Twitter followers—with time that could be spent reporting or writing for his employer rather than Twitter Inc.—many others face unemployment or a version of the old joke: "We lose money on every article, but we make it up in volume."

Nor are readers necessarily better off. Sure, we now have the ability to seek out independent reporters and the voices of the once-marginalized. We can read practically any newspaper in the world online, and that is a wonderful thing. While we can read far more broadly than ever before, though, we don't engage as deeply: 55 percent of Web users spend less than fifteen seconds on a page. With so much choice, publishers find themselves catering to the lowest common denominator or cultivating audiences that might never have to hear a political opinion with which they don't agree. Meanwhile, reporting on municipalities, city halls, and statehouses has decreased significantly, and many newspapers have closed international bureaus. Establishment journalists continue to maintain cozy relationships with those in power, knowing that access has its price and that the State Department or a communications consultancy might be their best options for future employment. In place of this difficult, expensive, necessary work, we've received an overabundance of cheap material—rote political commentary, pop culture detritus, insta-reaction pieces to the meme or faux-scandal of the moment. There's been a boom in aggregators, curators, linkers, and tastemakers, all glad to help you sort through the mess and tell you what to read (and to profit off of others' reporting).

The result is churnalism—cheap, disposable content repurposed from press releases, news reports, viral media, social networks, and elsewhere, all of it practically out-of-date and irrelevant as soon as someone clicks Publish. With no time to confirm stories, with being able to declare "first!" considered the path to page-view success and professional prestige (even though it's mostly colleagues and high-frequency trading algorithms who care about these things), errors appear regularly but carry few consequences. The definition of news itself is now fungible, as exemplified by the way the term "exclusive" is bandied about. We are far past Daniel J. Boorstin's concept of the pseudo-event—press releases, awards shows, press conferences, and other trumped-up events whose sole purpose for being is so that they can be reported on. (Consider the media fervor that erupts around the unveiling of a new product or feature by a big tech company.) Save the much-fetishized long-form genre—two-thousand-plus-word, thoroughly reported stories that have ascended to hallowed status in an industry in which only the most privileged are allowed to do such work—much of journalism is now pseudo. We have other names for it: clickbait, linkbait, trolling. These headlines might promise a revelation or an exclusive but reveal little at all or are written in the form of provocative questions (to which the answer is almost always "no"). The story may not even be true, but if it can gin up a scandal—if the credibility of the story itself becomes the story, or the poorly sourced story is deemed to have raised questions—then that's enough. Attention is the most sought-after commodity, and the motto of its purveyors might as well be, "Ask for Forgiveness, Not Permission" (which is not unlike Facebook's "Move Fast and Break Things").

Here's how the churnalistic cycle usually goes. On July 29, 2013, the *Daily Beast* tweeted what it claimed was a "scoop": Cory Booker, Newark's mayor at the time, would be visiting Iowa in the next month, presumably to lay the groundwork for a 2016 presidential campaign. Their source was a calendar on the University of Iowa's Web site, though the university "did not return a call asking for comment," a euphemism which, in such cases, often means that the call was placed shortly before the story was set to run. They also didn't contact Booker's office. Within hours the story was shot down. There was no planned Iowa visit, Booker's spokesperson said, and the event disappeared from the university calendar. No matter: the story, which after all was speculative, thinly sourced, and concerned a presidential election more than three years away, made a bit of a splash on social media. The *Daily Beast* issued a correction, allowing it to tweet out the link again. Then the site's reporter followed up with another story, this one with better reporting, including comments from spokespeople for the university and Booker. The new article ended by recapping well-worn criticism surrounding Booker's campaign—the implication being that this nonstory was but another example of a dysfunctional political operation, rather than a dysfunctional journalistic one. Two articles, a brief flurry of social-media attention, one major correction—and a solid helping of page views.

This kind of thing happens all the time; the pace and volume required by today's publishing environment ensure it. Still, the pattern is not necessarily native to social-media journalism; cable news has been dining out on nonstories, self-created narratives, fake controversy, and other such inventions for years. The point is to create the *appearance* of a story and then to continue attracting attention, to hook the audience along, by a process of perpetual teasing, revising, and updating. Errors are recast as a fast-changing story going through another iteration. Along the way, we readers lose track of why the story mattered in the first place, but that's immaterial, because we're already consumed by the narrative and are invested in the outcome. This type of reporting has a long lineage—more than a half century ago, media critics charged that presidential press conferences were but a way for the government to invent news stories that it could then control—but it's become even more pervasive in recent years, a principal feature of both newspaper journalism and more fast-and-loose digital media. And it couldn't have happened without the demands of speed and page views, the lucrative promise of virality, and the decline of editing, fact-checking, and copyediting.

Here's another good example. On August 4, 2013, the *Washington Post* published a column by Danny Hayes, a political scientist who, the WaPo's *Wonkblog* promised, would "offer an empirical perspective on the issues dominating Washington." With the dramatic headline

"Obama Is Wrong. Traditional Journalism Isn't Dead," Hayes issued a 1,200-word, statistically rich riposte to the president, who, Hayes said, had claimed in an Amazon Kindle interview that "those old times" aren't coming back for journalism. Hayes's readers might have nodded along as he built his rebuttal to President Obama; here was a solid, analytical take on the news business—the kind of data-driven political analysis for which *Wonkblog* had become known. There was just one problem: President Obama had never said that about the news business. Unlike the *Daily Beast*'s snafu, in which a correction notice was placed in the original headline, this one was confined to the very end of the article, where an anonymous editor noted, "This post originally stated that Obama said 'traditional journalism is dead' in his interview with Amazon. That was incorrect." The entire premise of the article was wrong. But either too embarrassed to admit it up front or afraid of robbing itself of needed user engagement, the *Washington Post* tucked away this correction at the end of the article. (It probably was no coincidence that Hayes's confusion had come from reading a Kindle Single—long pieces, generally 10,000 words or more, that take time to read and that are frequently strip-mined by third-party aggregators, who take the juiciest bits and present them as top 10 lists. Who has time to read all that?)

The rush to be first can produce some pretty obvious blunders. When *TechCrunch* published a story with an enticing headline— "Dispatch from the Future: Uber to Purchase 2,500 Driverless Cars from Google"—other publications went nuts. No matter that the story carried a dateline from the year 2023, or that *TechCrunch* had published cheeky, fictional stories before, or that no other outlet had reported the news. It was picked up everywhere, from small design blogs to the *Daily Mail*, the world's most popular online newspaper, which later deleted its story. *Wonkblog* even pulled its post making fun of the *Daily Mail* for falling for the hoax after Ezra Klein, then *Wonkblog*'s editor, sent a tweet touting the original *TechCrunch* story. As *Slate*'s Will Oremus argued, the incident showed that many aggregators don't really read the stories they rework (follow-up reporting is usually out of the question). "At a time in the media business when unique visitors and Facebook likes drive editorial agendas," Oremus said, "some outlets may find it more profitable to write up every wild claim that comes along than to go to the trouble of figuring out which are legitimate and which are bunk." It also added another bit of evidence to the growing consensus that most quickly spreading, viral stories are false—if not at inception, then through the game of high-speed telephone that occurs when aggregators, bloggers, social-media editors, and others pick up, repackage, and share them. The inaccuracy may be wholesale or a matter of degree, but it does seem as if the high-velocity, churnalistic atmosphere of today's digital media is more conducive to propagating this kind of story than vetting or questioning it.

And why not, when there aren't really any consequences for being wrong? In most of these cases, corrections are played down or not done at all. Stories may quietly disappear. Sometimes, as with the *Daily Beast*'s Booker nonstory, they become an opportunity to write another story in response to the initial one, thereby wringing multiple articles, and even more page views, out of the invented controversy. Around the same time that the driverless car story ran, *Forbes* published a story claiming that the Houston Astros, then mired in last place, were Major League Baseball's most profitable baseball team. But the story's main premise was wrong and it was filled with numerous other errors, causing another *Forbes* contributor to publish his own article refuting it. The original story's author then published *another* story, this one with comments from Astros officials, with the ultimate takeaway being that team finances are complicated and "it all depends on how you count it."

As with Booker, or a similar nonstory in which the tech reporters Jessica Lessin and Amir Efrati reported that Amazon was going to release a free smartphone, the bad reporting caused the principal players—Booker, Amazon, the Astros—to issue statements and engage with the reporter. The end result may be that more information was released; after all, these people would want to correct the record. Mistakes do happen in journalism, especially in a fast-paced environment. Stories surrounding secretive companies or complicated financial arrangements may contain errors, and journalists should be forgiven the occasional correction. But with this kind of reporting—a race to be first to be picked up by an influential Twitter feed or Google News—the process is messy, and only those who happened to catch all the updates, or who are keyed into these kinds of new-media mechanics, would get the full picture. Meanwhile, each story and counter-story arrives with the gale-force promotion once accorded to a major exposé. And in the end, publications are still rewarded for doing shoddy work.

Journalists aren't disembodied observers of social media. It's both a source and a medium for them; they are among its most active promoters, chroniclers, and users. In short, they alter what they touch. When they focus on social-media outrage (which one can find for practically any topic), that becomes the story, even when it might not be the most important part of the story. It fuels that day's news cycle, along with the typical meta-level navel-gazing about the strangeness of reporting in the social-media era. Journalists, and their editors and readers, should adopt a more critical attitude. They shouldn't presume that something trending is prima facie evidence of a story, or privilege it over other kinds of reporting and consulting other sources.

As I'm writing this very section, a shooting has been reported on Capitol Hill—outside the Capitol building or inside it, depending on which journalist's unconfirmed tweet you believe. As usual, there are many "conflicting reports"—a handy phrase that has come to justify immediately publishing whatever one hears. A female suspect might be dead, or alive at a hospital. A baby might've been pulled out of a car. There might've been a car chase that started near the White House. And already there are journalists warning one another to show restraint this time, while others plunge heedlessly into the informational morass and professional commentators take this story, only minutes old and with practically nothing known, as a reason to advance arguments about the ongoing government shutdown, gun control, or any other pet issue. News consumers have a role to play here, too, in part because they often provide raw material on social media that's passed on by reporters, and they share reports that may be wrong as well. We're all publishers now, according to the latest conventional wisdom. "Beware reflexive retweeting. Some of this is on you," WNYC's *On the Media* tells consumers in its "Breaking News Consumer's Handbook"—a handbook written in response to the shooting two weeks prior at the Washington Navy Yard, another frantically reported catastrophe rife with misinformation about multiple gunmen with multiple identities.*

What we're witnessing here is a real-time fetish, where news becomes more and more valuable—and perversely, both more raw and true—the more quickly it is reported. Temporal distance from the actual event begins to seem intolerable. We need to know *now*. We need to be bathed in the numinous feeling of "news as it happens." We won't care later.

But what do we need to know, and when do we really need to know it? What is news, and what is just entertainment for an audience conditioned to need constant feeds of information, however speculative or inaccurate? If you live or work in the area of an active shooting, or know someone there, of course you would want to gather information from television, radio, the Internet. But even then, most of that information is probably unreliable or not very useful. It will provide little certainty and less comfort. Painful as it might be, you'd probably be

---

* Later, it would be reported that the Capitol Hill shooting involved a driver who struck a security fence and a Secret Service officer in front of the White House, before being chased by police to Capitol Hill, where the driver, a woman named Miriam Carey, was shot to death by police in front of her infant child. Follow-up reports revealed that Carey had a history of mental illness. After the social-media frenzy, few commentators bothered to consider how Carey ended up in such a precarious position, hundreds of miles from her home state of Connecticut, or whether police had the right to use deadly force.

better served by staying indoors and waiting a few hours until you can get a more definitive report. We suffer, however, from an informational addiction, going into withdrawal when we know that others might be getting their fix. "When everyone is monitoring everyone else, no one can bear to be left out," James Gleick argued after the Boston Marathon bombing manhunt. A similar idea runs through the minds of cable news personalities: "It is devoutly felt at CNN and Fox News that prestige or viewership or both depend on being the first, even if only by seconds, to announce practically anything." That is why CNN anchors, reporting on an unfolding disaster, have taken to reading e-mails and tweets directly off their smartphones, as soon as they receive them. The real-time fetish now pervades digital media in all its various manifestations.

For journalists working online, the sentiment is expressed in more populist tones. How dare we deny readers information by waiting to vet a report or publish it in an article before posting it on Twitter? Information should be shared immediately; just add some caveats about its being unconfirmed. With this philosophy, reporters adopt their own version of the "information wants to be free" adage of which cyber-libertarians are so fond. It's the illusion of democratization, of just putting the information out there and letting the reader decide. Because the landscape, we are told, has shifted irrevocably. We are all supposed to be both consumers, creators, and distributors of the news, acting out these roles simultaneously. It's a convenient posture to take, for it allows for a more expansive conception of what is news and invites consumers to feel like they're part of the reporting process and should contribute material (photos, tweets, eyewitness reports) for free. CNN, the *Huffington Post*, *BuzzFeed*, Examiner.com, and other journalistic outfits have brought in a lot of traffic and money by giving readers perches from which to post as "iReporters" or unpaid bloggers or "community contributors."

In the chaotic aftermath of the Boston bombing, and the bungled reporting that accompanied it, *BuzzFeed*'s Ben Smith and John

Herrman wrote: "But now we should assume our readers and viewers see virtually everything that we see. We can no longer decide which rumors and scraps of information should be dignified with publication—a sufficiently compelling scrap of information, be it a picture of a man with a black backpack or an anonymous, single-sentence Reddit post from the scene of the crime, will become news on that merit alone." The two went on to write that some misleading, inflammatory, or false material will "inevitably" spread and that the media should both anticipate what will do so and respond to it, for example by adding context when there is none.

This strikes me as nobly intentioned—be responsive to your audience, they seem to be saying, as they have lots of information at their disposal—but wrongheaded. It's a version of the trending fallacy: the notion that just because something has become popular in one's timeline or has gone viral, it's significant and worth addressing. Certainly, false information should be countered with facts, and some outlets specialize in that more than they do in hard reporting. (After spending years building traffic by spreading dubious viral stories, *Gawker* introduced a weekly feature, called "Antiviral," in which a writer debunks the week's false stories.) And news consumers now have more information on hand and can challenge reporters. But the basic job of reporting, the essential fact of what constitutes news, and the responsibility of due diligence and circumspection have changed little. Readers should be respected, but they shouldn't be pandered to, nor should they give reporters their assignments (though no one turns down a good tip). And many readers aren't like *BuzzFeed* reporters: they can't afford or don't want to be online all day, sifting through masses of information, creating a news narrative for themselves. They expect reporters to use their expertise and resources to do this for them, and to be able to read the results at their leisure. They might not have all the information that reporters have or know where to get it. They don't have the experience of being on the front lines of a dramatic breaking story and figuring out the best way to present it. They don't have the phone numbers of government officials. They turn to paid experts to fill in these very gaps.

What *BuzzFeed* and a new generation of social news sites specialize in is a firehose approach to news. Everything is content, nothing more, and meant to be sharable, to have a chance at virality—from the listicle of cool turtle photos to the occasional rigorously reported, long-form story. These sites produce a tremendous amount of material, more than anyone could read in a day, but like a social network, they have to deal in volume, as that's the only way to make money off of advertising. Scale is everything. Because of how much material they publish, and how much time their reporters and editors spend on social media, *BuzzFeed* suffers from a sense that everything is important and amazing, which is the equivalent of saying that nothing is. This is a natural impulse when faced with a flood of information. Journalists who are active on social media, as well as the social-media editors who are paid to pretty much be just that, are perpetually faced with trying to describe and sell the importance of the latest thing to come down the pipe. They are constantly historicizing, telling us which is the definitive take, which is the must-read article, which just-published photos will surely be iconic, who is the breakout star or meme of this news event, who "won the day." It's not just about promoting the latest item; it's a reporter's own effort to stand in the stream of content and try to make sense of the thing surging around him. For the same reason, "I'm late to this, but . . ." has become a common phrase for journalists on Twitter. It's supposed to be a way of apologizing for not posting about an important article or event within a few minutes, or hours, of its appearance. The excuse works two ways: allowing the journalist to confess to not being on the bleeding edge of what's current while also implying that he or she might have had better things to do than kibbitzing on Twitter. It also reflects an anxiety borne out of the conflicting demands of remaining informationally up-to-date and being able to pull away long enough to concentrate, to do one's job.

Sometimes, though, there are real-world consequences to being wrong. After the Boston Marathon bombing, the *New York Post* splashed two young, brown-skinned men on its cover and said that the government was looking for them. The two men were never suspects and had nothing to do with the attack; later they sued the *Post* for libel. It's during major events such as the Boston bomber manhunt that the supposedly self-correcting tendencies of the Twitter hive mind are far less appealing.

Indeed, the Internet is filled with would-be armchair detectives and vigilantes, who operate in uneasy accordance with journalists and law enforcement. Police departments tweet out photographs and screencaps from CCTV cameras, asking the public for help in identifying suspects. The social news site Reddit became notorious for its role in incorrectly identifying suspects in the marathon bombing and other attacks, with some site moderators eventually trying to ban the practice. Newspapers trawl Reddit, police scanners, and social media feeds in order to pick up on the latest speculation and rumors, all in a rush to be first, but in the process often publicizing the names of wholly innocent people. Every rumor, blurry photo, and explosive tweet gets immediately passed on, perhaps with a question mark that's supposed to indicate doubt and prove exculpatory for any journalist whose conduct later comes in for criticism. These mistakes are excused as being par for the course in real-time reporting. Or the vague notion of virality is invoked: this person's name and photo had already "gone viral"; the newspaper was simply reporting what was already out there in the Twittersphere, without taking into account the various factors, technological and otherwise, that make something trend. As a result, they elevated a potentially innocent person's profile, while granting the imprimatur of legitimacy, and occasionally a measure of fame, to digital vigilantes. These incidents reveal that there are costs to being wrong—for people incorrectly reported as suspects, for those grieving, for citizens who are afraid, for news organizations' credibility, and also, though it matters less, for the stamina and mental clarity of reporters expected to be always on and uncovering every possible nano-scoop.

- - - - - - - - - - - - - - - - -

BUZZFEED AND THE PROBLEM OF SOCIAL NEWS

BuzzFeed began as an aggregator of memes and barrel-scraping viral content, much of it sourced from Reddit and other viral spawning grounds. *BuzzFeed*'s growth into a full-fledged media organization, complete with original reporting on politics, culture, technology, geopolitics, and other beats, has been aided by its pseudoscientific claim to have a grasp on what makes something go viral. (The fact that many of *BuzzFeed*'s popular listicles—e.g., "13 Cozy Cocktails to Warm You Up"—are taken, uncredited, from other sites and then glossed up with cheeky captions and large, high-resolution images has done nothing to damage its traffic or its reputation as an up-and-coming, zeitgeist-grasping media property.) *BuzzFeed* not only presents content that's "hot" (a favored in-house term) but also claims to know what generates a viral hit, which is to say what makes content social and shareable. Each article shows how many views that page has gotten and also its "viral lift," a concept that apparently refers to what portion of an article's views come from sharing. (The site uses "viral lift" and the similar "social lift" interchangeably.) That term's vagueness, though, is part of its power; viral lift sounds like a fancy metric, something worked out by statisticians and researchers, helping to give *BuzzFeed* the image of a company that knows what people want and how to make them share it. It's all a distraction from how irremediably superficial the site is, though its founder, Jonah Peretti, freely admits that *BuzzFeed* caters to the "bored at work" crowd, trusting them to wade through its endless listicles and mindlessly click to share them on the social network of choice. He has also said that "it is hard to make viral media especially for serious topics" and that "content is more viral if it helps people fully express their personality disorders," so publishers should "target crazy people."

Social is *BuzzFeed*'s alpha and omega. Comments, sharing widgets, and chartreuse buttons labeled "COOL," "OMG," "WTF," and so on encourage constant, reflexive responses from readers. The site isn't geared toward providing newsworthy, interesting, thoughtful material but rather content that has the capacity to go viral, racking up huge traffic numbers. That's what makes *BuzzFeed* both rather successful in the social-media age and an excellent avatar of its sins and structural defects. Like the *Huffington Post*, *BuzzFeed* engages in what Peretti refers to as a "mullet strategy," using a deluge of salacious, shallow, cute, and similarly disposable, Facebook-ready content to generate page views and subsidize more straightforward reporting. Unlike HuffPo, which Peretti helped found, *BuzzFeed* adopted this strategy after it decided to add an original reporting arm, but it's arguably perfected it. At the very least, being the newest kid on the block and with its obvious talent for drawing traffic, the site has earned a modicum of respectability, mostly because, with its ample venture-capital backing, it has the money to poach journalists and vacuum up those left unemployed by the industry's years-long attrition.

While *BuzzFeed* now reports original stories, it is still focused on creating content that can be socially shared. (Even the site's political journalists find themselves posting listicles of, say, photos of then U.S. ambassador to the UN Susan Rice as a high school student.) Granted, one might say that trying to create social or viral content is simply a means to the end of making money, something that any for-profit journalistic enterprise strives for. But most traditional journalistic outlets, at the very least, would claim that they strive to provide a public service and that quality earns readership—and, by extension, higher advertising rates, greater revenues, and institutional prestige. For *BuzzFeed*, quality is an accidental by-product, rather than a goal in and of itself.

As part of its relentless quest to find out what makes content shareable, *BuzzFeed* has adopted what it calls a "no haters" policy. Positivity is both a content strategy and a company philosophy. The phrase crops up anywhere from its job listings to interviews with its editor in chief. To that end, the site is defined by the same naïve, cheery tone that permeates much of viral media. Complexity is shunned in favor of easy narratives or simple lists that make weightier stories easily digestible. Through these techniques, a story about the aftermath of a shooting at Fort Hood, the United States' largest military base, is reduced to a stream of photos, each paired with a one-sentence caption explaining the base's amenities. As other publications ran stories featuring interviews with survivors and base residents, or explored the causes of the second mass shooting on the base in five years, *BuzzFeed*'s featured story was characterized by photos of Starbucks paired with captions such as "The coffee shop serves Starbucks." The story, for which the reporter traveled to Texas, received more than 120,000 views. (The article's writer, Benny Johnson, who held the title of viral politics editor, was later fired after being unmasked as a serial plagiarist.)

The resemblance of most *BuzzFeed* posts to advertising—in tone, style, and presentation—is likely intentional. And not only because they're considered highly shareable. One of *BuzzFeed*'s main revenue streams is the use of sponsored posts, a form of native advertising in which big companies pay for posts, written in the company style, promoting their products. Even in a cash-strapped industry, this practice is controversial and has gotten more than one media organization in trouble for seemingly giving their editorial brand over to commercial partners. (When the *Atlantic* published an advertorial from the Church of Scientology, the resulting outcry caused the company to pull the ad.) On *BuzzFeed*, the only distinction between these posts and regular content is that the listed author is a corporation—e.g., "Durex, Brand Publisher." It's not uncommon for *BuzzFeed* to publish similar posts by both in-house staffers and commercial partners, or for one to be easily mistaken for the other. A listicle titled "Undeniable Proof That Dunkin' Donuts Is the Best Place Ever" or a video called "Dazzling Facts about Dunkin' Donuts" are both original editorial material, as far as I can tell (though the listicle uses photos pulled mostly from other sources, likely in violation of copyright). But they *feel* like advertising, which is all the better if you hope to attract Dunkin' Donuts, or a similar company, as an advertising client.

This uncertain distinction between advertising and editorial has caused some problems, and not only in the eyes of confused readers or critics like me. Mark Duffy, whom *BuzzFeed* hired to be their ad critic, has claimed that he was asked to soften his work, and even to delete some posts, in order to appease potential advertisers. One of his posts, in which he blasted the misogynist ad campaigns of Axe body spray, was deleted a month after it appeared on the site "due to apparent pressure from Axe's owner Unilever," Duffy said. Duffy was eventually fired, in part, he claimed, because his work didn't jive with *BuzzFeed*'s "no haters," advertiser-friendly environment.

As the first major social news site, one wholly dedicated to grabbing users' attention within their social-network feeds, *BuzzFeed* represents both a success story and a cautionary tale of how a social-first mind-set can be corrupting. For many users, the site's rivers of mindless content are exactly what they're designed to be: enjoyable trifles to be idly digested and shared during slow periods in the white-collar workday. As one monolith in the new-media landscape, then, *BuzzFeed* isn't so menacing. But as a prominent example of how a news organization can be built, ground up, for the social web, it is troubling, all the more so because its occasional quality content is hidden in thickets of dreck. (*BuzzFeed*, like some other digital organizations, believes that home pages don't matter much anymore—referrals through the social web are much more important—which is one reason why the home page of BuzzFeed.com is a mass of links, none of them communicating their relative importance.) More than that, the site is based on a patronizing and infantilizing view of users, who are seen as easily manipulatable totems—target the crazies! *BuzzFeed* doesn't even bother engaging in the connectivity rhetoric of Facebook or Google+ or the civic spirit of a traditional news organization. It sees users as conduits for its own viral content, and as a consequence, the material it produces in such great quantities is largely fatuous, slickly cheery, disingenuously naïve, and eminently disposable. In other words, it represents all of the worst qualities of social media and online advertising.

The listicle, the signature element of *BuzzFeed*, perfectly encapsulates its strategy. It's good for any occasion, from breaking news to war coverage to generic nineties nostalgia. Combining a number of simple, numbered statements and stock photography (or photos taken, with varying amounts of credit and permission, from outside sources), listicles provide readers with something familiar. Easy to get through in a minute or two, they are a handful of M&M's amid the never-ending feast of Internet media. *BuzzFeed*'s listicles often go for cute or quirky, but many of their most successful lists practice a technique similar to Internet advertisers buying placement in your social-media feeds: they micro-target. With these "demolisticles," as they've been branded, *BuzzFeed* goes after small populations—Cornell alumni, sufferers of irritable bowel syndrome, people with Puerto Rican mothers—and targets them with information that only they understand. So you might not know what a "rowie" is, but if you grew up in Aberdeen, Scotland, you do, and you might click on and share "31 Signs You Grew Up in Aberdeen" (it has more than 12,000 Facebook likes). These lists are potentially evergreen—there will always be more people graduating from the University of Wisconsin or growing up in small towns—allowing them to continue drawing traffic well after they're published.

BuzzFeed has so many lists that PepsiCo sponsored a feature called ListiClock, a rotating clock that offers up a new list every second. No micro-demographic is immune; everyone can find something that appeals to what's been called "thisness"—a feeling of instinctual identification, of a general feeling or concept coalescing into something concrete and palpable. Suddenly, your hovering concerns about life in your twenties are made particular. They're all there, glibly summarized in a few sentences and funny photographs. There's seemingly no more honest response than to declare, with happy relief, "Yes! This!" It makes sense then, as Reuters's Chadwick Matlin pointed out, that "this!" has become a common affirmative response on social media, as embedded in the language as "like" or "+1." We look for things which remind us of ourselves and speak a private-yet-communal language.

The writer Alice Gregory, commenting on this kind of targeting, remarked: "We adore being targeted by art. We love getting nailed. Among those who write for a living, 'nailing it' is one of the most succinct and meaningful compliments. Implicit in the idiom is conclusiveness: nailing it *shut*. The phrase also usually implies a gimlet eye, the ability to articulate the ineffably obvious. As readers, we've grown addicted to it." Gregory was commenting on art that flatters certain specific, often culturally influential demographics, but she threw in some comments about *BuzzFeed* and similar sites. The problem with "thisness," or being targeted, is that "it's intellectually disingenuous to allow that recognition to masquerade as some higher order of feeling. Owing to the rise in niche media, specificity—of language, of dress, of eating habits—is taking the place of narrative empathy. People love thinking about themselves, and getting someone to like something—or to 'like' something—seldom requires much more than giving them the chance to celebrate their own personal history."

If social media has to answer to charges of narcissism, it's here, in this endless succession of lists designed to show readers a pleasing image of themselves. They don't ask us to be curious or imaginative or empathetic toward others, nor do they require us to view material that may be outside our comfort zones.

Listicles have in turn led to quizzes, which have also become a staple of the *BuzzFeed* viral arsenal. Some of the quizzes tend toward the absurd: "Which Sandwich Are You?" Others target people with specific interests; fans of cop shows will likely find something familiar in "Which TV Detective Are You?" A slice of these quizzes clearly target certain avid subcultures—the Internet is nothing if not a collection of tribal associations and expressions of fandom—but the larger point is that anyone can feel a sense of identification. Even though it's a lighthearted game, the quiz still produces a result: you are this person/sandwich/*Star Trek* character, or you should be drinking this type of wine/have this celebrity butt/identify with this chemical element. (These are all quizzes posted over a few days on the site.) These questionnaires *address* you. Whether using the voice of a comedian, an expert, or your best friend, they speak to you directly. After you answer some questions, they spit out an answer seemingly calculated on your behalf. In this way, they tap into an essential feature of contemporary media, what anthropologist Thomas de Zengotita calls, in his book *Mediated: How the Media Shapes Your World and the Way You Live in It,* "the flattery intrinsic to representation." We feel flattered by seeing ourselves represented, targeted, or addressed, which is why cable news anchors and advertising alike frequently use the second person and confront their audiences directly—*you* should have these jeans; *you* should keep watching CNN because it has a story about auto safety that might save *your* life. (It's also why some of us love seeing characters who remind us of ourselves or our friends on TV. The rise of the HBO series *Girls* owes much to the fact that it chronicles the lives of the same type of people who would cover it professionally—young writers and creative people living in Brooklyn.)

Thisness is the flattery of representation in concentrated form. Thisness allows us to feel like pieces of media were made for us. Thisness is your own personal filter bubble, showing you exactly what you want. (It also allows *BuzzFeed* to divide their readers into highly specific, personalized categories, aiding in future targeting efforts.) It's not just for entertainment. Outrage and grievance also play well, which is why talk radio and strident political blogging have boomed in the last fifteen years. It's an equally powerful technique in advertising, but it's more troubling when applied to journalism. It means that stories only matter so far as they can reflect your sensibilities or make you feel better about yourself, whether your appearance or your status as a socially conscious person.

This latter notion has taken hold with what is now one of the Web's biggest purveyors of viral media. Issues matter, but feeling good about paying attention to them—and being able to share that feeling—matters more.

THE SOCIAL NEWS MODEL SPREADS

Upworthy is what happens when a *BuzzFeed*-like emphasis on cracking the code of virality meets a half-baked social mission. The site mostly serves up popular videos and memes, all of a relentlessly upbeat, positive, inspiring, or otherwise cheery bent. Its motto is "Things that matter. Pass 'em on," and it carries itself as a sort of digital self-help outfit that is also infused with progressive politics. None of its content is created in-house, except its headlines, but it makes up for its lack of originality with a sense of mission. Upworthy wants to make people care about social and political issues, and feel like informed citizens by watching a four-minute video and then sharing the link on Facebook. (Facebook links provide the bulk of social referrals to Upworthy.)

With long, chatty headlines—"What Happens When an Old Woman Says 'NO' to How Fashion Orders Her to Be?"; "Scientists Discover One of the Greatest Contributing Factors to Happiness—You'll Thank Me Later"; "What Brings This Farmer to Tears Is Worth 3 Minutes of Your Time"—the site is your best friend here to share the coolest, most wonderful thing it just stumbled upon. Each post is called "a bit of Upworthiness," and even the saddest anecdotes can be rescued by a dose of treacle, such as a post titled "Ellen's Tragic Lesbian Love Story Has the Most Beautiful Ending." Unlike *BuzzFeed*, the site isn't dedicated to being naggingly up-to-the-moment; much of its content is material that has fallen through the cracks on YouTube or image-sharing sites. But it takes advantage of similar methods: harvesting others' content, gussying it up with evocative headlines and dramatic imagery, and constantly pushing its users to share (the site is barnacled with social widgets). Emblematic of the social web, the site presents sharing as a form of identity creation. By sharing Upworthy's posts, you are both doing a good deed and publicly demonstrating your social consciousness.

Upworthy was founded by Eli Pariser, who previously was the executive director of liberal advocacy organization MoveOn before going on to write *The Filter Bubble*, a book warning that search algorithms, by taking into account our preferences, browser histories, locations, and other personal information, limit our ability to access diverse points of view. This all became richly ironic when Pariser founded Upworthy in March 2012. Upworthy is a self-proclaimed liberal site trafficking only in positive and uplifting messages; by its very construction, it's dedicated to pushing one political point of view bound in a certain package. It's confined itself to its own filter bubble. Many media outlets and writers have political biases, acknowledged or otherwise, but Pariser's own history of writing about the dangers of being exposed to a narrow range of views makes this a curious enterprise, though his past career as a political operative goes some way toward explaining it.

Even so, these are precisely not the types of entities we should want to see succeeding on the social web. It's not just that they produce—or, in Upworthy's chosen term for aggregating, linking to, or otherwise stealing others' content, "curate"—a steady stream of serious-but-not-too-serious material, all of it designed to accrue as many views as possible. (Ideological mission or not, Upworthy is above all in the page views business.) It's that its success as a meme factory—sixteen months after its launch, Upworthy was declared "the fastest growing media site of all time"—sets an unfortunate example for the rest of the industry to emulate. Upworthy has spawned a gaggle of imitators, sites with names such as 22 Words, ViralNova, Distractify, and Independent Journal Review (this one a conservative viral curator).

The rise of Upworthy and other what's-happening-on-the-Internet aggregators encourages every news outlet to keep some sort of social-media editor, viral editor, viral news blogger, or similar position on staff in order to pump out its own steady stream of the stuff cresting on blogs and social-media feeds. This kind of work helps goose page views and bring in some more ad revenue, arguably subsidizing worthier parts of a news operation. But often it comes shrouded in a disingenuous play for relevance while simply introducing more noise. Even worse is when we see the tics and tropes of viral content start to enter more conventional reporting and blogging. So we see *Washington Post* journalists using cheekily titled listicles to explain the Syrian Civil War ("9 Questions About Syria You Were Too Embarrassed to Ask") or CNN.com explaining the fiscal cliff with animated GIFs.

Upworthy has partnerships with some journalistic entities, which syndicate its content. Some mainstream outfits apparently see it as a potential competitor. In October 2013, the *Washington Post*'s *Wonkblog*, which prides itself on its ability to explain complex policy issues in long, sober posts, unveiled its own viral media subsite, called *Know More*. The site features a tile-like arrangement of large images, each with an evocative headline in the deliberately naïve *BuzzFeed*/Upworthy style. These have become known as "curiosity gap" headlines, teasing at an essential or dramatic story without revealing too much. For example, in pushing a story about Apple's quarterly earnings, *Fortune* tweets to its followers, "How many Macs did Apple sell last quarter?" rather than just simply telling its readers how many computers the company sold. This style of headline has become widely imitated, even if, for many readers and critics, it feels a little underhanded, a way of convincing readers to just click. If you were to succumb and click on one of *Know More*'s curiosity-gap headlines, you'd be served up a small dose of shareable media—a video, an image macro—along with the opportunity to click in order to "know more"—that is, to learn the real backstory behind a congressman's crazy statements or the Higgs boson particle. The real story is another click away, but the site has already gotten your page view.

By this point, the language of viral media had already infiltrated *Wonkblog*. The day it unveiled *Know More*, it also published *Wonkblog* posts titled "The 13 Reasons Washington Is Failing" and "This One Photo Shows How Humiliating the Government Shutdown Is for the United States," the latter a textbook curiosity-gap headline. The site's posts were becoming facile, promoting the kind of simplistic explanations for tangled issues that fill the social web. It sounded rather transactional: take a couple of these posts and you'll be a hit at your dinner party. The success of *Wonkblog* and *Know More* was such that Ezra Klein, the sites' founder, was able to establish his own news enterprise, Vox, under the banner of Vox Media, the owner of several well-regarded digital properties, including *The Verge* and SB Nation. Vox's trademark form is the explainer, an article that claims to simplify a complex issue but which, in practice, often condescends toward readers, treating them like bumpkins for whom traditional news is too sophisticated to understand.

There's a patent sense of manipulation and hucksterism surrounding these sites, with their promises to crack the code of what makes something shareable, their shying away from uncertainty and difficult answers, and their folksy temperaments. Across the board, headlines tend to overpromise while the articles under-deliver, much of it from the too credulous, gee-whiz attitude they project. It's not uncommon to click on what one thinks might be a full-fledged news report only to find that it's a wire photo accompanied by a couple of sentences. Or that it's but the first, brief installment of what's being teased out as an exposé, requiring you to return again and again to learn more. And in the articles themselves, photos and quotes and other elements have been broken out into shareable modules, so that each has a chance to be shared and go viral, where it'll lose all context but might rack up reblogs or appear on a list of trending topics.

These sites also exhibit the overflowing confidence that comes from treating an audience as a collection of data points and mistaking that for knowledge. Like social networks, with their algorithmically sorted feeds and carefully targeted ads, they show people what they think they want to see. Readers of these outlets are nodes in a feedback loop where trending search terms, not newsworthiness, dictate what's written. In the same way, these sites constantly test their audiences, trying to get to know them better. Upworthy brags that it tries out as many as twenty-five different headlines for some of its articles, testing each on a subset of visitors, before deploying the one that seems most effective. We are simultaneously readers and behavioral test subjects.

Business Insider isn't much better. Although it occasionally features strong reported stories,* BI's governing ethos is a profound cynicism masked by an earnest credulousness, one perhaps best embodied by trollish pieces such as "Why Do People Hate Jews?"—which was written by Henry Blodget, the site's founder, whose previous notoriety was earned when he agreed to a lifetime ban from the securities industry after being charged with fraud. That post, which earned a tremendous backlash, caused Blodget to apologize and change the article's headline to something more circumspect: "What Are the Sources of Anti-Semitism?" Along with the usual fare of listicles, linkbait, and overaggressive aggregation, the site is known for pushing the kinds of micro-scoops and this-photo-explains-everything articles that don't add up to much. Its success lies in profligate filling of news feeds and timelines with the viral stories of the moment. While you could get better, smarter, or more authoritative journalism elsewhere, *Business Insider*, for whatever it's worth, is deeply tied to the pulse of digital media, chronicling each petty outrage, each invented controversy or nonstory, with speed and unfettered enthusiasm.

The same might be said of the sports site Bleacher Report, which takes the practices of social journalism and gustily advances them. Among its innovations, Bleacher Report is known for assigning stories with opposite points of view on the same issue. The message isn't that reasonable people can disagree, or even one of false balance. It's simply

---

* Good journalists do work for these sites and shouldn't necessarily be held accountable for the sites' ideological failings or corrupt practices. In this book, I sometimes quote from articles published on these outlets, which is to say that I'll take wit or wisdom wherever it might be found.

a blatant grab to get traffic for whatever people are searching for or whatever might flatter their already held views. They know that people are searching for "Tom Brady bad quarterback" and others for "Tom Brady good quarterback," so why not publish articles on both, while pumping up both of them with bravado and macho exaggeration? One article will tell you "Why Tom Brady Is the Most Overrated Quarterback in NFL History"—an opinion that the article's writer, in an interview, admitted he doesn't even hold—while another will tell you "3 Reasons Tom Brady Is the Greatest NFL QB of All Time." (*Buzz-Feed* has similar tendencies; the site has, for example, two dozen or so listicles about the movie *Clueless*.)

But that's just one example of Bleacher Report's essential nihilism, part of a range of practices that has it churning out tons of witless content—and becoming one of the Internet's most-read and most financially valuable sports sites in the process. The site began by luring thousands of sports fans to blog for it without any compensation, much like the *Huffington Post*. It also invested heavily in titilating, or just vile, listicles and slideshows, such as "Japan's Earthquake and Tsunami: The Worst Natural Disasters in Sports History" or "The 20 Most Boobtastic Athletes of All Time." Bleacher Report has made an effort to hire experienced, respected journalists, but most of its material is created by fans who compete for medals, rankings, and other virtual rewards. The site strings along these unpaid contributors by offering them better virtual perks—an inflated title, a slightly more prominent spot in its popular newsletter—and promises of one day being paid like the well-known sportswriters they admire.

Bleacher Report produces eight-hundred-plus articles each day, but don't expect to find any original reporting: company policy declares that writers are not allowed to break news. It's much more profitable to aggregate others' work and use search-engine optimization techniques to make sure that their version of the same story shows up at the top of Google results. Provocative opinions are highly shareable, too (remember Jonah Peretti's dictum about appealing to "crazies"). Bleacher

Report also relies on an analytics team that looks at what's trending or tries to anticipate the same; those topics are handed off to editors who craft the most shareable, search-ready headlines; finally, a writer is drafted to produce the actual words, which may just be a few sentences introducing a slideshow. "They'll make up whatever people search for," one tech columnist told *SF Weekly*. Not that this data-driven approach produces accuracy: a company handbook instructs writers to create "hyperbolic headlines" that "either overstate or understate your position." Even the site's paid veteran writers are handed prefab headlines and strict, computer-determined dictates about article length.

Bleacher Report is a perfect distillation of social ideology. It produces easily shareable, worthless viral content for very low cost, and it produces immense amounts of it, all while claiming total faith in its data. Its main constituency is advertisers, but it claims to be responsive to readers. As one critic remarked in *Columbia Journalism Review*, "Here's a labor force that creates the product for free or for very low pay, and the owners reap almost all the reward." (That happens to be the business model of nearly every social network.) Bleacher Report justifies this crooked structure by saying that amateur sports fans, the kinds of people who might've once called into talk radio or, a few years ago, started their own blogs, now have an outlet for their passions. This is their platform, just as Twitter might be for amateur comedians. Aspiring professional writers are granted exposure—that nebulous kind of visibility that may allow them one day to ascend the ranks of viral fame and finally start profiting on all their spent labor. It's the same promise made to writers and would-be journalists by many other sites, from content farms like Examiner.com to digital outposts of respected legacy publications like the *Atlantic*. One day this will pay off for you; just defer your payday for a while; build your brand, network, and share, share, share. In the meantime, you can sit securely knowing that some (likely well-compensated) data analysts are coming up with the kinds of statistically driven practices that guide your exploitation. Few writers who started at Bleacher Report have actually achieved much success or parlayed their experiences into stable, paying jobs. But some people at the top have done well. In August 2012, Turner Broadcasting System bought Bleacher Report for a rumored $200 million. One of the site's founders, Bryan Goldberg, has since gone on to found Bustle, a venture-capital-backed women's Web site, despite evincing little knowledge about women's media, feminism, or even the lives of the other sex.

The existential anxieties surrounding the journalism profession are real. But the forces of digital disruption don't justify the kinds of cynical plays for attention found here. Journalism's embrace of social media has been accompanied by the adoption of social media's data-driven feedback systems and its privileging of relevancy above all else. The result is a new-media landscape that prefers to give people not what they want but what will keep them on the platform and contentedly clicking. Should this trend continue, journalists may indeed—as many of us fear—find themselves replaced by automated algorithms that synthesize the latest press releases, box scores, earnings reports, and crime statistics into articles. If that were to happen, however, we would be forced to admit that the industry deserved some portion of its ill fortune by placing so little value on quality work, originality, and the intelligence of its readers. When a writer becomes simply a fabricator of content, one station in the digital production line, then it is only natural that he, too, will one day be made superfluous.

To Watch and Be Watched

The only thing more powerful than celebrity is to own the tool that makes it.

—Katherine Losse, *The Boy Kings*

We live in what the British theorist David Lyon calls an "emerging culture of surveillance." Broadly speaking, there are three types of surveillance that take place in this culture: surveillance by governments, surveillance by companies, and surveillance by us—that is, surveillance we perform on one another, particularly through social media. I'll get to how social media is a surveillance environment, but first it's important to consider the types of surveillance performed by governments and corporations, sometimes in league with one another, and how these acts help to create a culture in which to watch and be watched has become not just a matter of law enforcement or intelligence work but also a social practice. Through it, we become conditioned to want surveillance, not only for paternalistic protection but also for self-expression.

Over the last decade, the U.S. government has, by law and by policy, granted itself massive electronic surveillance capabilities. Due to vigorous reporting and important leaks from intelligence-industry whistleblowers, we now know that U.S. intelligence agencies, along with their foreign partners, have access to immense quantities of Internet traffic, some of it available in real time. "They quite literally can watch your ideas form as you type," NSA whistleblower Edward Snowden told the *Washington Post*. The NSA recently opened a $2 billion data center in Utah, a facility designed to help crack previously unbreakable encryption methods and to store and mine the billions of e-mails, phone calls, and other electronic communications that the NSA sucks up around the world. The data center is designed to handle a yottabyte—1 septillion bytes—or more of data. That's hundreds of times more information than passes through the Internet each year; it's vastly more than the sum of all human knowledge. In short, the data-capture and -mining ambitions of the U.S. intelligence apparatus are boundless. Already the NSA is able to record and sort all phone calls in several countries in real time. As storage costs decrease and analytical powers grow, it's not unreasonable to think that this capability will be extended to other targets, including, should the political environment allow it, the United States.

Some of the NSA's surveillance capacity derives from deals made with Internet firms—procedures for automating court-authorized information retrieval, direct access to central servers, and even (as in the case of Verizon) fiber optic cables piped from military bases into major Internet hubs. In the United States, the NSA uses the FBI to conduct surveillance authorized under the Patriot Act and to issue National Security Letters (NSLs)—subpoenas requiring recipients to turn over any information deemed relevant to an ongoing investigation. NSLs often come with gag orders preventing the recipient, such as an e-mail provider, from disclosing to anyone that they have been subpoenaed. The NSA and its partners in the Five Eyes—five English-speaking countries (USA, Canada, the United Kingdom, Australia, and New Zealand) that have agreed not to spy on one another—also hack into computer systems around the world, those of everyone from civilians to terrorists to heads of state to major cloud-storage systems and Internet backbones, the links between the major networks powering the Internet. Government agencies also benefit enormously from information gleaned by phone companies, market research firms, credit card companies, health records, travel documents, and numerous other digital records that are created by default as part of daily commerce. "Any transactional information that's available [is] accessible by the government in secret," NSA whistleblower Thomas Drake has said. And there are vast quantities of data, including much of what we do on social media, that are simply public, out there for the taking, whether by an intelligence agency or a small company conducting sentiment analysis.

These digital bread crumbs, the tiny leavings that we produce on a daily basis, have spawned a new term: dataveillance. Dataveillance is surveillance enabled by the data produced by credit cards, E-ZPass devices and toll booths, RFID pass cards, transit cards, browser histories, and on and on. Often, they are not important for the actual data they contain—such as what you bought with your credit card—but for their metadata—the time and location you made a purchase, where you went to next, the route you took. (Metadata is data about data, data that describes other data.) This information would be as useful for an intelligence agency as it would be for a smartphone app trying to develop a pattern of your daily commute and your buying habits, the better to target you with ads that fit your profile.

Dataveillance capabilities are often side effects of digital, networked technologies that are supposed to make our lives easier or enhance our communications. For instance, cell phones are light, relatively cheap, and highly portable. They let us talk on the phone almost anywhere and to practically anyone. But they also allow for unprecedented surveillance possibilities. Your cell phone needs to communicate with cell towers, meaning that your location can be triangulated by just collecting data from a few towers. Using the same technique, your movements could be tracked and a map made of your routines. Of course, now that most smartphones have GPS chips, this process is made even easier and more precise. GPS has made it possible to track stolen cars or to check in on Foursquare; it's also made it easier for people to spy on loved ones or for intelligence agencies to follow dissidents. And it's turned location-tracking into a booming business. Cell phone providers such as Verizon sell aggregated data on their customers' movements to businesses.

It's not only governments and intelligence agencies, of course, that have increased their surveillance programs. Social media has created opportunities for us to be tracked by potential employers, schools, insurance companies, and companies trying to market products to us. Although much of the data and Web sites these organizations are scanning is public, this kind of surveillance raises uncomfortable possibilities, particularly because much of it is punitive in nature. It is generally not meant to provide assistance or services, although advertising networks repeatedly claim that close tracking of users' browsing habits leads to more useful or relevant ads. Subjecting social media to surveillance-type scrutiny can produce some unwelcome outcomes. Consider, for example, companies such as Varsity Monitor, which some U.S. universities hire to monitor athletes' online activities for offensive behavior or for indications they've violated their scholarship agreements, such as by accepting gifts. "We look for things that could damage the school's brand and anything related to their eligibility," Varsity Monitor's CEO told the *New York Times*. Given the controversial labor conditions of Division I college sports, which generate billions of dollars for universities and the NCAA but not a cent for the athletes who fill the seats and sell jerseys and risk their health, these kinds of practices further the image of college athletes as indentured servants with few privacy rights. Nor do they seem to acknowledge that many of these athletes are teenagers, denied some of the freedoms and pleasures traditionally afforded to college students, and while in return they may receive free education and celebrity status on campus, policing an important means of expression deprives them of essential agency. It takes their social-media accounts and makes them an extension of the university's public relations apparatus. Despite having done nothing wrong, they are surveilled all the same.

Troubling because it's unnecessary, this kind of surveillance is based on the assumption that companies and organizations have suffered from *not* being able to monitor their workers or members. Workplace and social-media surveillance is here to satisfy a need you might not have known you had. But this surveillance can also easily become capricious, used to punish some employees instead of others, or to reinforce existing hierarchies. It's acutely problematic for students, who are encouraged to try out different identities, to express themselves, and to make and learn from mistakes. If a high school student fears that his principal or guidance counselor is monitoring his Facebook and Instagram accounts, will he abstain or censor himself? Will that compromise his social life or inhibit the kind of personal exploration and expression expected of teenagers? And what happens when it's not just high-profile athletes who are surveilled, but when this practice spreads, as it already has, to schools, churches, workplaces, and neighborhoods?

The result is communities turned into mini–surveillance states. In August 2013, the Glendale school district in California paid $40,500 to a company called Geo Listening to monitor its students' social-media feeds and issue school administrators daily reports. In typical fashion, the company says that it's in students' best interests to be part of this program. Geo Listening's Web site informs potential clients: "Your students are crying for help. We have heard these cries of despair, and for help and attention, loud and clear from students themselves via their public postings on social networks. Many feel as though no one is listening, and they are falling away from societal connections." One suspects that this program is less about protecting students than creating a lucrative new market for private contractors to track and monitor children well beyond the school environment. Administrators may appreciate this newfound power; they can tell concerned parents and supervisors that they're doing everything they can to secure school zones. These school officials may not recognize another potential downside: the money spent on Geo Listening and the time spent reading its daily reports could be better spent hiring social workers and spending time actually talking with students.

Social-media surveillance operates from a place of suspicion in which the surveilled are perpetually on the defensive, knowing that an authority figure is watching. It allows organizations to acquire a kind of sovereignty over an individual, in which he or she is always beholden to the organization, always its public representative—even if this kind of representation really matters only when the person in question does something wrong. Some states have passed laws forbidding employers from asking employees (or prospective ones) for passwords to their social-media accounts, but that offers little protection for employees' online behavior that may be public, or at least discoverable.

- - - - - - - - - - - - - - - - - -
WHY WE WATCH

The new surveillance possibilities provided by our consumer devices, as well as the revelations of widespread government eavesdropping and data collection, have been greeted in some quarters with a shrug. Perhaps your iPhone is too fun and useful for you to worry about how it may be used against you, whether by a local police force looking for a drug dealer or Google tracking your movements so it can show you a customized map filled with its advertising partners. Widespread government surveillance is often accepted as a suitable response to the threat of terrorism. "You can't have 100 percent security and also then have 100 percent privacy and zero inconvenience," President Obama has said. Others have surrendered to the notion that privacy is somehow passé or irrecoverable, a bygone luxury in a world saturated with recording devices. As I argue in the chapter on privacy, it most certainly is not. But in tracing the societal acceptance of corporate and governmental surveillance, we also have to look at the ways in which surveillance has become a cultural value deeply tied to our appetite for voyeurism, self-display, the public performance of identity, and confirmation of our own existences in the form of micro-affirmations from a disembodied audience.

Indeed, if we live in David Lyon's culture of surveillance, it's one that has been developing for decades and by the public's consent. Reality television and the democratization of celebrity—anyone can be famous now for simply being themselves—are intimately tied to the use of surveillance. Reality TV show contestants are perpetually watched and monitored. The bedroom cameras on *The Real World* are often perched in a corner near the ceiling and produce the kind of low-resolution, black-and-white or lurid green images that we associate with CCTV and night vision. There's an invitation to be a voyeur, to be titillated. The contestants are there willingly after all, and they perform for the cameras—acting histrionically, submitting to written scripts or ginned-up conflict, sitting before a confessional camera and responding to provocative questions from producers that are then edited out of the final product.

These shows, particularly the aptly named *Big Brother*, are premised on the unspoken notion that only by eradicating privacy can we see what is "real." (They deal in terms similar to that of social media, which promises authenticity and unmediated access to others while also undermining privacy.) There is nothing real about having no privacy, nor about being monitored at all hours of the day and having the edited results broadcast to millions. Nor is there some sudden authentic moment, when the mask slips and we see the real desperate person behind the televised persona, but like porn, we might watch with some part of our brains tuned to this possibility. Perhaps we'll discover, embedded deep in the artifice, signs of raw humanity. Occasionally, too, reality show contestants try to rebel against their own surveillance. They hide from camera crews (or make us think they do), they obscure mounted cameras in an act of defiance, they whisper to one another. Any escape is a temporary one—conveniently, the camera's mic still works well—but each small revolt heightens the sense of something illicit going on here, that we are peeking into someone's tender moments. Privacy is never so precious as when someone tries to recover a semblance of it under difficult conditions.

Part of what makes reality TV an indelible symbol of our

surveillance culture is the way in which participants are constantly torn between competing states of self-consciousness. Reality TV stars know they're being watched but also try to act natural, as if no cameras are there. A similar tension occurs in social media, in which each broadcast or gesture is for some imagined public but is also supposed to seem spontaneous, unpracticed, a true reflection of some in-the-moment epiphany, experience, or scene. Acting unself-consciously becomes its own kind of self-conscious exercise. It's something practiced, performed, so that we become like the famous actor who dresses up for the paparazzi, who have already been told where to expect the star to appear. The star himself might act harried or surprised, but this is part of the act, part of how premeditated performance becomes packaged into something spontaneous and authentic. That we as savvy consumers—whether of celebrity gossip or Twitter—are now wise to these rituals only lends them a more delicious theatricality. They are garlanded with irony and we are granted a sense of intimacy, of being given specialized knowledge. We know how this kind of thing works, and yet we submit to it anyway, in part because there doesn't seem to be any way out. How can you act unself-consciously on social media when focusing on that very idea? The question inevitably leads us to a paradox, one common to living in a world filled with the flattery of self-representation.

These practices have led in turn to the techniques of surveillance being adopted by political insurgents, nonviolent and violent alike. The rise of wearable computing and cheap, portable digital cameras and smartphones has contributed to a phenomenon known as sousveillance, in which people participating in an activity, such as a protest, film it. That footage may prove useful in a lawsuit that claims police brutality, or it may simply help a group to analyze an event after the fact. In many cases, the video is then shared on social media, helping to catalyze political awareness. Sousveillance became particularly popular during the Occupy movement, in which activists filmed and photographed police and made heavy use of streaming video in order to transmit their message to a wider audience. Syrian rebels, from secular Kurdish groups to foreign jihadists, use video to publicize their exploits, as propaganda against Bashar al-Assad, as a plea for foreign aid, or as a way to inform the world of a government attack.

Variations of sousveillance have become central to the quantified self movement, in which people use smartphones, cameras, sensors, and other devices to record and analyze data about anything from their eating habits to their daily movements. Frequently hailing from the tech industry, quantified selfers believe that every human problem—whether it's your slow metabolism or trouble sleeping—is a bug amenable to a technological fix. It only requires enough data, the right algorithms, and a bit of self-discipline (which itself can surely be enhanced with the proper app or device). And you might share this data on social media, perhaps through regular, automated updates, in order to brag about your progress and prompt your friends to encourage you to do better. The various fitness trackers that track joggers and then post their run times—*Jacob just ran 2.2 miles in 30 minutes. He's slow.*—on Facebook are a prominent example, but committed quantified selfers go much further, following the belief of intelligence agencies and Facebook alike, that all data is potentially useful.

The collection of data might be its own reward. By lifelogging, as the practice is also called, you can create a vast trove of data on your own life, one that both expands upon and substitutes for your own fallible human memory. In order to satisfy this need, a range of tools has appeared—fitness trackers, wearable cameras, diet apps, automatic check-ins, wearable computers. Some of these devices operate independently of their operators, making the act of self-surveillance autonomous. The Memoto Mini Camera, for instance, clips onto a user's shirt and automatically takes a photo every thirty seconds. When synced with a computer, the device pulls out what it thinks are the best photos.

The strange logic of lifelogging is that devices such as Memoto are somehow empowering, even though they tend to take choices out of the hands of users. Instead of deciding when to take a photo, a device does it for you. Instead of choosing your favorite photos for safekeeping, Memoto does. Memoto's manufacturer describes its device a "searchable and shareable photographic memory." (Panasonic calls its A100 wearable camcorder a "point-of-view lifestyle" device.) At the same time, an incredibly detailed chronicle of your life is created—one that companies with access to this data can sell to commercial partners or mine to sell you products and that you must manage in order to extract any value from it.

The quantified-self movement may argue that building up such a cache of data about oneself allows for self-improvement, for changing bad habits, for improving health and quality of life. But an equally important part of the quantified self seems to be sharing that data, turning the act of self-surveillance into a social occasion. Data becomes something to be proud of, something to brag about, not only for what it shows about how one lives his life (look how often I exercise! look at the photos my Memoto took in Tahiti!) but also for the sense that one is being bold by exposing so much of himself for public consumption. Disclosure becomes seen as a good in and of itself, a potentially brave act of radical transparency.

One pitfall of lifelogging is how the collection of this data can become normalized, even expected. That data can in turn be used against us by insurance providers or mortgage firms. A June 2012 story from the *Economist* recounts how Rigi Capital Partners, a Swiss insurance company, decided not to purchase the life insurance policy of an elderly woman with dementia. The reason? Her Facebook profile "suggested she had a vibrant social life, not dementia." Perhaps the insurance company employee who examined the woman's profile was correct. Or perhaps, as Jaron Lanier proposed, "information underrepresents reality," and the Facebook profile was inaccurately interpreted. Who knows if the woman controls her own profile? Maybe she does tinker with her own Facebook page, but her grandson tagged her in some photos and checked her in at some events. Regardless of other possible explanations, the incident emphasizes the foolishness of trusting social-media data as a definitive depiction of someone's life and of using that data to inform important decisions. It also points to the precarious state of a lifelogger: that once you make self-tracking and self-surveillance a value unto itself, you may begin to feel that you can never have enough. You may not want to have your feces' microbial content analyzed, as some quantified selfers do, but think about how *useful* that data could be. Your spouse might be annoyed that you photograph every meal and date and label each photograph, but something extraordinary will happen at one of these dinners and you'll want to look back at it and remember. You can always collect more, spend more time shaping your life around the collection, storage, analysis, and curation of personal data. The possibilities are endless, if not, to my jaundiced eye, very fulfilling.

Your data is also unlikely to remain your own for very long. Many of the apps and products that life loggers use have rather porous privacy policies. These companies, such as Nike, which produces the popular Nike+ Running app, share user data with third parties. That may end up being a health foods company looking to target users who want to lose weight, or, more ominously, a health insurer who decides that a client's fitness tracker data indicates someone woefully out of shape whose premiums should go up. Of course, the client would likely never know why his premiums increased.

Surveillance is always part of a power relationship, with one party trying to monitor, protect, or control another. The kind of self-surveillance practiced by the quantified-self movement could be empowering, but it also requires a tremendous amount of work and vigilance. We've accepted corporate surveillance as a price to pay for free services—just as we might accept governmental surveillance as the price of security, despite several prominent studies, including one funded by the Department of Homeland Security, showing that data-mining is ineffective for predicting terrorist attacks. But this exchange of privacy for services (or the promise of security) is accompanied by an exchange of power as well.

While corporate surveillance takes place because many Internet firms don't have any business model besides advertising, which depends on using surveillance data for targeting users, there are occasions where it is part of the product itself. The mapping app Waze asks users for data so that they can develop real-time traffic models, which in turn are sent back to users. You give up information about your route and receive useful traffic information. Some cities have created similar apps that use smartphones' accelerometers to track potholes, which then allow cities to better understand where roads need improvement. Citizens get to feel like they're helping their communities, but they're also helped by making the city's pothole crews more responsive, which may save their cars' tires and suspensions from damage.

Even in this more utilitarian model of surveillance, there are downsides. Mapping apps such as Waze are able to build up extensive profiles of your movements—information that can be used to target you with ads from Google, Waze's parent company. The ads themselves may be an acceptable annoyance, but some might find them more troubling, not least because they impart a sense of being watched. If you're not comfortable seeing the same ads follow you all over the Internet, you might also be troubled by seeing ads on your phone for a certain restaurant every time you approach it.

Another problem with this kind of surveillance is data's tendency to be fluid. It ends up in unexpected places and it can be put to unwanted uses. A few years ago, a Web site called PleaseRobMe appeared. Using publicly available data from Foursquare, the site showed when certain Foursquare users weren't at home, as indicated by their check-ins at restaurants, movie theaters, and the like. The site's creator presented it as a paternalistic exercise in reminding people about the value of privacy, but it arguably undermined some people's privacy, or at least their sense of security, in the process of making this point. An app called Girls Around Me used Foursquare check-ins and public information from Facebook profiles to track women. The idea was to alert heterosexual men to the presence of women at bars and other venues, but this already sexist premise was quickly deemed creepy because it used personal data in a way that Foursquare and Facebook users never intended; it could easily facilitate stalking. That's why Foursquare blocked the app's access to the company's API—essentially rendering the app unable to access data from the location service. (The app's makers later pulled the program from the iTunes App Store.) In some respects, Girls Around Me wasn't that much different from location-based dating apps such as Grindr and Tinder. But the important distinction is one of informed consent. (Packaging matters too: Girls Around Me employed the crude silhouetted-women iconography found on truck mud flaps.) People on Tinder and Grindr want to be found and may not mind having their dating profiles connected to their social-media profiles. But the creepy line is crossed when our data is used against us in ways we didn't intend or approve of. As these kinds of technologies become ubiquitous, consumers have some obligation to inform themselves and to avoid certain services or implement privacy features. Yet much of the power resides in the hands of app makers and platform owners who build this surveillance infrastructure. Our emerging culture of surveillance provides opportunities to keep us safe and informed, and even to fashion new identities, but it doesn't take much for us to become prisoners of these same systems.

- - - - - - - - - - - - - - - - -

FROM WATCHING OURSELVES TO WATCHING ONE ANOTHER

Social media has ushered into the culture a remarkable level of mutual surveillance. On social media, we are constantly watching one another, indulging in the sense of being watched, making judgments and being judged, forming mental dossiers of our social-media contacts, and knowing that they are doing the same about us—all the varied patterns of thought that make up ambient awareness. We become a mass of little brothers, thousands and millions of surveillance artists, practicing our craft upon one another. We learn not only what people are doing, but also *how* to watch them, intuiting their habits, discovering that a best friend tends to go on drunken Facebook rants on Saturday nights or that a colleague has left unlocked his Instagram account.

Social media allows us to be surveilled but also gives us more control over how we are surveilled, though the nature of that control varies among networks and apps. Still, despite whatever privacy controls we employ or choices we make to limit our number of followers, it's often difficult to tell who is watching us. (Unlike many dating sites, social networks generally don't alert us when someone looks at our profiles.) We rely on feedback: a like or retweet, a reply or e-mail, a comment in person to indicate that our update was seen. In the absence of that, social media is defined by a sense of a lurking, disembodied audience, some mass of watchers. Their watching is tolerated, invited even, though it still retains an air of the illicit. For that reason, we often speak of "stalking" people on social media. The usage is lighthearted, always surrounded by scare quotes, but all the same there is something sly and subversive about it. We invite people to be voyeurs and—particularly when we're over-sharing, posting photos of our young children, our vacations, our every mood and daily routine—to take some vicarious pleasure from our lives.*

For some, the prospect of mutual surveillance is a cause for concern. Invoking Orwell's *1984*, Walter Kirn writes that Winston Smith "hid from the telescreens whenever possible and understood that the price of personhood was ceaseless self-censorship and vigilance." We may have to consider doing the same, he says, not knowing when someone is watching or chronicling us. He cites the story of Tyler Clementi, a Rutgers student whose roommate secretly broadcast footage of Clementi in a sexual encounter with an older man to other students in

---

* Discussing her experiences as an early Facebook employee, Katherine Losse writes in *The Boy Kings*, "We even had an internal tool, called appropriately, Facebook Stalker, that showed who had looked at our profile, which revealed fascinating insights."

their dorm. Clementi, who wasn't openly gay, later committed suicide. Clementi's roommate, Dharun Ravi, used mutual surveillance tools—Twitter, a webcam—but by doing so without his roommate's knowledge or permission, the uninvited surveillance became something despicable, a lurid voyeurism designed to embarrass.

The uncanny valley of mutual surveillance appears when we have to face our surveillers, whom we prefer to think of as disembodied and remote. The prospect of a stranger holding a smartphone, its camera pointed in your direction, has become almost routine, but it can also be cause to stop and wonder, *Is she photographing me? Why would someone want to photograph me? And do I want to do something to encourage or discourage that?* Google Glass is an ostensibly social surveillance device which, in practice, places the medium, a bit too literally, between the wearer and the world. Its red recording light is a vivid signal to onlookers: I'm watching you. Wearing it becomes a desocializing act, as some early adopters found when people asked them to take them off in social interactions or stared at them skeptically. In response, some coffee shops, casinos, and bars banned the devices. Even so, some early adopters also found that while many people were puzzled or repulsed by the devices, others flocked to them, fascinated. Disgust and curiosity can exist here in equal, if opposite, measure, two poles of the magnet.

From the Rodney King beating to the "47 percent" video that helped doom Mitt Romney's 2012 presidential campaign, the covert filming and sharing of footage has become a central part of the new mutual surveillance. We film out of curiosity or prurience but also, not unlike the quantified selfers, out of a belief that this footage may become important. The documentary impulse is its own reward, so we submit, or demand, to be documented. Maybe there's something we're missing in the moment—beyond the fact that filming, staring into that small screen rather than at the thing itself, can make us feel like we're not *in* the moment. Maybe the two people arguing in the public plaza will soon come to blows; better film it, so that you can give the footage to the police or post it on Facebook, where it'll get a lot of likes. As this mind-set develops, the sense of being recorded becomes omnipresent but ambient, not a specific awareness but diffused like a gas. The average Briton is caught on CCTV seventy times per day. How many of those are noticed? How many times, in turn, have we been caught on some tourist's video camera, or been snapped by an iPhone because a person took an interest in us, or, walking past a group of college students, ended up photobombing their shot, discovered later when it was posted on Instagram? This is a post-panoptic kind of suspicion: we know that people and devices might be watching, but there are gaps in the coverage, bouts of inattention and fortuitous accidents. We live in a constant state of possible surveillance.

It's an anxiety that should only become more urgent in coming years, as recording devices continue to proliferate, get smaller, and develop new capabilities. We used to have a better sense of how we were monitored and who was doing the surveillance. "Now it's more random, unpredictable," Jared Keller explained to me. "Like, should I really be talking about this, or is somebody on Amtrak behind me live-tweeting everything I say?"

He was alluding to an incident in October 2013, when former NSA chief Michael Hayden, traveling on a New York–bound Amtrak train, was giving an interview on background to a reporter from *Time.* Another passenger overheard Hayden's conversation and recognized him and tweeted some of Hayden's remarks, outing him as the anonymous official "giving reporters disparaging quotes about admin," as one tweet stated. It was sloppy tradecraft by the career spy, especially on a train frequented by government officials, journalists, and other nosy types. It was also emblematic of the new surveillance possibilities—unexpected, covert, widely distributed, but perhaps, if only occasionally, in the public interest. The Hayden incident showed how the sausage of national security reporting gets made, with powerful former and current government officials tending to dole out quotes and intelligence behind the shield of anonymity. It did not, on its own, lead to a change in the practice.

- - - - - - - - - - - - - - - - -
MICRO-FAME AND THE CELEBRITY IDEAL

In social surveillance, being watched isn't just the price of a free service; it's native to the culture, which couldn't exist without it. It invites us to participate, to become shapers of our own surveillance practices, and to provide material for others to surveil (some theorists accordingly use the term "participatory surveillance"). Social surveillance allows us to feel as if we have an audience at any time, waiting to be summoned. The individuals we imagine as seeing our updates may not actually be there; they may not be online at all, but the amorphousness of our public is part of its appeal. It's potentially infinite, if only we satisfy them, causing them to spread our words through the network.

Social surveillance fulfills our hunger to see and be seen. It offers a finishing school for the self, in which our public performances complete and complement our private identity construction. If you want to thrive in the attention economy, then you want to be watched—on your own terms. Unwanted surveillance—a sext that appears on a popular Tumblr; a vacation photo that you didn't mean to share with an ex; or, in the worst scenarios, the total violation of the sort experienced by Tyler Clementi—impinges on the sense of control that allows us to continue in the surveillance economy. Without this sense that you can open yourself to surveillance when you want to, however illusory or contingent that feeling might be, surveillance would feel suffocating and unsustainable.

John Steinbeck said that poor Americans see themselves as "temporarily embarrassed millionaires." In the same way, today's Americans tend to see themselves as unrecognized famous people. And so we submit to social surveillance because it helps us to *feel* famous. Facebook and Instagram are a thousand paparazzi, ready to train their cameras on us. (Or each of us is his own paparazzo and Facebook his press agent.) The reality TV metaphor only seems clichéd because it, and its attendant fantasies, has so thoroughly penetrated our culture. Social media thrives on the promise that anyone can be a celebrity, that anything can go viral. This promise ostensibly reflects the democratization of celebrity, creating a merit-based rubric for fame, but virality and social-media success often have nothing to do with the worth of an idea. And the particular form of micro-celebrity endemic to social media is a volatile one indeed.

It's no coincidence, though, that celebrities and brands have flourished on social networks. In some cases, this has been a result of a concerted push by the networks' owners. The highly publicized race to one million Twitter followers between Ashton Kutcher and CNN—a news outfit that indulges in as much pop culture and personality-driven coverage as any entertainment organization—highlighted the burgeoning popularity of Twitter while also holding out the unspoken promise that you, the average user, might scale similar heights of popularity. Perhaps you might not reach the million followers of a TV star, but he still might see your tweet about him. Here the democratizing promise of Twitter was not that everyone was on the same footing, that people could communicate with one another regardless of political, social, geographic, or other boundaries, but that you could communicate with someone famous. You could have *access* to fame. You could stand under its shade. A little of that fame might, in turn, rub off on you—which explains why ordinary Twitter users don't ask celebrities to reply to them but instead beg to be retweeted. It isn't conversation that's desired. It's to be granted the spotlight, an unreliable one, yes, but which says that this person *recognized* me. He or she saw me as a peer, deemed me cool or beautiful or just lucky enough to appear in his timeline, alongside tweets promoting his new movie, his cologne, the camera-friendly charity project in the Congo. We don't want to connect with celebrities. We want to dispatch them and take their place, although we'd settle for just a little dusting of celebrity attention. Their attention is the real commodity, and so is the victory that it represents: if Snoop Dogg or Beyoncé retweets you, it means that they passed over perhaps thousands of others. This unacknowledged mass of fans in turn becomes the audience, as the retweeted party, having been pulled onstage for just one song, rises above the rabble and can look down on them.

No celebrity represents both this ideal, as well as this ability to keep his audience in a tight orbit by regularly recognizing one of its members, better than Justin Bieber. The young pop star is the most popular Twitter user, with more than forty-one million followers. According to one report, 3 percent of Twitter's servers are dedicated to serving up Bieber's tweets (and the thousands who retweet and reply to any utterance that appears under his name). His Twitter account is filled with the usual celebrity pablum—promos for friends' work, updates about his latest tour, behind-the-scenes photos, tossed-off sentimentalities about how grateful he is for his success. All of it is dispensed with the usual all-lowercase textspeak and playful hashtags that help make a carefully constructed celebrity seem like a person, rather than a massive cash-generating machine for various corporate masters. Among Bieber's most savvy behaviors is to periodically retweet or reply to one of the thousands of fans who tweets at him daily. Within minutes, these tweets are retweeted and favorited thousands of times by Bieber's followers. Bieber also follows more than 120,000 accounts, meaning that he uses this ability much in the same way he doles out retweets and @-replies: to grant recognition. (No one would reasonably follow so many accounts expecting to be part of a community or to keep up with their updates.) His fans who have been recognized by @-replies or follows often mention that fact in their Twitter bios. It becomes incorporated into their Twitter identities, raising their status in the Bieber fan community, particularly among the vast subgroup of diehard Bieber fans whose whole profiles—from avatars to account names to the contents of their tweets—are devoted to chronicling Bieber's every move. This particular community of digitally committed Beliebers (as some of them call themselves) is a key part of spreading the Bieber phenomenon. Some of these fans have thousands of followers of their own, and their continual activity ensures that even if Bieber might be momentarily inactive on Twitter, a flourishing community of tribute and media (songs, photos, links, outbursts of praise) remains. They are his doppelgängers, simulating Bieber, bot-like, substituting for him in his absence.

Bieber's Twitter presence isn't just for his cult of Beliebers. It also is an easy way to earn mainstream media coverage, producing a volume of media impressions that normally costs millions in advertising and PR dollars. And many journalists are happy to play along. Social media has become an incredibly fruitful source of material for reporters, whether data journalists mining public sentiment or bloggers charged with producing a half dozen posts daily, or those for whom celebrity social media is a news beat in and of itself. It's not just usual suspects like *People*, TMZ, the *Huffington Post*, or the UK's *Daily Mail*; celebrity social media is the rare growth segment in journalism today. Whether it's coverage of a supposedly accidentally leaked nude photo or a soberly reported trend story (a sampling of *New York Times* headlines: "Twittering Celebrities Take Fans Backstage in Their Lives"; "The Celebrity Twitter Ecosystem"; "Celebrity Spats Thrive on Twitter"; "Celebrities Are Leaving Twitter"), social media can be tapped for content practically on demand. And as digital journalism, following its cable TV counterparts, has increasingly become a twenty-four-hour, always-on operation, driven by the insatiable need for page views above all, its practitioners have turned to celebrities' Twitter and Instagram accounts to provide the kind of easily constructed clickbait that earns traffic.

Unfortunately, these two parties need each other. Celebrities know that even an unplanned, perhaps regrettable Twitter outburst will lead to useful media coverage. The cycle is well established: a public faux pas; the deletion of offending tweets/Instagrams/posts or the entire social-media account; a press release accompanied by a public apology, perhaps issued through the same Twitter account; and some genu-flecting before a cultural authority (the Anti-Defamation League or a late-night talk-show host) and a promise to do better. Many journal-ists don't like this state of affairs, but they answer to a higher power, whether it's page views, an editor's edicts, or the tempting possibility of seeing a half-baked trend story about celebrities' Instagrams go viral. A similar concordance exists with politicians—that other form of celebrity—who have learned to use social media as a bully pulpit and as a way to provoke media coverage.

We follow in their wake. "Merely being on social media is sufficient to make users feel micro-famous," Rob Horning writes, "regardless of the particular number of likes or reblogs attracted." Managing an audience, managing a self for an audience, speaking in the language of branding and PR—we play with the tools and machinery of fame, even if we are not, by any standard, famous. We position ourselves to be recognized as famous. Lifestyle blogs, Tumblrs, and other acts of self-documentation prepare us to catch a crest of fame or the attention of some industry doyenne or tastemaker. These are the shortcuts to fame for its own sake, or to the kind of micro-fame that might secure one a job or a book deal, with a cooking blog serving as the new screen-play, something everyone has in the drawer, ready to show to someone more influential than we are.

I experienced some of this process when I won a few episodes of *Jeopardy!* in 2012. I chose to post a note on Facebook about each win, in part to tell friends and family when they could watch me. Maybe I wanted to brag a bit and secure a little fame of my own, to allow my fleeting *Jeopardy!* success to be stretched out a bit longer and cata-logued in Facebook's database-cum-scrapbook. The episodes had been taped months earlier, the outcome already known to me and a few friends and family members; in some regard, it was all over and the broadcast and Facebook postings allowed me to relive a victory that had felt all too brief (who doesn't like free money?). Those few days ended up being the most active of my time as a Facebook user, with dozens of likes and ecstatic comments on every post. I had tapped into one of those rare veins of pop culture about which nearly everyone has an opinion or a couple of questions ready at hand. Facebook friends with whom I'd long fallen out of touch chimed in, some more excited that I was on TV than anything else. My roommate from freshman year of college, whom I hadn't spoken to in years, reposted a photo of me from the show on his own Facebook page, adding an exuberant caption about his old roommate's game-show victory. There was no direct contact between us, no private exchange of pleasantries, but the gesture allowed him to become associated with my micro-fame, taking some of it on for himself, and submitting it to his public, to be approved by likes and comments and shares.

It's not easy trying to become a micro-celebrity. Living in an environment of mutual surveillance produces its own anxieties. Surveillance, after all, is predicated on vigilance, discernment, and judgment. It's an environment of suspicion, the definition of which has become elastic in an age of ubiquitous surveillance, when intelligence services, Facebook, and market research firms insist on collecting data about everyone. Edward Snowden helped popularize the term "suspicionless surveillance." By this, he meant surveillance without cause, which shifts the presumption of innocence toward guilt. The mutual surveillance of social media still has the skeptical and judgmental qualities of classic surveillance, but now we must prove ourselves not as innocent but as worthy of interest, of being followed and celebritized.

When viewed through the lens of surveillance culture, the rhetoric of tech executives isn't surprising. Just as government officials claim that widespread digital surveillance is in Americans' best interests—it's targeted, it receives oversight, it keeps us safe, and so on—Zuckerberg and company claim that the surveillance of the social network also promises security—of identity, of your stuff in the cloud, of your relationships and data self. The ads and suggestions you receive will be tuned to you, good for you, because we are watching out for you. The best way to accommodate yourself to social-media surveillance is to give into it entirely. Don't think, just share. After all, given how insubstantial and evanescent each update is, how soon it is likely to be forgotten, you're incentivized not to dwell on any one update. Reformat your routine and your mental metabolism around snap-twitch documentation, followed by a blast out to your networks. It's supposed to be instinctual, frictionless. Eventually, it'll be automatic, as we become more bot-like or choose a bot to substitute for us. The progressive response to social surveillance becomes to submit to it. Avoiding it is the Luddite's path, and that kind of work is far more boring and self-defeating (a succession of untagging, unsubscribing, deleting, and requests for photos to be taken down) than the perpetual labor of tending to one's social-media profiles. The victors in this community are those who prove themselves worthy of being surveilled most and whose metrics rise to the top of the heap. If you go unwatched, you've done something wrong.

MANAGING VISIBILITY

Celebrity and surveillance are united by one chief quality: visibility. Visibility is variable and fickle. One day, you might appreciate having your witty tweets widely retweeted, followed by funny and appreciative responses from throughout the Twittersphere. The next day's tweets—now available to a slightly wider audience, thanks to the followers gleaned from your briefly viral jokes—may earn derision. Perhaps someone misinterpreted a stray remark and now you're faced with an angry commentariat, demanding that you explain yourself. Now you must work again to regain the favor of your audience and, consequently, their help in keeping you visible.

Visibility is primarily an asset for those who have the resources and support systems to deal with celebrity-style fame. The relatively anonymous individuals who are briefly thrust in the viral spotlight rarely seem to do well—not unlike in previous generations, when a local hero might have had trouble reacclimating to civilian life after some heavily chronicled event. (Robert O'Donnell, the paramedic who helped save Baby Jessica in 1987 during a widely watched telecast that became an archetype for these kinds of stories, later saw his life fall apart and committed suicide; many participants in high-profile rescues have suffered similar patterns of sudden fame, dislocation, trauma, and suicide.) Faced with this possibility, it is traditional celebrities and those who openly pursue celebrity who may be most suited to thriving in a world of persistent mutual surveillance. Both groups may be mocked, but the former exists above quotidian concerns (all virality is good virality; money absolves the pain of abusive @-replies), while the latter has made a devil's bargain, knowing that becoming a social-media power user may lead to both Internet infamy and money making opportunities in traditional media. Notoriety has become relatively indistinguishable from more prestigious forms of fame. Only some violent felony convictions, perhaps, would make a major brand or mainstream media outlet shy away from someone with a million Twitter followers.

Acting as if one is famous, or adopting the pose of micro-fame, is also a way to cope with the visibility of social media. It allows us to make it work more for us, to retain some control over it, because we're contending with it through the lens of our own perceived renown. Instead of hiding from it, of spending too much time removing things from our Facebook walls or asking our friends not to post on Instagram, we encourage it, knowing that there's value to be wrought from living publicly and without self-consciousness. This attitude allows us to give into the values of social-media culture, to adopt them uncritically. We might not be famous (or micro-famous) but we can feel that way, at least among our own networks. Still, the network's constantly visible metrics remind us of how others are always doing better in their pursuit of fame. Our follower counts and retweets and shares are insufficient, whether compared to a more famous friend or someone listed as a trending topic. You can always be more famous.

Pursuing this kind of social-media-driven fame, then, allows us to shed some of the anxieties associated with social-media use. We're buying into the system, saying that yes, we can do better, we can be better consumers by treating ourselves as we would Justin Bieber or the Lakers or Dow Chemical. All is branding, all is PR. We shed the neuroticism of sharing and public scrutiny by treating network fame with utter seriousness. We exist alongside corporations and movie stars in the attention economy, and if we must pursue this—if social media, indeed, feels so intrinsic not only to the culture but to anything from getting a job to being allowed to sign up for Tinder or Airbnb—then we'll grab some of that attention for ourselves. We deserve our visibility. Our friends become our press agents, and we return the favor for them, knowing that it grants us social capital. And we'll present our lives as precious and perfect and worthy of being known, because there is no trite saying that better embodies celebrity than "fake it till you make it." Your public awaits; it just might not know it yet.

In rare cases, visibility serves as a form of security. Edward Snowden outed himself as the source of NSA leaks because it ensured that he couldn't be spirited back to the United States and kept incommunicado. He pursued and embraced viral fame because there was safety in having his name known. To use a term native to digital culture, one that I'll explore more deeply in the next chapter, he "doxed" himself—he revealed his identity. By shedding the cloak of anonymity, he hoped to protect himself and gain sympathy from a wider public. If he hadn't been outed to U.S. intelligence already, he likely was about to be—a vigorous investigation was then under way—so he dramatically expanded his circle of visibility. Viral fame, for Snowden, held the promise that he could become too famous to secretly arrest or kill. The numerous public opinion polls that followed in the weeks and months after Snowden's self-doxing demonstrated the wisdom of his choice. It also reflected his canniness. These polls and constant debates over Snowden's personality, his fate, and practically every knowable aspect of his life may have distracted from debates over his revelations (though that was inevitable, with responsibility lying with celebrity-obsessed media organizations and not Snowden himself). But they also showed that one can play the instruments of celebrity, surveillance, virality, and visibility for personal, even noble gain. His relative success was far from assured. An array of forces, from administration-friendly media outlets to senators calling Snowden a traitor, soon coalesced to counter the former NSA contractor's media campaign. There also was no guarantee that the public might not eventually turn against Snowden. As soon as he became a public name and face, a horde of journalists, well-wishers, security officials, and others began tracking Snowden's every move, swarming Hong Kong hotels and, later, the Moscow airport. As any Hollywood star might claim and any paparazzo might confirm, it's exceedingly difficult to embrace some parts of fame while evading others. Managing visibility is a full-time job, and Snowden's reputation will continue to be litigated in the court of public opinion, perhaps for years.

The War Against Identity

Anonymity is a shield from the tyranny of the majority . . . It thus exemplifies the purpose behind the Bill of Rights, and of the First Amendment in particular: to protect unpopular individuals from retaliation—and their ideas from suppression—at the hand of an intolerant society.

—Majority opinion in Supreme Court case *McIntyre v. Ohio Elections Commission*

Whether you take your cues from postmodernism (it's all a performance) or your parents (you can be anything you want, dear), most of us are made to think that identity is mutable. Your identity can change, sometimes as easily as buying new clothes or finding a new watering hole, with people who know you not as a banker but as the guy who likes to go bowling and drink old-fashioneds on Friday nights. Many of us experience this sense of possibility most poignantly at the beginning of college, that much prophesied transition that's supposed to be all about starting over, becoming the person you couldn't be in high school.

In the social-media age, all of this is changing—perhaps irrevocably—and particularly for the college set. Arrive on campus now and all of your new friends will be able to pore over your Facebook profile, ingesting the CliffsNotes version of your teens. For many, this is an uncomfortable realization. The clinical psychologist and technology researcher Sherry Turkle says that "this sense of the Facebook identity as something that follows you all your life is something that many adolescents feel is a burden." Identities are no longer toyed with, tried on and cast off, adopted for various settings or as a method of exploration. No, they're cloud-based, filtered through a standardized profile that never forgets. As Turkle says, "Now there's one identity that counts—it's the Facebook identity." It must be carefully tended to and managed, because it's the only one you have.

The relative decline of Facebook usage among young people may be attributed, at least in part, to this growing feeling of stasis. (The influx of older Facebook users, who render the network uncool and easily monitored by parents and other authority figures, also doesn't help.) Private and ephemeral messaging apps such as Snapchat, Kik, and WhatsApp offer young people—who are already used to cleverly managing their privacy when dealing with prying parents at home—an opportunity to communicate creatively with less fear of repercussion. Like e-mail, these apps aren't immune to eavesdropping, but they help return communication to a more protected space. Messaging apps are, however, illusory in the measure of privacy they offer. Just as Google did with Gmail—scanning private e-mails to serve up ads and contribute to the records the company holds for each user—there is little reason to think that messaging companies won't submit to similar tactics. Surveillance and advertising remain the industry-standard business models. And though Facebook, which purchased WhatsApp in a $19 billion deal, has promised to respect that company's privacy-friendly policies, it's hard to believe that one of the world's biggest data-mining firms won't make some use of all the juicy consumer information passing through its networks.

This is the problem when communication becomes inextricable from surveillance, data permanence, and publicity. From a social networking profile to one's Google search results, one's identity is increasingly a matter of public consumption. In some sense, each of us is now a public figure, thanks to the development of digital systems designed to make sure that Internet users are always locatable and identifiable by their real names, all so that they can be connected to a digital profile that reflects their tastes and habits. When these systems are combined with smartphone GPS data and the proliferation of advertising screens, sensors, cameras, and facial recognition throughout our urban environments, we are looking at a future where we will never be anonymous, even when walking down the street. The local barista may not remember your face or your order, but the sensor in the coffee shop's doorframe will, and it'll tell him to get started on that double espresso as soon as you pass by it with your smartphone. Advertisements will follow you throughout your day, using billboard cameras to recognize your face or a sensor in a bus stop to identify your phone. Once you submit and buy the new video game they're pushing, they'll harness your social graph and move on to your friends, imploring them, "Jacob bought this game this morning. Don't you want to play with him?" Based on their demographic data, your friends may be offered a higher price, but they won't know that.

This scenario raises some uncomfortable questions. What does it mean to be anonymous, beyond the ability to say something without attaching our names to it? Are we on the path to trading the freedom and flexibility of anonymity for the conformity of the named? Who really benefits from making social-media users employ real names and fixed, stable identities? Is it online communities, or is it the managers of social networks, the ad purveyors, the data brokers, and the intelligence agencies? Is anonymity a right worth fighting for, or has it been ruined by a host of bad actors?

- - - - - - - - - - - - - - - - - -

TELL US YOUR REAL NAME

Google's Eric Schmidt has cast himself as a philosopher-king of digital networking, assuring us that he has a clear eye of where these technologies are headed and how his company can avoid crossing the "creepy" line. He also has a tendency to sound somewhat detached from the very societal transformations his company is helping to foment. Consider some of his comments about the changing nature of online identity.

"For citizens, coming online comes to mean living with multiple identities; your online identity becomes your real identity," Schmidt once said. "The absence of a delete button on the Internet will be a big challenge. Not just what you say and write, but also the Web sites you visit, and do or say or share online. For anyone in the public eye, they will have to account for their past."

In the same interview, Schmidt raised concerns about online behavior, explaining that parents will have to talk to their kids about what he called "digital footprints" as much as they will about sex. Some parents, he speculated, may give their kids unusual or, alternatively, common names, depending on how they want them to show up in Internet search results—essentially practicing search-engine optimization, known in industry circles as SEO, from birth. (A few years earlier, Schmidt said that perhaps young people, in order to shed their digital trails, should be given the right to change their names upon turning eighteen. Of course, people already have that right.) He said that fake digital identities, complete with concocted records of online shopping, may become equally valuable to dissidents and drug dealers.

Schmidt was mostly speculating, riffing on recent history and trying to predict where we might soon be headed. His remarks carry some truth and his predictions seem possible, but what is stranger about it all is how removed Schmidt sounds from the very concerns he's presenting. This is, after all, one of the most powerful people in the technology industry, the executive chairman of Google, a company that has done as much as any other to push for the use of fixed online identities and established widespread Internet surveillance, and he's apparently concerned about the consequences of these same practices. If only someone could tell Eric Schmidt this! He might actually do something about it.

Demonstrating a shift in rhetoric, if not in practice, Facebook has been far more paternalistic in telling us why we must always be identifiable online. It is apparently for our own good. In an interview with Charlie Rose, Sheryl Sandberg, the company's COO, said, "The social Web can't exist until you are your real self online. I have to be me, you have to be Charlie Rose." Here is the airy rhetoric of authenticity, though what represents a "real self"? If I use the Tor software—favored by activists, hackers, and cyber-criminals alike to anonymize their Web browsing—am I being inauthentic? If I change my Facebook name so that an ex can't find me, am I being insincere? Company founder Mark Zuckerberg went so far as to impugn his users' character, explaining: "Having two identities for yourself is an example of a lack of integrity." His sister Randi apparently agrees. In July 2011, Randi Zuckerberg, who at the time was Facebook's marketing director, said: "I think anonymity on the Internet has to go away." She claimed that "People behave a lot better when they have their real names down."

Regulating behavior is an odd goal for a company devoted to connecting people. Such a policy can easily lead to measures to chip away at users' freedom of expression or to coerce them into certain actions. (Facebook's history of secret experimentation on users, along with its interest in boosting ad click-through rates, suggests that they are already deeply involved in the behavior modification business.) And there's little evidence that these sorts of real-name policies accomplish much. In the last decade, South Korea experimented with requiring real names to post comments on many Web sites, eventually requiring them on all sites that received more than 100,000 visitors per year. But so-called malicious comments only decreased by less than 1 percent, while people who posted frequent harsh comments appeared undeterred.

The more important question is not whether these policies work to reduce rudeness or antisocial behavior (the definition of which may vary widely not only between cultures but also among individuals), but whether companies should be allowed to impose such requirements on users. The U.S. Postal Service, for example, doesn't require you to use your legal name to mail a letter; why should digital media be any different? While many companies claim to worry about civility online, they in fact have financial incentives in establishing real names for their users. A user who is always browsing and posting under her real name is easier to track, monetize, and keep within certain bounds of approved behavior. And with U.S. social-media firms directly tied to the country's surveillance programs, your complete digital dossier is potentially available to the U.S. government. Seen this way, anonymity becomes closely linked to privacy, to control over who knows your identity and when they're allowed to know it. An assault on anonymity is an assault on privacy.

Facebook's anti-anonymity rhetoric is wrongheaded; it's also hypocritical. In May 2011, a report surfaced explaining that Facebook employed a public relations firm to urge journalists to air privacy concerns about Social Circle, a Google search feature that allows users to see search results that draw on their friends' social-media feeds. The PR firm, Burson-Marsteller, didn't reveal who its client was, but the relationship was exposed by a journalist for the *Daily Beast*. This type of mudslinging is common, although in this case, it represented a particular embarrassment for a company that preaches values of openness and transparency. Facebook—itself notorious for its fickle and confusing privacy policies, with each frequent change inevitably exposing more user information—was secretly using a PR firm as a front to drum up criticism of a competing company's privacy practices. Add to that Facebook's creation of what have been called "dark profiles" for people who have never signed up for the service, along with its habit of retaining information that users believed they had deleted, and one gets the sense that the promulgation of a real-names policy is but another element to gather as much information as possible, to make us transparent first and foremost to Facebook and its advertising platform.

A SINGLE LOG-IN

Not long ago, Web users had more options about how they conducted themselves online. Chat rooms, message boards, and online games invited us to employ whatever name we wanted. An e-mail provider would give you a mailbox and that was it; it wouldn't scan the contents of your messages in order to provide more relevant advertisements. Even early social networks offered some degree of flexibility, and at the very least, these sites stood alone. Your Friendster or Myspace identity, for whatever it was worth, wasn't connected to a range of other services. These services didn't spread widgets throughout the Internet that allowed them to monitor the habits of millions of people. In contrast, life online was about finding what you wanted and, occasionally, establishing a persona for a particular online community. We trawled the Web relatively unmolested; now the Web watches us and invites, or forces, us to identify ourselves at every opportunity.

Sites such as Quora, a question-and-answer forum where people can come together to solicit expertise or ponder life's big questions, exemplify this shift. As soon as you go to Quora.com, the site asks you to log in with Google, Facebook, or Twitter; if you sign up with your e-mail address, you are expected to abide by the site's real names policy. If you are linked directly to a question, a pop-up message might obscure your view as it prompts you to log in. Click your way past that (it's not easy to find the small link to dismiss the dialogue box) and you might be able to read the first response to a question, but likely no further. The remaining answers will be obscured, unless, of course, you choose to log in with one of your other accounts (or set one up with your e-mail address). The enticement here is that it's easy and that it seamlessly connects to other services you use all the time. For Facebook, Google, Twitter, and LinkedIn, all of which offer these kinds of open graphs or social log-ins, the appeal is obvious: it's one more way for them to spread their power beyond their walled gardens as they follow you wherever you go and collect more information on what you do and who you know.

All of this is a shame. It flies in the face of the more open-ended, freewheeling Web that many of us first alighted upon in the late 1990s. The Web then was perhaps even less civil than what we see now. It was easy to encounter shady characters in a chat room, or stumble upon a malware site promising free downloads of expensive software. There was also a degree of openness—of a sort totally different from that found in Zuckerberg's remarks, for whom openness is a way for customers to expose their lives to his company—and of intellectual freedom that seems on the verge of being snuffed out, if not subordinated to the sensitivities of Facebook's advertisers. It was a lot like an alternative newspaper (growing up in Los Angeles, I discovered *LA Weekly* around the same time I started poking around the Internet). You could read fascinating dispatches about culture and politics; you could also flip a page and end up smack in the middle of thinly disguised ads for drugs or prostitution. The Web and the alt-weekly were both anonymous, while the latter was free and the former rather cheap, provided you weren't being charged by the minute. Both of these media showed me information that was, at times, a bit beyond my understanding, but I turned out all right. And thankfully, many of my early explorations—through malware, porn, chat rooms, gaming, and elsewhere—didn't contribute to a permanent digital profile, nor were they syndicated in real-time feeds viewable by my friends and colleagues. I was allowed to explore what I wanted to without declaring myself or leaving a trail behind. Looking at something didn't automatically declare my interest in it, or allow a corporation to classify me accordingly and promise to serve me up similar content and ads. I went where my curiosity took me.

"We went from a Web that was interest-driven, and then we transitioned into a Web where the connections were in-person, real-life friendship relationships," said Christopher Poole, the creator of 4chan, the raucous, at times repulsive, but immensely popular online message board, where anonymity is treasured as an absolute right. "Individuals are multifaceted," Poole continued. "Identity is prismatic, and communities like 4chan exist as a holdover from the interest-driven Web."

I would go a step further than Poole. The social web treats everything, every personal encounter or article you read or thing you buy, as if it were a transaction between friends. Everything is perceived to reflect a deliberate intent—when you're shopping for new shoes, posting on someone's wall, or, whether for research or on a lark, you decide to read *Dabiq*, the Islamic State's English-language magazine. It all is supposed to be part of you, which is why it must be tracked. And yet even this process of tracking has difficulty measuring intent. There is plenty that I do for reasons that I couldn't articulate or that I don't tell my friends or my family, either because I choose not to or I don't think it'd be interesting to them. (And there is much more that I would prefer not to tell companies monitoring my clickstream.) I sometimes act differently in front of my parents than I do in front of my partner or best friend or a police officer. This kind of "prismatic identity" might shock Zuckerberg, who would accuse me of inauthenticity. The truth is that we all do things like this. I have a couple of friends who are comedy writers, and when I'm with them, I become a little more eager in my jokes, looking for anything to riff off of, enjoying the sense that everything is material and that we are all trying to entertain one another. I doubt anyone who knows me would mistake this for insincerity; it's a performance, as one's identity often is, and quite deliberately so in this case.

On an identity-driven, persistently surveilled Web, discrete bits of information matter more for what they say about us and how they inform our public demonstrations of identity. As the Danish academic Anders Colding-Jørgensen argues: "We should no longer see the Internet as a post office where information is sent back and forth, but rather as an open arena for our identity and self-promotion—an arena that is a legitimate part of reality, just like our homes, workplaces and other social arenas in our society." We've moved, he explains, from an information economy to an identity economy. This is a bit self-serving—commentators have developed no shortage of dubious new types of "economy," from the "attention economy" to the "knowledge economy"—but Colding-Jørgensen is onto something. Our consumption of information online has shifted from purely utilitarian to an expression of the self. This is the paradigm of "Pics or it didn't happen," where every incident is worthless without shareable documentation, because our experiences are made fuller by being shared. Even what we might think of as plainly utilitarian—a recipe, for instance—becomes an object for sharing and identity-crafting. Whereas a decade ago, you might've downloaded or printed a recipe and cooked from it, now you might find a recipe, ask others whether they've used it or have comments, cook the dish, photograph and share the dish on Instagram before eating it, and finally offer a rating or comment on the site where you found it. The relatively straightforward act of finding a recipe and preparing it becomes bound up in complex questions of identity and self-image—Do you want to seem domestic? Do you share this on Facebook or your more exposed Twitter account? Do you take an elegant, well-lighted photo of the prepared meal, or one of your date happily chowing down?

It's these calculations that show how illusory the notion of authenticity is. We can be deliberate in shaping our public presentation, but that doesn't make these gestures insincere. Each of us is engaging in practiced, sometimes Machiavellian calculations about how we want to present ourselves and what we might want to get out of it, and there's no inherent shame in that. Our motivations are complicated, our identities multifaceted. Some of Japan's biggest social networks allow pseudonyms, and yet the country is awash in all sorts of digital interactions and eruptions of new cultural phenomena, from cell phone novels to virtual pets. A person might value his online pseudonym—I still have a soft spot for the one I used for many years in various online role-playing and action games; he exists as a distinct character in my mind—precisely because it is a form of expression, bound to certain experiences. And indeed, handles, avatars, and the other raw ingredients of online identity have long been treated as types of expression and play, things to be tried on and cast off, manipulated and customized. Markus Persson, the creator of the enormously popular game Minecraft, is widely known as Notch, and the nickname is no less real or authentic because it originated online. His continued use of it, both online and off, only shows how much he values it.

Our digital and offline lives are more intertwined than ever, and in some respects, that's a good thing. These two worlds have never been fully separate. Actions in one arena can easily affect us in another, and the notion that the digital is all illusory has often been employed as a justification for trollish behavior online. A conversation on Facebook is no less real than one on the phone, though each medium offers different possibilities of interaction and may produce varying complications. I might prefer one to the other, but they both exist and whatever I learn in one happened to me as surely as an in-person encounter. What is important is that I have the freedom to do these things and that I am not forced to tote around my Facebook identity just to access other services. Identity shouldn't become an unshakeable shadow.

The ultimate irony of an identity-driven Web where one is pressured to use a single log-in across many sites and apps is that it actually makes us less secure, in more ways than one. Knowing that every interaction is linked to our real-name accounts, we find it easy to become neurotic about what might become part of our digital records or what might be shared, without our consent, on the home platform. Surveillance is nothing if not a form of pressure, in its capacity to cause us to preempt our usual habits, knowing that we're being watched and recorded. It may also cause us to share more in order to alleviate that anxiety, in pursuit of the same nebulous degree of authenticity promoted by Facebook. We feel the need to post more in order to demonstrate our real selves, to overcome the strictures of Facebook's rigid environment.

This instinct also emerges on LinkedIn, where the site features pop-up messages and alerts telling users that they should fill out their profiles in order to make them more complete, to have a better experience, or to "quickly grow your professional network." Information sharing will improve your LinkedIn experience, which will, according to the site's mission, boost your value in the world. Similarly, Facebook

sometimes prompts me to input my phone number; this is for security purposes, the site tells me, so that they have another method of
verifying my identity. But in the same dialogue box, I'm offered the
option to show my phone number to my friends. That giving Facebook
my phone number makes my experience there more secure is, on its
surface, somewhat dubious, though the site uses text messages to verify potentially compromised accounts. At the same time, the overlap
here of promising security while also encouraging disclosure of one's
phone number to friends in the interest of openness or authenticity is
revealing of Facebook's motives: the more personal data they can get,
the better.[*]

- - - - - - - - - - - - - - - - -

NAME AND SHAME

In certain quarters, digital anonymity has become a precious
commodity—for dissidents, activists, journalists, and as a cultural
value in and of itself. On the social news platform Reddit; in the madcap, all-anonymous message board 4chan; in the hacker collective
Anonymous (whose roots trace to 4chan)—in these and other online
communities, anonymity is something to be treasured and protected.
Chalk it up to scarcity, perhaps. Here an assault on one's anonymity is
considered a grave act.

The act of unmasking an anonymous Internet user is often called
doxing. Doxing isn't always done on purpose or with the intention of
harming someone. Doxing can be accidental or out of the belief that
someone deserves to be publicly recognized. It's this very mutability

---

[*] And if my account were to be compromised, would I then want my phone number
accessible to some hacker? While working on this book, someone logged into my
Facebook account from a Ukrainian IP address. I have no idea how it was done,
but Facebook locked down my account and notified me via e-mail. I was glad that
Facebook acted promptly, but I was also glad that I hadn't given them my phone
number or address.

that means that the ethics and eventual consequences may not always be clear. (Whoever doxed J. K. Rowling as the pseudonymous author of the novel *The Cuckoo's Calling* was likely interested in getting more recognition for the book but may not have anticipated how much this act would anger Rowling herself.) Anonymity can be a tool of power or a way to fight against it; it can also be relatively benign. But many of the most notorious cases of doxing are tied to a desire for revenge over some perceived slight. And there are still other instances in which the wrong person has been doxed, leading to harassment. This kind of doxing doesn't differ much from, say, the *New York Post* rushing to name a suspect—recklessly and wrongly, as it turned out—in the Boston Marathon bombing. The goal is the same: make someone infamous, so that they can suffer the consequences. That's why doxing can seem freighted with grandiosity and self-righteousness.

The treasuring of real names, of names as a private thing, brings to mind the use of secret names in traditional oral cultures. Anthropologist Claude Lévi-Strauss famously manipulated members of the Nambikwara, a preliterate tribe in Brazil, into revealing their proper names to him. Among the Nambikwara, proper names were forbidden, so Lévi-Strauss and his colleagues tried to assign what he called "arbitrary appellations," or nicknames, to members of the tribe. (In our own culture, we might think of Internet screen names or the call signs granted to fighter pilots.) In *Tristes Tropiques*, Lévi-Strauss recounts an incident in which he was playing with some children when one girl came up to him and began whispering in his ear: "Out of revenge, the first little girl had come to tell me the name of her enemy, and the latter, on becoming aware of this, had retaliated by confiding to me the other's name. From then on, it was very easy, although rather unscrupulous, to incite the children against each other and get to know all their names. After which . . . I had little difficulty in getting them to tell me the names of the adults."

Essentially, Lévi-Strauss was engaging in what hackers call social engineering, cajoling and tricking his subjects into sharing privileged information. He got them to dox one another and to think that it was to their advantage. The girl who initially revealed her enemy's name was doing much the same thing. When names are private, when they reveal something fundamental about a person, there's power in revealing them—or threatening to do so.

Like privacy, anonymity is about preserving control over what someone knows about you—in this case, that most fundamental of identifiers: your name. In a networked, data-rich society, knowing someone's name is potentially a way to know all kinds of other things about her. Imagine if you were to walk down the street at all times with a sign above your head telling everyone your name, how to contact you, and other information about your background. That's how we appear to trackers, ad networks, and other companies online.

Doxing is closely tied to the concept of public shaming, which has found new forms on social media. Shaming and viral villainy are made all the easier by the use of real names and the ways in which data travels between social networks. The practice is flexible, as easily applied to an airline that's mistreated a passenger as it is to a relatively unknown Twitter user from Nevada guilty of tweeting a racist epithet. Shaming remains problematic because of its close association with vigilantism and because, in its leveraging of viral channels, it can spin out of control, producing a disproportionate response. The hive mind may respond with a dozen people tut-tutting, only to then melt away, or it may be ten thousand people issuing death threats, publicizing the target's address, calling his employer, and ensuring a permanent data trail of shame and embarrassment—what has been termed "SEO-shaming," after the practice of gaming Internet search results.

When is someone taking the initiative to dox or shame another person a courageous act, or, at the very least, an effort to defend an injured party, and when is it self-righteous or malicious? Such standards aren't clear, in part because, as Danah Boyd notes, "the same tactic that trolls use to target people is the same tactic that people use to out trolls." Both sides in a conflict may be engaging in similar behaviors but toward very different ends.

It's easy for a shamer to come across as a bully, particularly when the shaming is directed at someone with little renown or power of his own. Like satire, shaming seems less effective, and less conscionable, when someone punches down rather than up. In one notable incident, the feminism and pop culture blog *Jezebel* publicly called out a dozen teenagers who tweeted racist remarks after Barack Obama's reelection. The site went beyond posting the tweets by researching the students, writing short bios for each, and contacting their schools. While the students' conduct was abhorrent, they were minors, and the manner in which *Jezebel* went about publicizing their own behavior offered the impression that the act was more about allowing *Jezebel* to grandstand as a moral authority and to rack up page views based on the resulting controversy. *Jezebel* could as easily have contacted the students' schools—the kind of institution of authority that might be able to positively influence the children's behavior, or, perhaps, enact some punishment in concert with the children's families—and written a story about the experience while also keeping the students anonymous. Instead, the site ensured that, for many of these students, they would spend years trying to scrub the Internet of their bad behavior, while likely nursing a (perhaps understandable) grievance toward *Jezebel*, rather than reforming their own racist attitudes. It's easy to forgo self-examination when you, too, feel like a victim.

There's a self-aggrandizing element to public shaming—the unearned self-regard of the mob leader. It tends to privilege dramatic gestures of pique and knee-jerk outrage over quieter or private efforts to engage, educate, and criticize. It can allow one party to call out another's bad behavior, while also overlooking complicating issues of class, power, and influence.

But it's not always like this. At their best, public shamers find common ground with activists whose goal is to show that, contrary to the conventional wisdom that we live in a post-racial or socially progressive society, racism, sexism, and other forms of discrimination are still endemic. It's toward this kind of end that public shamers should dedicate themselves: surfacing examples of abuse and injustice. Twitter accounts such as @YesYoureRacist and @EverydaySexism or the "Public Shaming" Tumblr are most useful at showing that these phenomena are still very much alive. They allow a wide public to see that discrimination is still often expressed with extraordinary callousness and casualness. They can also provide spaces for people to come forward and share their experiences. Many people are unaware how cruelly women are treated online, especially when they try to speak out on controversial issues. These platforms can be sites of ongoing conversation, where alliances can be made and important issues aren't allowed to recede.

I talked to a couple of people involved in public shaming, partly as a way of working through my own ambivalence toward the practice. I wanted to hear what some of these people had to say for themselves and how they viewed their behavior. Matt Binder runs the "Public Shaming" Twitter and Tumblr accounts. A producer for a political radio talk show, Binder began highlighting racist comments on Twitter in the run-up to the 2012 presidential election. Binder specializes in emphasizing the hypocrisy of some of his targets—for example, he's found young Middle Americans who complain about the laziness of food stamp recipients, only to discover that in the past, these same people have tweeted about being unable to find work. The implication is that these young people are, if not lazy themselves, then victims of the same economic system as food stamp recipients, but their racism and classism leave them blind to this fact. From this kind of myopia comes the site's tagline: "Tweets of Privilege." Sometimes, Binder will pick out racist responses to a news event, retweeting them with an added bit of commentary or arranging a dozen on a Tumblr post and contributing his own sardonic remarks. His efforts have found him a wide audience: when we talked in 2013, @MattBinder had about 11,000 followers, @PublicShaming had 3,000, and his Tumblr, with more than 60,000 followers and surges of traffic around pop-culture mini-events such as Marc Anthony's MLB All-Star Game performance, was sometimes listed among the top 10 most popular pages on the social network—an impressive distinction for one of the world's most-visited blogging platforms. (Binder also shares his Public Shaming posts on a Facebook page of the same name.) Mainstream news outlets have picked up on his posts, sometimes borrowing them wholesale, and a few ads appear on the Tumblr but only enough, Binder says, to buy lunch once a month.

Binder's presentation mixes righteous outrageous with acid sarcasm. He's happy to mock his targets. "It definitely needs to be entertaining—otherwise people aren't going to pay attention to it," he said. "Social justice blogs and sites like that have been around a long time. Not every one of them has blown up as big as this site has."

At the same time, he recognizes that some online opprobrium is unlikely to provoke contrition: "A couple people telling them online that they're idiots isn't going to change their outlook." Instead, he sees his role as surfacing incidences of hate speech and making them visible to a wider audience, even if the person making those remarks doesn't understand their impact. "A lot of people don't even realize what they're saying," he said, "and even when they do, they don't seem to have a problem with it, so they double down on it. It's kind of shocking." Binder's site then exists more as an example to others, particularly liberals who may have developed some sense of complacency about social progress. Many of his readers, he said, also aren't politically attuned and may not be aware of the kinds of opinions spouted on social media. "The point of the blog is sort of to show people that these opinions exist and these types of people still exist."

I asked Binder how he responded to accusations that his site was self-righteous or damaging to those he targets. He was unconcerned, explaining, "I don't really care about bullying people who are assholes." He added: "If you saw a kid getting bullied and you went and stood up to [the bully], would you be considered a bully because you stopped" it?

Binder doesn't believe in censoring the vile remarks of others, but he does have some limits. He won't go after anyone who "looks especially young." He doesn't pick out anonymous or pseudonymous accounts, because he wants to show people who are willing to broadcast their awful opinions under their own names (and, often, with a photo of themselves attached, along with other personal information).

Logan Smith, who runs the @YesYoureRacist Twitter account, has shown more leniency. He retweeted someone's racist comments only to backtrack after seeing that the man's account contained some remarks about suicide. "I really don't want to be responsible for pushing someone over the edge," Smith told me. Another man was in basic training in the military; he apologized to Smith, and though Smith doubted his contrition, he removed the tweets because he didn't want to jeopardize the man's career.

Smith went to college in South Carolina and lived there afterward for another four years. Every day when he drove to work, he said, he drove by the state capitol, where the Confederate flag still flew. A local barbecue restaurant chain also flew the flag and distributed segregationist literature. Later, he moved to Raleigh, North Carolina, where he's found work in progressive politics. The racial climate is somewhat better there, but he still lamented the state's passage of a restrictive voter suppression law. There's something in common between Smith's everyday activities—his political activism and the institutional racism he's observed in these communities—and his Twitter project. (When we spoke, Smith said he was working on a book project that would also examine the history of racism in public policy in the South.)

His account is particularly interesting because he seeks out people who start by hedging their comments—"I'm not racist, but . . ."—only to spill out remarkably prejudiced comments. Each of these tweets arrives with a sort of cognitive dissonance baked into it—a prophylactic denial of being racist followed by a clear example of that very sin. Smith explained how widespread he's found this phenomenon to be: "It's really opened my eyes to how many people, especially young people, don't seem to understand the concept of racism. It seems like they think that unless you're out there lynching somebody or burning a cross in someone's yard, then you're not a racist." Smith's not as gleefully belligerent as Binder (the two aren't well-acquainted but spoke respectfully of each other; they've also had contact with other people using social media to highlight homophobia and racism); but he has a similar attitude toward the cause. "I'm not at the forefront of the civil rights movement," Smith said. "I do not have any illusions about that. I simply found a simple method of publicizing racism."

Talking with Binder and Smith left me feeling more approving of their efforts, though the former's enthusiasm, as well as his self-identified status as a bully for a good cause, left me a bit uneasy. Perhaps it's that they operate from a position of security—liberal white men upholding respectable values by pointing to the buffoonery of others. They're not risking much, and they are mostly preaching to the converted. It's possible to detect a halo of sanctimony, though Smith's experience in progressive politics showed him to be a thoughtful activist. But there's no doubt that the comments that they seize on are vile, and there's something to the claims that many people, especially young people, seem to think that their online racism is somehow hidden or doesn't count as racism, particularly if it's presented with a caveat. Exposing racism is, on its own, a laudable goal, and there should be some social cost to being a bigot, no matter the form in which it's presented.

Shaming can be effective if it's directed toward worthy targets—corporations guilty of discriminatory behavior, powerful figures who deserve to be called to account, a pattern of destructive behavior by a community leader that has received insufficient attention. A random Twitter user with a few hundred followers, unused to being in the spotlight or interacting with traditional media, will likely recoil in the face of a public shaming. He'll adopt a defensive posture, as his friends rally around him or laugh at his sudden exposure. He might delete his Twitter account, as did many of the students shamed by *Jezebel*. And he might be more guarded in his public postings, which by some measures would constitute an improvement. But he's unlikely to embark on any real soul-searching. Shaming people in positions of influence strikes me as more useful in fomenting social change.

The court of public opinion, however, is likely to be an increasingly busy place in the coming years. Going viral can be seen as a threat, as Taylor Chapman thought when she filmed her tirade at Dunkin' Donuts. In an online environment in which we are always visible and named, reputation is an increasingly valuable commodity. Damage to one's good name can seem equally perilous, which is why so many online disputes escalate so quickly and why their private, muted resolutions receive less attention than their explosive beginnings. Given a traditional legal system that often seems rigged for the wealthy and the powerful, online speech, despite its quirks and limitations, feels like a more honest, democratic place in which to litigate one's problems. Our ability to exact justice or defend ourselves, it can seem, is only limited by our eloquence and appeals to reason and emotional honesty. Unfortunately, that's not always true; the fallacy of social media as an inherently meritocratic, democratic space cuts both ways. Corporations are making their own efforts to game social media to their advantage, employing sentiment analysis, consultants, and always-on PR and social-media representatives to nip any crisis in the bud. Your complaints about mistreatment by an airline may mushroom into a mainstream news story, or they may be snuffed out by an attentive social-media marketing officer, who responds to your tweets, monitors your mentions, and strategically buys sponsored tweets to appear in your followers' feeds. A campaign against an offensive newspaper columnist may fall apart as his powerful colleagues fall into step behind him, leaving the protesters less powerful than when they began. And once made, accusations can't be withdrawn. They can only be revisited or corrected, followed up upon in the same way we often over-share to make up for some crappy joke or faux pas we suddenly regret. Meanwhile, some archival record lives on: in search results, in someone's screenshots, in the disembodied audience to whom we've made our appeals.

What's the end game? What kind of consequences do we want for the shamed? More speech is usually a good thing; bad speech can be countered with good speech, our liberal impulses tell us. But that is again to assume a level playing field. As it is, the cyclonic effect of social media, its tendency to act as a perpetual outrage machine, can be wearying. Depending on the size of your networks, you may be used to seeing another scandal, another villain, every day. These overheated campaigns tend to run together, even as each provokes a need to comment, showing that we care and that we are on the right side of—well, not history, but some progressive sensibility shared by others in our timelines. There's certainly a place for shaming and declaring your anger, as there is for other forms of protest. Sometimes we need to point at something and say, this is it, this is the thing itself, we must do something about it. But in a mediascape where attention is scarce and valuable, there is power in refusing to grant it.

THE MERITS OF ANONYMITY

The safety of a pseudonymous Twitter account might encourage some people to be trollish, but it also allows for the ability to speak freely without fear of consequence. A future political commentator might find his footing by starting out under a pseudonym. A woman used to being harassed online might find a respite by shedding her female identity for a while or adopting a new name. It allows us to determine who we are on our own terms. To that end, given the increasing recognition of gender as fluid and gender identity as something personally defined and mutable, it is surprising that it took Facebook a decade to add gender options besides male and female. Its introduction of a few dozen different gender options is an improvement but far from the ideal, and simplest, solution: a blank box, in which users can decide what they want to write, if anything at all.

Anonymity need not be seen as only a form of digital refuge. In a society besotted with publicity and granting credit for everything, it can be liberating to reject all of this. Anonymous expression has a rich tradition, from the Federalist Papers to graffiti. An anonymous publication can also have the feel of being a stunt, as evidenced by the speculative furor surrounding the identity of the author of *Primary Colors*, before it was revealed to be the work of *Time* magazine journalist Joe Klein—someone surely familiar with the machinations of publicity and media fame. That said, writers such as Stephen King and Doris Lessing have published novels under pseudonyms in order to see how the works might be judged. And in a surveillance society, where power is known to act capriciously, the right to anonymity, and anonymous expression, should be treasured. Anonymous protest is both a prudent tactic and a savvy way of keeping attention on an issue, rather than the people engaged in some act of subversion. In the process, the markers of anonymity—such as the Guy Fawkes masks worn by the Anonymous collective and other loosely associated leftist protest groups—help to bind individuals together as part of a community devoted to a larger purpose.

Anonymity can also improve digital security, especially when your digital persona is networked across so many platforms. A hacker might gain access to one of your social-media accounts only to find that they can use it to log into numerous services, essentially gaining control over your entire digital life. Do you know how many apps you've authorized on Facebook or Twitter? Do you know what each of them is allowed to access, and what each app's data-sharing policies are? Probably not, but don't be hard on yourself: these permissions settings have become like the terms of service agreements that we all consent to every day and that almost none of us reads. And even these agreements are often broad and intentionally vague, so that an app or site claiming that it will only share your data in ways meant to provide you with more relevant services can easily translate to: We'll sell your data to whoever comes calling. One study examined fifty health and fitness apps and found that the free apps were more likely to sell personal data—information that would be less valuable, and less harmful, if it's not tied to your real name and social-media profile. Consequently, it's important to keep a close eye on these apps, particularly on ones you haven't looked at in a long time, because you don't know what they might do with your data. Like the popular Facebook groups that suddenly and surreptitiously change from supporting some charitable cause to promoting a credit card, third-party apps and sites may find themselves under new, less scrupulous ownership. By accepting the terms of service agreement for Facebook's own Messenger smartphone app, you authorize the app to make calls, send texts, take photos, record audio, read your contacts, read your call log, and look at personal information in your device settings. (The app also requires users to opt out if they don't want the app to automatically append their location to each message they send.)

The battle between real names and anonymity need not be zero sum. Unfortunately, Facebook and Google have done their mightiest to make it so. But remaining anonymous and presenting oneself publicly should be practices that can coexist. As in the physical world, one should be able to move between digital spaces that are anonymized and others in which one's real identity is needed and useful. But the trend is clear: anonymity is under assault. If the positions were reversed—if, perhaps, Facebook had a financial interest in preserving anonymity as a value and technological capacity—then it might be the single-identity faction who would be demonized and would have to defend their practices. Theirs, after all, is the more stringent ideology. But that is not how the powers are arrayed. Anonymity certainly has a public relations problem, and some defenders of anonymity haven't done enough to recognize how anonymity can be a facilitator of some of the worst behaviors of online abusers: death threats, misogyny, stalking and verbal abuse, racism, and so on. But it's also important to distinguish between that which anonymity enables and that which would be otherwise impossible without the ability to be anonymous. Most of these horrible behaviors are not solely a product of anonymity, much less a shared value of those who do believe in defending anonymity. Some people choose to act in such a way because they know they can hide behind the protection of a pseudonymous Twitter handle or a 4chan message board in which no one is ever identified. But these actions often take place online anyway, just as they do in the analog world. Anonymity may occasionally be a shield for sexual harassment, but it's not a cause. To stamp out anonymity with the intention of making a more civil or humane online environment is to choose a technological solution that merely papers over the underlying social and political problems. Rape culture, misogyny, the marginalization of minorities—all these and more can't be fixed by making people register for Facebook under their real names. Plenty of real-name social-media accounts are the bearers of despicable messages, as are talk-radio hosts and mainstream politicians. Features to report abuse and block users are therefore necessary and helpful in policing bad actors. But the major social networks have become victims of their own rhetoric. In promoting Facebook as an electronic agora of limitless connection, Facebook has created a sanitized version of human life, one that bears little version to the physical world it claims to represent. It has also helped to create a digital culture which, rather than working to tackle existing social issues, often devolves into a cacophony of anger and recrimination whenever an unsavory person finds himself with a megaphone—a cycle that can lead to calls for further clamping down on digital speech. Facebook and its users would benefit from recognizing that, if the company hopes to link people together in massive numbers, a range of behaviors will inevitably appear, because that is how human beings act. Trying too hard to guide these behaviors is likely to result in manipulation and a crackdown on anything but the most prudish forms of speech.

THE POWER OF REAL NAMES

It's important to understand the motivations of those calling for real names online and the potential future implications of these practices.

When Arianna Huffington said that anonymous comments must be abolished on the *Huffington Post*, the enormously popular news site and aggregator she founded, she was calling for a change in policy that would serve her own business practices. If HuffPo were to adopt real names, say, by replacing its commenting system with one provided by Facebook, it'd immediately become both a partner to the social network and a node in its data-collection apparatus. HuffPo would then be able to build up even more data about its millions of monthly readers. It would know even more than it already does about who they are, what they read, what they buy, and what they think. This information would be helpful for the site's targeted advertising efforts, and it'd be quite valuable to commercial partners. It also would offer another tool for banning unruly commenters.

Banning anonymity is, in short, a strategy of the powerful. At minimum, it allows for greater data collection or control over a communications network. At its worst, it's a tool for authoritarian governments to monitor and track their citizens. That's not to say that governments don't like anonymity—when it serves them. For decades, U.S. presidential administrations have employed anonymous leaks in order to plant stories, guide public opinion, or stave off controversy. Despite occasional harrumphing from critical journalists and media observers, the practice of anonymous sourcing continues because national security journalists fear losing access or being scooped on a story. Tor, the free anonymizing software that allows for covert Internet browsing, was originally sponsored by the U.S. Naval Research Laboratory to help democracy activists communicate overseas. Another U.S. government agency, the NSA, has been trying to break the Tor network ever since.

There are, of course, many other instances in which we'd like to preserve anonymity: voting, visits to doctors or lawyers, double-blind medical studies, buying porn or a gift for a friend. Anonymity can also be a way of removing certain motivations—greed, ego, vanity—from practices that might be better served by a form of silent cooperation or unacknowledged altruism. Maimonides, the twelfth-century Jewish polymath, ranked various levels of tzedakah, or charity. Among the highest levels was a donation where neither the giver nor the recipient knew the identity of the other party; they're anonymous to each other and so no one can claim excessive pride or credit. The recipient also retains some dignity by not having to know the donor. The expression of charity can stand alone, just as an anonymously published essay or a pseudonymous Twitter account can be judged on its own merits, without worrying about the confounding roles of follower counts or influence metrics. Anonymous online speech mitigates some of the obligations that come with digital publishing: incessant promotion, worries about audience composition, appeals to the whims of advertisers and members of one's peer group or professional network.

In a time of precarity—widespread unemployment, record income inequality, rapid technological change, looming environmental calamities—identity has become ever more fixed. One would think that it should be the opposite—that with society, and job markets in particular, in such a state of flux, people should have more flexibility to define themselves as they wish. Perhaps you are a Muslim immigrant living in a midsize American city and wouldn't mind posting background information on your Facebook profile. But then your neighborhood begins to feel the effects of a recession: houses foreclose; crime goes up; a neighborhood watch group forms and soon starts spouting noxious anti-immigrant rhetoric. You might decide, with some reluctance and sense of inner conflict, to change your name and avatar on Facebook to something that won't mark you as Muslim. Again, you may not feel good about it; but perhaps that's what you choose to do so long as your neighborhood feels unsafe, or before you can move, or before you can try to rally some neighbors to support you or talk with law enforcement. Under Facebook policy, you would be allowed to do no such thing. What's more is that if you attempted to make a similar change with other social-media accounts and the various online services to which they're connected (at the time, having your Facebook log-in as a near-universal Internet ID seemed so convenient), you might run into a host of competing policies, with most of them tending to push you give them as much information as possible.

Consider another scenario in which you were fired from a job doing data entry for twelve dollars an hour. Your boss was actually a real pain—among other things, he was always pressuring you to have drinks with him after work—and it led to so much friction that eventually you were let go. Perhaps you want to be able to post an anonymous review of him on a job board, without fear of retaliation, or simply to vent anonymously (and without naming your ex-boss) on your Google+ account. It was also the kind of experience that you don't want to explain in future interviews; it was only a couple of months of low-wage work, and you believe that it's your right to withhold this information from another employer. Yet by the logic of LinkedIn, you would disclose all of your employment information, along with your educational history, your entire career network, and every skill in your quiver. Your account also would be viewable by anyone who happens to search for you on LinkedIn. Sure, it's a truism that people fudge their résumés, but wouldn't a little obfuscation, some strategic elision, be understandable here? And wouldn't you also want to look for a new career without your imperious ex-boss looking over your shoulder? On LinkedIn, you can't: the site tells you who's been looking at your page. It doesn't even include a block feature—a design choice which, along with the site's tendency toward high-volume spamming and too-intrusive "you might know" reminders, has earned it the label of "the creepiest social network."

It's in conditions very similar to these that Internet entrepreneurs in recent years have been touting the virtues of public identities that serve as advertisements for yourself. Developing a personal brand, selling yourself, maintaining public profiles, monetizing the things you already have, be they underutilized expertise or a spare room in your house—the good neoliberal subject is someone who is always hustling and available. Among the most popular products in this "sharing economy," as its partisans call it, is Airbnb, a service that allows people to rent rooms, houses, and apartments directly to each other. The idea is that the supposed inefficiencies and extra costs—taxes, insurance, maintenance—of hotels can be bypassed. Your house is going to be empty for a week anyway while you're visiting your in-laws; why not rent it out? There are problems with this kind of economic philosophy, which I'll get to later. But first, consider the unintended consequences of these kinds of transactions, which rather than being anonymous, cull information from social-media profiles in order to bring people together.

One of the nice things about booking a hotel over the phone or online is that they don't know much about who you are. Yes, hotels may be collecting some data behind the scenes, perhaps to market to you better in the future, but they have no interest in turning you down as a patron, nor do they have the legal right to do so based on your identity.

Those rules don't quite apply with Airbnb, which encourages users to log in with their Facebook accounts and provide detailed profiles. Some Airbnb users have reported being discriminated against because of their appearances. Franklin Leonard, who owns the Black List, a script discovery and reading service in Hollywood, recounted to me a story about trying to book Airbnb lodging for a business trip. Leonard travels frequently and had had success with Airbnb in the past. Six months before a film festival, he tried to rent a house in Austin for himself and several employees, but the owner refused, saying he wouldn't rent anything more than three months out. Leonard offered 20 percent above the list price but was rebuffed. Three months later, he tried to rent the same house and was rejected several more times, despite offering to pay well above the listed rate. Leonard was baffled. He e-mailed the owner asking what the problem was. "He responded by saying that he wasn't not renting it to me because I was black and that he had rented it to a member of Jurassic 5 only recently," Leonard told me.

It was the e-commerce version of "some of my best friends are black." The homeowner was conscious that his behavior *appeared* racist, so he tried to preempt it by saying that of course his decision wasn't racially motivated. But the message had already been communicated: Leonard and his employees weren't welcome there, and the reasoning was fairly clear.

Leonard began changing how he styled his Airbnb profile. "As it is, I do everything I can to make clear that I'm a responsible tenant," he said, "typically either mentioning that I'm in town with the company that I founded and run or that I'm traveling with my fiancée.

"After this event, I actually changed my profile photo on Airbnb to one with my fiancée [who isn't black] and I together. Sad but true. Got the idea from another black man on Twitter who had had similar experiences."

Despite tweeting a complaint at Airbnb, which encourages Twitter feedback from users, Leonard never heard from the company. He had little recourse but to take precautions in how he presented himself—to, oddly, show less of himself on a site that encouraged him to be himself. And his experience isn't unique: one study by researchers at Harvard Business School found that black hosts charge on average 12 percent less than nonblack hosts for comparable rentals. The study concluded that black hosts have to charge lower prices in order to overcome the racism of some renters. The authors also recommend that Airbnb adopt measures similar to those of online marketplaces such as eBay, where sellers and buyers don't need to share their photos and names.

Racism disappears no more with anonymity than it might with face-to-face encounters. But stories such as Leonard's provide a reminder that the promised bonhomie of transparency can be elusive. Rather than bringing people closer together, insisting on furnishing full identities in situations such as these can lead to discrimination, abuse, or other fractious social behavior. A person's authentic self may be a racist one. Sometimes people don't want or need to be seen, and there are good reasons for that, which should be respected. Anonymity can be a refuge. It can also just make life easier, which is why booking a hotel online is more frictionless than negotiating the personalized stalls of Airbnb.

The German film director Werner Herzog once said in an interview, "We have to have our dark corners and the unexplained. We will become uninhabitable in a way an apartment will become uninhabitable if you illuminate every single dark corner and under the table and wherever—you cannot live in a house like this anymore." Herzog, a delightful eccentric, was railing against psychotherapy. But while his antipathy toward Freud's science might be overblown, his point is equally applicable to digital life. The social web combines a confessional society—with its privileging of openness and self-revelation for its own sake; its equating of brute honesty with virtue—with the totalizing demands of a surveillance state. When we are always identified, we step further down the path of the totally illuminated world that Herzog fears. We don't need to revere subterfuge, being withholding, or even lying, but we don't need to eliminate these things either, whether by custom or a programmer's directive. We need to allow for ambiguity, the freedom to do wrong, the freedom to be not yourself but some other self. Shade gives texture to life's landscape.

In the same interview, Herzog also expressed some appreciation that there are many false versions of him on the Internet—people imitating his distinctively dour German accent on YouTube, fake Herzogs planting flags on deceptively official-looking Facebook pages. "What it's about I don't know," he said, "but I welcome it, because I see them as some kind of protectors around me. As though they were bodyguards." Herzog is speaking from a position of privilege—he's a successful director of art films who's never cared about the expectations of others. But there's something wonderful about his philosophical bent. He's unintentionally offering a defense for the practice known as deliberate obscurity, putting out false traces and concocted data trails to give would-be surveillers a misleading sense of who you are. His words also call us to tolerate uncertainty, to find that there can be something weird and interesting about not knowing who we're dealing with or who even might claim to be us. This sentiment is the animating joy behind the verve for strange bots—performance art projects, moody spam bots, automated accounts that mimic our tweets upon request. Of course, there can be consequences of such practices—identity theft comes to mind—but we might also discover a measure of freedom by surrendering some concern. We would replace the cynical, verify-yourself skepticism of the current moment with something more creative, unbounded. Along the way, we might even regain a measure of security, for acknowledging identity as fluid and self-directed would make for a more interesting, and trusting, culture.

To reach such a state would require that our actions online go untracked, much less be used against us in ways we don't expect. Unfortunately, an entire industry has built up around doing just that.

The Reputation Racket

They're becoming increasingly wary that their lives are going to be no longer their own.

—John Pezold, Georgia state representative

If you've ever used Uber, the ride-sharing service and smartphone app, you're probably familiar with rating drivers. After each trip, a passenger is expected to rate his or her driver on a score from 1 to 5. Too many low scores and the driver's job is in jeopardy. What many people don't know is that drivers rate their passengers as well, and these scores are usually kept hidden from passengers. There are only a few ways to get them. You could ask an Uber driver to reveal your score, or you could call up Uber's corporate office and see if they'll share it with you. Or if you happened to be online on June 27, 2014, you might have found the method publicized by Aaron Landy, a software engineer. Writing on the Web site Medium, Landy showed that a person can find his Uber score by going to Uber's mobile Web site, logging in, bringing up the browser's console (essentially, a command line for Web browsers, where one can look at and tinker with a Web page's code), and executing six lines of Javascript. No one had discovered this method before in part because it was somewhat elaborate, but also because Uber didn't *want* anyone to find it. (It's also usually not a good idea to run Javascript code with which you aren't familiar; it can be a quick ticket to allowing some unscrupulous operator to take over your machine.)

If you correctly followed Landy's method—which was soon reported across dozens of tech sites—a pop-up window would show you your score. But you had to be quick on the uptake because by the next day, Uber had already modified its site to stop Landy's hack. Passengers' scores became hidden once again.

These ratings matter, though. It's just that, because Uber keeps much of this information secret, it's difficult to know the exact role they play. In that way, they're much like the algorithms that sort your Facebook News Feed or provide you Google search results or decide who you are a good match with on OkCupid. We might know some of the factors that these algorithms take into account, or that there's some guiding intelligence here, but we can't understand them fully. We do know, however, that riders with lower scores will have a worse time getting rides. Uber drivers have some autonomy in this regard; they're free to pass over a lower-rated passenger requesting a pickup in favor of a higher-rated one. And according to another report, a passenger who receives a rating of 3 or less from a driver will no longer be visible on the driver's app, making the customer effectively invisible to him. (It's still possible that other drivers will see the low-rated passenger and choose whether or not to pick him up.) This mutual rating between driver and passenger is much more than a standard review; it's intrinsic to the system and determining who gets service, the quality of that service, and, in the case of the driver, whether he gets to keep his job.

The Uber rating model is increasingly becoming standard practice. There's almost nothing we aren't asked to review now. And why not, when the individual's voice is supposed to have the power to start a movement or sink a political campaign? The opinion is the prototypical expression of the social web: everyone has them, and yet there are never enough. Web sites, apps, purchases, restaurants, books, TV shows and films streamed on Netflix—everything and every place is open to being summarized in a few sentences and a one-to-five star rating. Even prisons are now reviewed, however cheekily, on Yelp. Sometimes we review as a favor to a friend, in return for a discount, or because our disappointment with the last season of *True Blood* forced us to act (strong opinions play well on the social web). Despite the widespread belief that many ratings sites are inaccurate or skewed—about 14 percent are fake, according to a 2012 Gartner study—reviewing continues apace. The collected wisdom of millions of customers has congealed into pabulum, with the average review on Yelp being about 3.8. (For this very reason, YouTube switched from a star system to the simple thumbs up/down.) There is also a known herding effect with positive reviews, with one study finding that, when comments on a social news site received an up vote—i.e., a positive review—as their first rating, a greater proportion of up votes would follow than if they started with a negative or neutral vote. Just like traffic and followers, reviews are often bought and sold and faked by restaurant owners, authors and their friends, or automated bots. Negative reviews proliferate as acts of revenge against scorned rivals or as ways to push one's own rating ahead of a competitor. Even so, companies remain extraordinarily reliant on these reviews. A 2011 Harvard Business School study found that, on Yelp, "an extra star is worth an extra 5 to 9 percent in revenue."

The result of all this reviewing has been the atrophying of the critical culture, with professional critics seen as dispensable, nothing more than recommendation engines who can be replaced with algorithms and free, crowdsourced reviews. (Even so, some prominent cultural critics remain, though with less influence than they used to hold, and a smattering of publications, from the actuarially precise *Consumer Reports* to the liberal humanist *New York Review of Books*, continue to thrive.) It's also expanded the idea of what should be reviewed, with everything now potentially susceptible to, if not a star rating, then the kind of up-or-down judgment we perform all the time when we choose to like things. Our judgments aren't just expressed in comments or reviews now; in the metric-rich world of social media, number of likes is a valuable form of review, doing for social content what linking once did for individual Web pages. This time, though, giving your endorsement may land your avatar in an ad on Google+ or Facebook.

The obvious next step in this reviewing culture is to start reviewing people. Many review sites offer spaces to review professionals—doctors, lawyers, mechanics, plumbers—while employers, teachers, and vacations are now increasingly reviewed on sites such as Glassdoor, RateMyProfessors, and TripAdvisor. But reviewing people who sell us goods or services, or treat our illnesses, is to be expected. How would you feel to be rated as a person, perhaps as a friend or lover, and to have that information be publicly available, linked to your real name and social-media identities?

Yelp helped inaugurate this movement, but soon the reviewing bug spread widely. For a time, Honestly.com offered a place to anonymously review bosses and coworkers. Honestly was previously called Unvarnished, and the site trumpeted that its reviews were candid (they could also never be removed, only replied to). The site used a few of the nefarious tricks common to the social web. It would e-mail potential users saying that someone had created a profile for them, implying that they had been reviewed. Then, to sign up, they'd be asked to sign in with Facebook and LinkedIn *and* submit their e-mail address book. This was a blatant grab for data and e-mail addresses, but it also offered people the opportunity to review everyone in their social networks—as if these people were just dying to know what their friends, colleagues, acquaintances, and scattered Internet friends thought about them. To even see reviews of yourself—which were, alluringly, grayed out behind a series of pop-up messages—you had to like the page on Facebook or leave a review yourself. When you finally gave in and clicked Like, you would find that you hadn't been reviewed at all. But Honestly had successfully recruited a new member and grabbed information about hundreds of your contacts. Perhaps it's not a surprise that Honestly eventually folded, reinventing itself as TalentBin, a search engine for what it calls "passives," people who aren't actively looking for jobs but may be highly sought after by companies. Despite Honestly's transformation, the tactics it used, and the philosophy it embodied, are now endemic to social media, the nexus for rating people.

The first act of rating comes when we decide to follow someone on any one of a number of social networks. Then we sit in judgment, parsing their tweets or postings, moving them up or down in a mental hierarchy, deciding if they're still worthy of following. We decide whether to heart Tumblr posts or repin their image, with each decision point serving as a critical judgment, a de facto review, one that might improve their ranking on Klout or Favstar. We rate their ideas on Kickstarter, and if we deem them worthy, we donate and share the listing. We endorse colleagues on LinkedIn and request endorsements from others. On dating apps such as Tinder, we rate on a simple binary, swiping away those we don't want to meet and hearting those we do. On many commenting platforms, we rate other comments and give stars or other plaudits to individual commenters. Lulu, a smartphone app, and ReportYourEx.com allow women to warn others about deadbeat ex-boyfriends. And a range of products and services—from eBay to Lyft to Sidecar to TaskRabbit, many of them falling under the umbrella of the sharing economy—allow us to review independent contractors who have only a tenuous connection to the app or Web site that connected us with them. When we rate an Uber driver, who doesn't technically work for Uber, we are, in essence, rating him as an individual, adjudicating his personal value to us.

Robert Moran, head of the Brunswick Group, a communications consultancy, sees what he calls the "rateocracy" as an opportunity for transparency, when good corporations and citizens will be rewarded for acting ethically and in others' best interests. It will be integrated with augmented reality apps, so that you can activate your Google Glass or pull out your smartphone and see ratings for people, businesses, and places all around you. Facial recognition will likely play a role: imagine being able to access information—social-media profiles, Google searches, biographical information, ratings from friends, colleagues, lovers—on anyone you see, without even talking to them. A universal ratings service might appear, or ratings services will become more deeply intertwined, with shared log-ins and metrics in the manner of some social networks. The term "rateocracy" is supposed to be a combination of "rate" and "democracy," but the word sounds eerily like "autocracy." A world in which everything and everyone you see is judged and those ratings distributed in real-time could lead to some horrific outcomes. As one critic put it, "think about the power to manipulate and control that algorithms and organizations which provide ratings will have, how ratings will lead to stigmatization and cumulative disadvantage, and how fear of automated social judgment will have a chilling effect on behavior."

Unfortunately, in many important respects, that world is already here.

- - - - - - - - - - - - - - - - - -

A NEW KIND OF RÉSUMÉ

Reviewing people may sound crass, but it's just a more honest description for the $5 billion field of online reputation-tracking and reputation management. Unlike the credit-reporting industry, which is dominated by four agencies and regulated according to the rules laid out in the Fair Credit Reporting Act, online reputation, and its kindred industry of consumer scoring, is largely unregulated. But as our personal information, reviews, and social-media data accrue online, along with a range of Web sites scraping this data and selling it back to us, private companies, and governments in the form of proprietary reports and ratings, the importance of one's digital reputation will increase. Everyone from health-insurance companies to potential employers is already looking at this information, whether in the course of normal background checks or improving risk models. Even worse, we have little control over how this information is used. And unfortunately, many people, particularly those cashing in on this industry, think this is for the best. In the vision of Moran's rateocracy, reputation, influence, and reliability should be entrusted to the free market, as third parties seek to profit from our personal identities and habits of consumption. In this scenario, we will have less sense of what goes into our reputation scores than our credit scores, and we will have even less ability to influence the opaque algorithms that determine our worth and afford us opportunities in the marketplace.

The importance of online reputation has proved to be a useful weapon for public officials looking to deter bad behavior and less scrupulous actors searching for a quick profit. Some U.S. municipal police forces have TV shows in which they air the names and photos of Johns; others buy billboards. Joe Arpaio, the fiercely anti-immigrant sheriff of Maricopa County, Arizona, has forced prisoners to live outdoors and wear pink jumpsuits. Several towns have taken to engaging in what one writer termed "scarlet SEO," shaming Johns for purchasing sessions with sex workers. This kind of shaming isn't original to social media. In the Middle Ages, many Jews had to wear yellow badges (a forerunner to the same insignia later mandated by the Nazi regime), while paupers and debtors often had to wear colored sleeves or other badges. Tattoos and brands were used to mark criminals, slaves, and blasphemers—a style of punishment in tune with societies that tended to treat moral failings as criminal offenses and resorted to public humiliation in the form of the stockade and widely attended executions. These marks could be debilitating not only for their permanence but also because the people marked with them often had nowhere else to go. Lives were short and transportation difficult, and few people traveled far beyond their home village or town. To be marked as a debtor or philanderer could ruin someone's life. Now, while we might be able to move, our reputations follow us digitally, linked to our names.

The problems with this kind of reputation are overlooked in Silicon Valley. There are only, in the parlance of the industry, "influencers"— people with large social networks who trust their expertise and recommendations—and people who have not yet become influencers. No one suffers from a bad digital reputation, or scarlet SEO, or lack of a digital footprint. And if they were to, the thinking goes, their reputation could be rehabilitated by pouring more time and energy into social networks. Reputation services promise to return value for the unpaid labor of tending to social-media profiles. Klout, which has emerged as the most visible, if frequently mocked, arbiter of online reputation, casts all social-media users as unacknowledged public figures waiting for their personal spotlight. "Earn recognition for your influence," the site says. The implication is that your influence is already there, quietly pervading your social network. You just need to let Klout help you tap into it. "The Internet has democratized influence," the company's CEO told a reporter, "and we want to help everyone share in the rewards."

Klout parses users' social-media profiles, examining what they talk about and the size of their social graphs, and then grants them influence scores, ranging from 1 to 100. It also provides a list of topics about which it thinks that user is influential. Since its launch in 2009, Klout has been frequently criticized for concocting a rating of perceived influence that doesn't mean much and that leverages users' anxieties about their own popularity, thereby pressuring them to do all they can to boost their scores. What's more is that Klout's algorithm, which draws data from a half dozen or so social networks, as well as *Wikipedia*, can seem rather arbitrary. It's rated people such as Barack Obama as less influential than Justin Bieber. (To be fair, Klout did later update its algorithm, moving Obama past Bieber.) Klout also has rated some Twitter bots as quite influential, despite the fact that its algorithm is supposed to account for people's fake followers. It also tends to mistake which topics users are influential on—if I made some jokes about NASA's Mars rover (always a crowd pleaser) that got a lot of retweets, Klout might suddenly decide I'm influential about planets, but a few tweets on the subject don't make me an astrophysicist. And of course, the very idea of being "influential" on certain subjects is a rather vague claim. What does that mean? That I'm knowledgable? That people listen to and trust me? Given that influence, however one defines it, is also a matter of how we socialize in the physical world, our relationships with friends and family and colleagues, our education, our purchasing power, and so forth, Klout is looking at its subjects through a pinhole view.

Despite these criticisms, Klout has stuck around and claims to have scored at least 100 million people and to be used by 200,000 businesses. The company has also found some willing partners. For a time, American Airlines allowed anyone with a Klout score greater than 55 to access its Admirals Club lounges. Klout's also given away Cirque du Soleil tickets, gift cards, hotel upgrades, and other rewards, such as access to Spotify before it launched in the United States. Marketing firms have used Klout scores in making hiring decisions. Microsoft, an investor in Klout, modified its Bing search engine so that information from some of Klout's designated experts would appear prominently in search results. The information came from questions that Klout fed to high-scoring users, who were solicited for recommendations about gadgets, makeup, music festivals, and other pricey products. Notably, the answers weren't vetted for accuracy or genuine expertise before appearing on Bing; rather, other Klout users chose to vote answers up or down, and those that received enough up votes appeared on Bing.

Notwithstanding its black-sheep status in the industry, Klout is an exemplary social-media site. Relying on superficial metrics, it creates its own hierarchies or simply reinforces existing ones, since interacting with high-scoring users can raise your score. It claims an unearned authority over human affairs and promises to know its users better than they do. It sucks up users' data while offering little in return. Because Klout would like to rank the influence of every person with an Internet connection, it is also an opt-out site, meaning that you have to choose not to have your data considered by Klout. (The site got into trouble, and later apologized, for automatically creating profiles for minors.) You don't ask to be a part of it; you have to ask to be left alone—and to log in via one of your social-media accounts in order to do so. The site judges you, and with little knowledge of how its algorithm works, you can only hope to figure out how to make it like you better. And despite not coming up with a consistent revenue stream, Klout still managed to sell itself to Lithium Technologies, a start-up providing social-media services to businesses, for a deal valued at about $200 million.

The role of a reputation service in your life may be as inscrutable and surreptitious as the algorithm powering it. Some hotels have experimented with providing upgrades to guests with higher Klout scores—without telling the guests. The software companies Salesforce.com and Genesys have integrated Klout into customer-service programs, allowing their enterprise customers to check consumers' Klout scores when they tweet complaints or call a service number. The implication is that consumers with higher Klout scores will receive more attention and more perks, in the hopes that they won't use their online influence to badmouth the company. Similarly, some e-commerce sites have used Klout scores to offer influential customers greater discounts. Next time you wait an hour on a customer service line, only to receive bad help from a harried staff member, you may wonder if your low Klout score is to blame. Or, perhaps, it was a combination of your Klout score and the vocal analysis the company's software performed when you articulated your reason for calling. Either way, you have been sorted, a computer determining how much attention and service you deserve.

Klout isn't the only site of its type. A range of other influence-measuring companies have emerged. The company Reppify offers a "job fit score" and promises to use social-media data to help your company generate business leads from people in their social graphs. (As might be expected, Klout also considers the Klout scores of prospective employees.) Favstar keeps track of how often your tweets are favorited and doles out digital badges to popular users. Some of these companies offer users rewards based on their influence or for their ability to sell items through their social-media postings. Building on the dream of democratizing influence, they use Internet consumers as vectors in their own marketing, while selling them on an image of being micro-celebrities, paid for their social-media use in the same way that A-list Hollywood stars are paid handsomely to tweet about products. For aspiring influencers, it's a full-time job: one graphic designer with a Klout score of 74 tweets about forty-five times daily and worries about going on vacation because, he told *Wired*, "Brands couldn't get in touch with me. It's easy for them to forget about you."

These scores are largely arbitrary and useless, and they do little more than encourage users to spend more time on the platform, where they must hand over more data. But for less skeptical users, these services can be, well, influential. One study found that a "mock Twitter page with a high Klout score was perceived as higher in dimensions of credibility than the identical mock Twitter page with a moderate or low Klout score." Simply having a higher score was perceived as a mark of greater credibility, even though 80 percent of the study's participants didn't know what Klout was. The study's authors had an explanation: "Individuals tend to believe information given to them by a machine or computer due to its lack of thoughts, feelings, personal biases, and the like." It *seems* scientific, with the kind of certainty and precision that only a computer can offer. Why not believe it?

Klout is easily gamed if you suck up to famous people, buy followers, post frequently, and engage in various other attention-grabbing exercises. But the social web is suffused with so-called credibility markers, which play into the existing Silicon Valley fervor for "influencers" and "thought leaders," as well as résumés so padded you could rest your head on them. From verified accounts to listicles of essential Twitter users, power, authenticity, influence, and reliability are accorded every day on social media. These qualities often accrue based on traditional hierarchies—successful people in an industry promote their friends and colleagues; those wishing to join them are forced to appeal to them, whether on social media, in the office, at a conference, or elsewhere. Having a lot of followers is frequently interpreted as a measure of success—a Web 2.0 version of the hit count featured prominently on many late-nineties personal Web pages, which told users how many people visited the site. Similarly, noting how many times an article or video has been shared serves as a shorthand for, "This is important and worth looking at." For advertisers or venture capitalists, number of shares can determine how much money changes hands.

None of this seems very authoritative. Metrics are ever shifting, and each contains its own implied set of values.[*] Klout provoked an outcry from its users when it adjusted its algorithm and many saw their scores drop precipitously. It can seem shallow to judge someone simply by how many people choose to follow her (or by analyzing her followers:follows ratio). Ironically, given how much data the social web produces and its promises of quantifying practically any human endeavor, influence and reputation seem to be among the most difficult things to pin down. This kind of epistemological uncertainty sits uneasily with an Internet that privileges rankings and status and the accumulation of hard facts, but many people don't realize that even basic Web traffic metrics are frequently unreliable. Yet the success of Google Search is very much based on the idea that it can instantly show us the most authoritative, useful, and accurate sites for a given query. Fans of Google have come to accept this as axiomatic, and in this way Google delegates influence to millions of Web sites every day. It is constantly choosing favorites, algorithmically; failing to show up on the first page of Google search results can sink a fledgling business. It only follows that a social or people-based search engine would promise to do the same. While both Google's and Klout's algorithms remain trade secrets, we do know that Google has long favored pages that are frequently linked to, hence the prominent perch of *Wikipedia* in so many search results. It would follow that Klout would similarly privilege people with a lot of followers or shares—the social web equivalent of being linked to—despite the knowledge that these metrics can be manipulated.

There remains an inherent absurdity in delegating influence—and by extension, the perceived value of a person and his or her opinion—to an algorithm. But many start-ups take a "use the whole animal" approach to publicly available data—that is, this information is

---

[*] The researchers studying Klout scores were wrong: computers only appear to lack biases; software is designed by human beings, who choose, for example, which qualities are important for quantifying influence and which aren't.

out there; it shouldn't go to waste. It's hard to escape the idea that value might be leveraged from the digital exhaust so prevalent online.

Some other companies in the reputation racket have gone for a less deterministic approach. They still deal in the language of celebrity, thought leaders, and reputation, but they lean less on the faux objectivity of algorithms parsing hundreds of signals. Quora promises to be "your best source of knowledge" with "real answers from people with firsthand experience." LinkedIn wraps itself in the gauze of net jargon: its featured members are capital-I "Influencers," which really is to say that they're milquetoast celebrities from the business world, here to dispense some wisdom. The career network now publishes plenty of content, much of it rehashed business proposals, valedictory conference speeches given by industry moguls, business-oriented self-help, and other entries in the genres of How I Got Here and How to Succeed Like Me. For the millions of less influential members browsing the site, these missives represent an aspirational ideal. Accrue enough endorsements and recommendations, and you may be asked to contribute, too.

None of these environments seems to match up with the democratizing, everyone-is-influential rhetoric that Silicon Valley CEOs prefer. They instead uphold an ethos in which one is always auditioning and always seeking to rise up the status chain. We have internalized the logic of advertising and branding, forsaking a sense of play, unpredictability, humor, or subversion. It's become possible, we're told, to earn a job through social media—recruiters are always watching—but it's far easier to lose one when you don't hold up the mask of bland respectability that every would-be influencer wears. One misfired joke or incorrect additive in the heady mixture of the personal and professional can destroy the whole enterprise, earning you a pink slip and a lower reputation score. When everything is measurable and is tracked and mined for insights, it becomes a social crime to be inactive, invisible, or not forthcoming. Don't be anything but a good digital citizen, concerned about the issues of the day and sincerely engaged with others, while also pushing the restaurants and designer jeans you love. (They, too, might be watching, waiting to reward you for your free advertising. Plus, spreading taste is spreading influence.) Klout measures your constancy, recommending to users that they increase their "cadence" of tweets and their positivity in order to improve their scores. It may be boring and soulless, but how else to succeed in the information economy except by turning your persona into a more rarified form of data?

DOES THIS ALGORITHM LIKE ME?

Reputation and influence services play on our anxiety about being invisible on social networks or about being visible in the wrong ways. We don't want to think that all that work is for nothing. As Taina Bucher writes, social-media associates "visibility" with "empowerment." The more you are creating content, the more you have a chance to distinguish yourself from the masses. But it's not enough to simply post updates or share funny articles. You have to learn how to work within the system. On a network such as Twitter, in which most updates are treated equally,* the challenge is more about appealing to your audience. Still, you might get lost in the endless stream of updates. On an algorithmically sorted timeline, such as Facebook's News Feed, you might get lost entirely, because you aren't deemed interesting enough or you haven't satisfied the algorithm's unknown parameters. Klout, along with a range of consultants, reputation enhancement services, and other social-media aggregators, promises to vanquish the threat of invisibility. If your social-media "moments," as Klout calls them, don't receive enough recognition in the first place, here's another service to help you get the attention you deserve.

But algorithms aren't neutral arbiters. They contain artifacts from their creators, including their errors and biases, and they aren't always

---

* This is changing on Twitter. With sponsored tweets, verified users able to segregate @-replies from non-verified users, and other replies causing conversations to reappear in feeds, the once-chronological timeline is increasingly being sorted.

put to noble uses. Although most algorithms are kept secret, sometimes we can decode their processes. We learn to figure out what they want. We might discover that pressing 0 repeatedly on a customer service line allows us to skip the menu system and finally (finally!) connect with a real human being. We learn how to phrase queries in Facebook Graph Search to show us more information about the people around us.

Algorithms require us to bend our actions toward them. If we want to enhance our Google search results, we read up on the latest SEO practices. If we want to earn more attention on Facebook, we consult the news feed optimization industry. As Bucher says, "In order to appear, to become visible, one needs to follow a certain platform logic embedded in the architecture of Facebook." Each platform has its own logic, guiding algorithms, and user customs, along with consulting companies that develop in response to the threat of invisibility. Every time we tag a Facebook photo or a Tumblr post, we're submitting to the network's logic, asking it to accept our catalogued, machine-readable data and fold us deeper into the network's fabric.

This is the mostly unexamined, unchecked power that algorithms have, the influence that algorithms exert in our lives as they become decision-makers and forces unto themselves. Increasingly, algorithms guide our daily lives and sort us into different categories of experience—whether it's a city's automated traffic system or a police force's facial recognition software. They make judgments on the data available and of their own accord. A red-light camera knows only if you ran a red light; it doesn't know that you did it because you were speeding to the hospital.

Recommendations may be the type of algorithm with which we're most familiar. We've made comedy hay out of Netflix's ultra-specific categories ("Race against Time Sci-Fi & Fantasy Movies"; "Fight-the-System Political Love Triangle Mysteries") or the strange juxtapositions of Amazon's "Customers Who Bought This Item Also Bought" listings. Twitter, Facebook, and LinkedIn constantly propose that we follow, friend, or connect with people who surprise us or seem down-right creepy (*How do they know I know that person?*). Receiving less attention is the way that, as we delegate more power to social networks and digital media services, algorithms rise in importance as cultural tastemakers—far more influential, to the chagrin of this occasional book reviewer, than any cultural critic. Spotify, Netflix, Hulu, and other services—particularly those, like Pandora, that automatically serve us media based on what they think we like—dictate our cultural consumption. As we link them to our social-media accounts and they automatically broadcast our media choices, algorithmic taste spreads through our networks. Software begins to make choices for us, with the software itself influenced by the whims of the marketers and pro-grammers who created it.

It's hard to chart exactly how this influence spreads. We can't ex-pect Facebook to suddenly reveal how its algorithms work or what it shows to each person in our network. For years, companies and small businesses wondered whether Facebook's News Feed algorithm was "broken on purpose," as some called it, limiting the amount of expo-sure unpaid posts got. In December 2013, Facebook acknowledged that that's essentially the case. Companies would have to pay up to guarantee better visibility. Facebook didn't want every post appear-ing in a user's feed, possibly overwhelming him. As a consequence, they said that organic reach for a brand page would decline to 1 to 2 percent, meaning that the Los Angeles Dodgers, with about 2 million likes on their Facebook page, would have their updates pushed out to 20,000 to 40,000 fans. That was sour news for large companies that had spent a lot of time and resources trying to boost their Facebook followings, but it was potentially even worse for local stores, activist groups, and other small operations who might only have a few thou-sand followers. For those smaller entities, a post might end up being seen by only a few dozen people.

Facebook didn't make these changes just because they wanted to lessen the amount of material that appears in people's News Feeds.

Now a public company, with investors demanding greater revenues and high profit margins, Facebook has an important constituency to please. By limiting the organic reach of brand pages, Facebook creates a reason to encourage businesses to pay for sponsored posts, increasing their visibility and bringing more cash to Facebook's coffers. Money, in the end, is the surest way to get the News Feed algorithm to like your post.

So that's one signal—a big one, but more useful if you run a business or organization. Facebook claims that its News Feed algorithm, which used to be known as EdgeRank, has about "100,000 individual weights." We're unlikely to be given a full understanding of it. That only means that the more that Facebook provides us news, alerts, social invitations, and cultural material, the more power its algorithms have over us. And even if you embrace Facebook's role in your life, it should give you pause that it claims to know you better than you know yourself. By choosing to organize your feed based on its proprietary criteria, Facebook is deciding what's important to you, who are your good friends and who less so, what news or advertisements most interest you, what your mood might be, why you're logging in from a different device or location. Remember also that Facebook's goals aren't the same as yours. You might go on Facebook to see what your friends are up to, find out about upcoming birthdays, or check on some blogs you follow. Facebook's only goals are to keep you on the platform and to get you to click ads, and its News Feed algorithms ultimately exist to support those efforts.

That's why some commentators responded with a shrug when, in the summer of 2014, Facebook revealed that it had experimented on almost six hundred thousand users by subtly increasing or decreasing the amount of positive material these people were seeing in their feeds. These insiders have long known what is only now becoming apparent: that Facebook, like every other big digital platform, freely and secretly experiments on its customers, moving the dials this and that in an attempt to coax the kind of behaviors out of them that will help boost bottom lines. OkCupid, the dating Web site, responded a few weeks later with its own proud boast: "We Experiment on Human Beings!" read the title to a blog post. What followed was a devil-may-care tour through several ways in which OkCupid experiments on users. In one instance, the post's author, OkCupid cofounder Christian Rudder, emphasized the power that OkCupid has over the behaviors of its members—even when it's intentionally messing with them. "When we *tell* people they are a good match," Rudder wrote, "they act as if they are. Even when they should be wrong for each other." In other words, people believe what the platform tells them—perhaps because people are inclined to trust the accuracy of computers, or perhaps because OkCupid has established a reputation for performing sophisticated analytics to help bring people together. No matter the case, Rudder was unruffled about the ethical consequences of manipulating people into believing they should date each other. As he blithely explained, "Guess what, everybody: if you use the Internet, you're the subject of hundreds of experiments at any given time, on every site. That's how Web sites work."

Rudder is correct on the facts if not the ethics. This kind of algorithmic experimentation is indeed widespread. And it's not only culture and social life that are being subjected to this process. Whether in airport security lines or on e-commerce sites, our data is being run through the decision-making mill. We are judged on the basis of our personal data and our social-media presence, with little opportunity to dispute its accuracy or confront a real human being. Like demographic, medical, or credit data, information gleaned from social media is increasingly being taken up with the promise that it can tell companies and governments about who people are and predict their actions. Lenders, from payday loan companies to mortgage giants, have taken to looking at customers' social-media accounts. Their reasoning is that strong social networks may reveal a customer's reliability, particularly if he lacks a credit history, and that this information may be more up-to-date than what's contained in a credit report. Others are pairing the social graph with even more information: GPS coordinates, the applicant's behavior on the phone or Web site while submitting an application, online purchases, friends' credit histories, hardware and software used. This information may help in risk assessment, but it could just as easily be used to target people with loans they can't afford or to get them to reveal information they may otherwise want—and have a legal right—to keep private. One start-up executive told the *New York Times*, "We feel like all data is credit data, we just don't know how to use it yet." In this executive's world, "More data is always better," helping customers and clients alike, with no thought toward possible conflicts of interest or privacy violations. As Evgeny Morozov warned, "Given how much they know about their clients, these companies can perfect the art of hidden persuasion and manipulation in ways that Madison Avenue could never even dream of." One arm of a company can glean as much as it can about you—when you get paid, how much, where you live, how big your family is, when you're most likely to buy something online—while the other can push products at you, whether a high-interest loan or a new computer to replace your six-year-old clunker. And they'll do it under the conditions that their analysis shows makes you most vulnerable to saying yes. MediaBrix, a company which has partnered with Microsoft to work on new tracking technology, says that it targets "game players at natural, critical points in game play where they are most receptive to brand messages."

This is the kind of power—hidden, diffuse, strangely influential, operating according to its own logic—that we need to contend with on social networks. And as social networks become plugged into other systems of commerce, insurance, and the like, we can expect the role of our social-media data to increase. With few consumer protections for the use of personal data, companies will also be free to collect, buy, and sell data and to use it to skirt around consumer protections. Your bank isn't legally allowed to ask you about your race, but if they've got a data broker as a partner or asked you to log in with your Facebook account in order to receive special offers, they may already know. This is part of the cost of visibility, of trying to rise in the world of online reputation. What you get in return may be far less than what you are giving away. The more authoritative your score seems, the more information they have on you.

- - - - - - - - - - - - - - - - - -

DIGITAL HOSTAGES

Visibility is complemented by the threat of disappearing, of becoming irrelevant—which in turn makes visibility a valuable, exclusive state of being. That is, we don't all have the time, knowledge, or resources to preserve our visibility, to try to achieve micro-fame. It's a constant process of cultivation, like bailing water from a leaky boat. It's a luxury of people who can spend the time needed to stay present and available on their networks. If you don't maintain this performance, if you don't keep bailing water, you'll sink out of view. This is best proven by a simple test: open a browser tab and go to Facebook. Take note of the first few stories in your feed. Close the tab, open a new one, and go to Facebook again. The stories you saw before are probably gone (almost certainly so if you clicked on a link before closing Facebook). Facebook is in a constant state of flux, trying to show its users exactly what they think they want. Not clicking or leaving the site is a sign of refusal. Those stories might surface again, though, if they accrue enough responses and Facebook determines you're interested in them. But in all likelihood, if you didn't respond to them that first time, they're gone. You'd have to seek them out by visiting the pages of friends or brands that shared them in the first place or go scrolling through your News Feed, hunting for them.

News Feed tends to promote posts that produce a lot of likes and comments, which in turn prompts us to respond to material that our network has already deemed worthy of attention. In what amounts to an amplifying effect, popularity accrues to those who can impart the impression of being popular, which is one reason why offline hierarchies often translate well to social media and why celebrities receive hundreds of responses for even the briefest or most banal post. Fame is self-reinforcing and algorithmically desirable.

If your job requires you to be online all day, or you have particular information about how Klout or Facebook's algorithms work, or you happen to have a vibrant social network, then you're going to do better in this new economy of reputation and visibility. You may still find yourself on the hamster wheel of online influence, always producing more content to stay visible and relevant, but you'll receive more rewards than those who can't keep up, don't know how, or prefer to withdraw altogether.

But what happens to these types of people? What happens when, for whatever set of reasons, your digital reputation is less than sterling? I found out when I went searching for a friend I hadn't heard from in years.

We had once been close, college friends and confidantes, but we hadn't talked in some time. I had his e-mail address and could've gotten in touch with him that way, but I was also curious to see what I could learn about him by searching online. His social-media presence was sparse—no Twitter, no Instagram, no Tumblr that I could find; only a bare-bones Facebook page, with few photos or updates. I did a simple Google search, made easier because he has an unusual last name. On the first page of results, there were links to a few skeletal social-media profiles, his personal Web site, a business he started. Farther down the page, though still, crucially, on that first page of results, was a mugshot: he had been arrested a few months earlier for a DUI. I was disappointed and saddened, though fortunately, it seemed that no one had gotten hurt. The Web site, an aggregator of mugshot photographs, promised that the photo would disappear within six months. That seemed somehow reasonable. My friend had committed a crime, but there was no reason that this information should be immediately available to anyone who searched for his name. That was between him and the law, and between him and any future employers who might require this kind of information.

But more than six months later, I discovered that the mugshot wasn't gone. At least, it was gone from that site, but it had popped up on another, justmugshots.com. This site was pretty cheaply designed—it didn't seem to be built on top of anything more sophisticated than a basic WordPress template. But Just Mugshots had thousands of mugshots from around the United States. My old friend's page included his full name, his height and weight, the date and location of his arrest, his birth date, and, naturally, his booking photograph. The photograph had watermarks from the law enforcement agency that had taken it, but in the top left and bottom right corners, Just Mugshots had added even more prominent watermarks of their own, essentially claiming the photograph as their property. And below the photograph was a bright red button, with the promising words "Remove This Mugshot." Clicking the button led me to a page offering two possibilities to remove the mugshot. For about $145, they'd simply remove the page; for about $200, they'd remove the page and "INSTANTLY put in six Google Removal Requests on your behalf which requests that the page(s) and image(s) are removed from the Google Search index ASAP." The site will only remove the photo for free if the person is under eighteen, the charges are dropped, he's found not guilty, or he dies. That's it. Yet Just Mugshots's rather brief legal page explained, "We are not in the extortion or defamation business, we simply provide a high level of service to our satisfied customers." They also brag that they've never lost a lawsuit.

Just Mugshots, and the wider industry of mugshot Web sites, exist at the intersection of public government data, search, and e-commerce. Trading on the language of public disclosure and information freedom—the site notes that its information was obtained from publicly available government records, adding, "Many government agencies publish the exact same information online"—they adopt a who-me? attitude when accused of any wrongdoing. This information is out there, they claim; we're just making it easy to find. By utilizing public-records laws and providing information on possible criminals, they are, some owners say, doing a public service. (This reasoning breaks down when you consider that violent felons, if they have the money, can pay to take their mugshots down.) They make their money from people paying to remove mugshots but also from Google Ad-Words revenue and, in some cases, deals with reputation-management sites.

There have been some efforts to counter mugshot-removal sites: a class-action lawsuit, legislation at the state level, intervention by MasterCard, Discover, Visa, and PayPal, which process payments for many of these sites. After being contacted by a reporter for the *New York Times*, Google introduced a change to its search algorithm that lowered mugshots in some search results. But despite years of reporting on the industry, these sites endure. And their owners are tenacious, finding new payment methods when others dry up and threatening rivals. One lawyer who filed a class-action lawsuit against mugshot sites received death threats and had crude reviews about himself appear on a consumer complaint Web site. The site said that the postings could be removed—if he paid $499.

Sometimes deals to remove information come through a third party, offering to work on your behalf. Remove Slander presents itself as a friend who will "end your humiliation" by helping get mugshots taken off of Web sites or, they promise, your money back. Remove-Arrest, which is part of a consortium of similar sites that includes one now-defunct operation called MonitorMyTeen, describes itself as "much like a document service which assists you in obtaining a passport, birth certificate, or other vital records." But on other sites, the parallels with viral media and social-media-driven humiliation are quite obvious. Florida.arrests.org allows users to browse photos by crude tags ("hotties," "grills," "transgender," "handicap," "WTF") and to leave comments. You can also tag the photos yourself. In what's either a bizarre irony or the logical culmination of an insider's experience, the site is run by a man who served three years in prison for credit card–related crimes. (He does not have his own mugshot on the site.)

The revenge porn site MyEx.com allows users to post the full names, ages, hometowns, and photos of women. The postings are all anonymous and feature misogynistic diatribes against the women. Anonymous commenters chime in below each post, often making fun of the people depicted within. Each posting has a "remove my name" option, which for a time led to a site called ReputationGuard.co, which offered to delete the page in question for $499. MyEx also has an active section of posts about men, but it's predominantly, like many sites of its ilk, a place to shame women.

These sites show that availability is not the same as visibility. Some information is harder to find, or downright obscure, for a reason. Nor is "the information was already out there" a reasonable excuse. You might know that a book is in a library, but you won't find it unless you look for it or a librarian helps you. And if you don't know that such a book exists, then you won't learn about it unless someone tells you or you stumble upon it in a featured display or on a shelf. Information, even in the ostensibly boundless, horizontally structured commons of the Internet, does not exist equally. When someone with a large follower count shames someone for racist tweets, that's a deliberate effort to raise the profile of that information. The racist tweeter might very well deserve such treatment, but the shamer isn't just a passive transmitter of the information. There are real power dynamics going on, with information—and the promise of making it more prominent and, perhaps, more closely linked to that person's search results and online identities—being used as a weapon.

Mugshot sites are clearly dedicated to weaponizing information. Their motive is evident in how the sites are designed. Most mugshot sites scrape data (that is, collect data en masse) from public law enforcement databases, of which there are dozens. These databases are easily searchable once you get onto the site, but their information lies behind CGI scripts, meaning that Google doesn't catalog it. Instead, you have to go to the site and type in a name to find someone. You have to have some sense of whom you're looking for. Information is accessible but not completely open; it doesn't spread easily. When mugshot sites scrape this information, structure it, make it easily searchable, and add various tricks (SEO, user comments) to drum up engagement, their intentions are obvious: they want this information to be the first thing you see when you Google someone, because that will push the person to pay to have the mugshot removed. In the meantime, they make money from ads sold against curious visitors who drop into the site and then linger, clicking through the most bizarre or infamous photos.

Here's another useful comparison, this one from the world of celebrity. In recent years, one of the most famous paparazzos in Italy was a man named Fabrizio Corona. Corona didn't take pictures himself, but he owned a powerful paparazzi agency that he named after himself (Corona's exploits and ego are well chronicled in the documentary *Videocracy*). Corona cultivated relationships with celebrities, right-wing politicians, and TV executives, including people close to Silvio Berlusconi, the on-and-off again prime minister known for his media empire, sex parties, and corruption. Corona's agency took photographs of Italy's celebrities, from politicians to actors to soccer players, but he rarely sold the photos to newspapers. Instead, he offered the photos right back to the people in the photographs—for a hefty fee. In this way, he became rich and famous, while celebrities could stay close to a powerful paparazzo and help manage their public image. It's easy to see why this behavior could be considered reprehensible and even illegal. After becoming a TV personality in his own right, Corona was found guilty of extortion but only had to spend eleven weeks in prison—an experience he managed to leverage into a well-publicized tell-all book. But Corona's jail time didn't deter him, and he is now serving his second stint in prison, this time for tax evasion and fraud.

Corona's practices were little different from those of people who extort others online to keep compromising information secret. He played on people's fears about their reputation, about being made famous for the wrong reasons, about losing control of their image. Although his customers were mostly celebrities, who rarely receive much sympathy for their choice of lifestyle, and although he worked in cooperation with

some of these same people, his methods were still dishonest and manipulative. The reputation industry reveals itself to be made up of a range
of actors operating across a continuum of morality. Klout may not be as
malevolent as a mugshot site or a revenge-porn site, but it manipulates
people in a similar way. It takes what has become a small, if emergent,
need—one's digital footprint—and inflates its importance, telling people
that it is essential and should be cultivated. Reputation-management services take the process one step further by institutionalizing this behavior,
telling people that their online reputation is as important as a credit score
and must be monitored accordingly. Pay us to take care of it for you, they
say. Let us remove the worry that we just instilled within you.

Reputation management comes in many forms and degrees of le-
gality. Astroturfing, or hiding the source of advertising or potentially
false information, is a crime in some states. It's a tool of the wealthy and
connected—big-time bankers were reportedly using reputation services
after their industry helped crash the global economy in 2008—that has
increasingly filtered down to the middle class. The services offered include
tracking what people say about you online, pumping out positive press
releases, cleaning up social-media profiles, creating Web sites, removing
embarrassing photos, and other manner of digital hygiene. A common
program includes setting up profiles on popular Web sites, particularly
jobs sites, social networks, and blogging platforms, which would be likely
to show up on the first page of search results. Reputation managers serve
essentially as a combination of digital PR agents and corporate fixers. They
do their work clandestinely, but the product is supposed to be highly visible. Among the most important goals are making sure that the first page
of Google results are clean and that Google's suggested search results aren't
harmful or defamatory. For example, when I type my name into Google,
the four suggested search terms are "jacob silverman duff and phelps" (an
executive at Duff & Phelps, a financial services firm, shares my name),
"jacob silverman sxsw," "jacob silverman twitter," and "jacob silverman
jeopardy," none of which is cause for worry. If one of them had been "jacob
silverman arrest," I might have reason to look up a reputation defender.

Some disgruntled Internet users have tried suing Google for these suggested search terms, but a more effective, cheaper method may just be searching many times for more positive terms in order to crowd out the ugly suggested terms. Sometimes reputation defense is less about technical savvy than brute effort.

Claiming more than one million clients, Reputation.com charges $9.95 per month for its entry-level, personal plan but also has more expansive offerings, including for small businesses and enterprise clients. The site Brand Yourself goes for a more populist bent, encouraging users to sign up with their social-media profiles and begin working on their reputations themselves (LinkedIn operates on similar principles, particularly in its recommendation to high school students that they work on their "personal brand"). A small firm called Metal Rabbit Media caters to rich clients, charging them $10,000 per month for basic reputation-enhancement services. Metal Rabbit Media creates fake online personas for some clients but using their same names, with the idea being that it might be more useful to obscure the client's biography than to enhance it with fluffy material. When challenged, the client could always say that the Web sites must be describing *another* Jacob Silverman who served in two wars, published a sonnet cycle, and sits on the alumni board at Yale. Similarly, all those blog posts attacking Jacob Silverman for being an asshole and a tax cheat must be describing that other Jacob Silverman or still yet, a third Jacob Silverman. The effect is, as the journalist Graeme Wood put it, like Saddam Hussein's body doubles: create multiple versions of the same person, so the real one will always be in doubt. Werner Herzog, with his profusion of digital doppelgängers, would probably appreciate this method.

- - - - - - - - - - - - - - - - - -

THE TRADE IN REPUTATION DATA

In an environment of frictionless sharing, "data has a life of its own," as Matt Waite discovered when he built a mugshot Web site for the

Tampa Bay Times. A few months into the site's existence, he found that some other would-be mugshot kingpin was scraping *his* site, sucking down all the arrest data that he had collected. Although Waite had prevented Google from listing his site and made it so that the mugshots expired after sixty days, his site was still constructed in a way so as to be easier to pull from than the cheaply designed databases on sheriffs' Web sites. This other mugshot site was linking directly to Waite's images—and running up his bandwidth bills—in order to play a game called Pick the Perp. The game had gone viral on some social bookmarking forums, deluging the *Times*'s mugshot site with traffic. Waite and his codeveloper eventually fiddled with their code to prevent Pick the Perp from linking directly to their material. But in the process, he learned some important lessons about the consequences of open data policies and the ethical complications of putting information freely online. "Because you can put data on the Web," he concluded, "does not mean you should put data on the Web."

Indeed, once you make data available in an easily accessible, networked environment, others are going to find it and repurpose it to their ends. Search engines will pick it up, catalog it, and sell ads against it. Google Alerts will inform people that information about them is out there. Numerous social-media aggregators, search engines, and analytics firms, such as Inagist and Topsy, suck up all the social-media data they can find, using it to churn out reports, search results, or trending topics. Our personal data, the blog posts we write, and the photos we upload spread and become syndicated in weird ways. Our stuff shows up in online white pages and obscure blogs that scrape articles from paywalled newspaper sites. It goes into the cavernous databases of powerful data brokers such as Acxiom, Epsilon, Datalogix, and Experian, which offer detailed, though sometimes inaccurate, profiles of hundreds of millions of people. From there, who knows where it might end up? Governments, corporate marketing departments, research scientists, prospective employers, advertising firms, insurers—practically anyone interested in large amounts of consumer data has reason to make deals with these companies. Sometimes the information flows both ways. State DMVs sell some personal information to data brokers—to be used, the agencies claim, for "identity verification." The TSA has said that its Transportation Security Enforcement Record System may pass data along to debt collectors.

Experian, one of the biggest credit-rating services, allegedly sold data to an identity-theft service, according to an investigation by KrebsonSecurity. The deal involved an Experian subsidiary, Court Ventures, and another company, USInfoSearch.com, that shared data with Court Ventures. Through this chain of operators, identity thieves in Vietnam managed to access Experian's databases by claiming to be private investigators in the United States, while sending their payments from Singapore. The peculiarly convoluted deal spawned a congressional investigation and highlighted the weak controls that many data brokers have. In this industry, data and money move fluidly, with few questions asked. And consumers have little right to know where their information is going or even, in the case of Experian and its role as one of the country's main credit bureaus, to opt out. Whether you like it or not, Experian, Equifax, and TransUnion—the three largest credit-reporting agencies—have information about you and are capable of selling it for tremendous profit. (For its fiscal year ending March 31, 2014, Experian reported $4.8 billion in revenue.)

Some regulatory bodies have attempted to intervene, with a few states passing data protection laws, but they remain largely toothless and unsuited toward what is an international industry. Perhaps they're too late, and the trade in personal data, influence, and reputation has already been mainstreamed. Forty years ago, few residents of capitalist countries knew what a credit score was, much less requested regular reports or paid companies to monitor their credit. Now, such practices are standard, considered in keeping with being a responsible steward of one's own finances. Credit and debt are, after all, among the most ancient forms of reputation. Who owes whom? Who is good at borrowing and repaying promptly? Who's trustworthy?

Reputation management might become similarly standardized. We will be as concerned about our declining Klout scores or our shaky Google results as we are about our bad credit, or our Social Security number being stolen by identity thieves. (Preventing identity theft—another bourgeois digital-age concern that is no longer so novel—is one of the selling points of Reputation.com.) Necessary or not, these services remain divided along class lines, and some of the people least able to afford them may be doubly hurt by not knowing that their mugshots are available online.

It's unlikely that reputation and influence management will stop where they are, not when venture capital money and personal data are sloshing around in such abundance, both accessible to those with the right connections. The next generation of these businesses won't just be about holding people hostage for mugshots or helping them manage their digital footprints. Nor will they be so overt in their fraudulent activities or try to entice people into accepting dubious metrics for rewards at hotels and fancy restaurants. Instead, they will fall under the label of risk management, market intelligence, or predictive analytics, catering to large businesses who want to profile customers and employees. They will go beyond collecting data and sorting people into categories for market research or presenting background reports on potential employees. For a fee, they will offer thorough reports on who people are, what they do, and how they should be handled. (As of now, this kind of work is largely the province of intelligence agencies.) According to the scholars Woodrow Hartzog and Evan Selinger, "Besides presenting new, unsettling detail about behavior and proclivities, they might even display predictive inferences couched within litigation-buttressing weasel wording—e.g., 'correlations between x and y have been known to indicate z.'" For mitigating corporate liability, this kind of work carries great promise. But it will serve to narrow the scope of socially acceptable behavior, to stigmatize and re-criminalize those who have had any encounter with the law, and to allow people to be manipulated without their knowledge.

This is the problematic result when you combine fluid data, a culture in which the public disclosure of personal information is now a matter of course, an ideology of information freedom, and powerful search and data-analysis technologies. The masters of these systems gain the ability to review people at their very core: judging them not only for who they are, but for their actuarial value and possible future actions, all dictated by a computer. The morality or accuracy of the system is secondary to getting people to buy into it, to care about it, to feed it with data and cash and their fears at being on the wrong side of it.

As for my old friend whose mugshot I found online, he spent more than $700 removing his mugshot from different Web sites, he told me. He was trying to get some consulting work and was worried about how potential clients might perceive him. He also learned something about how these sites operate:

"I noticed a pattern of several more listings appearing after I removed the first round, so I think some people either (A) run multiple sites and used my payment as an indicator that they can probably extract more money by posting me somewhere else or (B) sold my info and the fact that I paid to another similar Web site."

Paying up did not seem to have gotten him very far. More than two and a half years after his arrest, some mugshots of him still lingered online—one site had even turned his photo into a video, perhaps hoping to create a more permanent artifact. His court case still had not gone to trial. After it was resolved, he said, he would probably take another pass at getting some photos taken down. In the meantime, in the eyes of some shadowy Internet entrepreneurs, his guilt was all but assured.

Life and Work in the Sharing Economy

> What happens when everybody is a brand? When everybody has a reputation? It means every person can become an entrepreneur.
>
> — Brian Chesky, CEO of Airbnb

Nandini Balial was seven years old when she decided she wanted to live in New York City. Visiting from India, she and her family took a drive through Manhattan, her gaze drawn to the towering buildings. What she saw stunned her: it was one of those moments acted out a thousand different times through the years, when an outsider comes to New York and understands, from some feeling of mysterious but strong conviction, that this place—this metropolis that her parents, in line with the city's mythology, had told her was the world's greatest city—was meant for her. "Why don't we live here?" she asked her mom and dad. A spell had taken hold.

Not long after that formative trip, Nandini and her family (two parents, a brother) made the long trek from New Delhi to their new home of Fort Worth, Texas. She was nine years old. It wasn't New York, and it was difficult leaving her grandparents and other family behind in India, but for all of them, it was the start of an exciting new American life—one which, her parents hoped, would lead to better opportunities, particularly for higher education. By the time that Nandini was a high school senior, she knew that Texas wasn't for her, so she arranged to visit the city she had fallen in love with a decade earlier. She toured NYU, an experience that culminated with a trip to the top floor of the Kimmel Center, one of the university's lavishly appointed, modern buildings off of Washington Square Park. From that vantage, she could take in an inspiring view of lower Manhattan. The message, she would later recall, was that "this could be yours"—this city, this life, all the fabled magic of being a young person on the make in New York. The next fall, she happily matriculated into the school of her dreams.

More than five years later, the dream has curdled. The economy crashed before Nandini even finished high school, and its tepid recovery has done little to improve her prospects. After finishing her film degree at NYU, Nandini and a roommate moved uptown to a rent-stabilized, fourth-floor walkup. Despite a fine résumé that includes a number of internships in her field of interest, she's been unable to find any work. If not for her relatively favorable lease arrangement, she'd have difficulty surviving at all in New York and might be forced to make the lonely trek back to Texas.

When we met at a café that Nandini used to frequent as a student, she told me about the challenges she's had in finding work and how the promises of a professional career in the city she loves seemed to be slipping out of reach. For eight months, she estimates, she applied to six to twelve jobs per day—first in TV and film, then expanding in an anxious, ever-widening search. "I hit apply on everything I saw that wasn't surgery or corporate accounting," she said. These were entry-level positions, including some as assistants, the kind of work one would normally expect to find after college. She had a half-dozen interviews, none of which panned out. One company interviewed her several times and didn't tell her that she was no longer a candidate for the position until several months later. Other jobs seemed to fill faster than she could apply for them.

"I'm not screwing around," she said. "I'm not applying for jobs because I have time that I need to fill." Her position is very clear: "I'm doing it because I'm desperate," she said, a mournful note creeping into her voice.

Nandini is a child of the Great Recession; its long shadow continues to circumscribe her life. She saw the economy melt down before she even picked up her high school diploma, and though it might have deflated her expectations for the future, she couldn't have anticipated exactly how bad it would get. Soon, she felt the damage at home: her father was laid off during her junior year in college. The only job he could find was in Mumbai, so he moved there, ten thousand miles from his wife, who remains in Fort Worth, where she's an assistant principal in the local school district while also working toward a graduate degree. Because the family is stretched so thin, "there's no extra money," Nandini said. She accumulated six figures of debt in order to pay for NYU, and for the foreseeable future, she has no way to bring that amount down.

Ironically, the job that Nandini's father got was in television, the very field she hoped to get into. He works in international operations and content strategy for the Sony Entertainment Network. It would be a fortuitous coincidence, except that not even nepotism, Nandini has found, can get her a job, at least not in New York. Perhaps it's reflective of the peculiar nature of our global economy, the same arrangement that forces an Indian émigré to return home—to Mumbai this time—to work for a Los Angeles–based division of a Japanese conglomerate. The forces at work here are buffeting and bewildering in equal measure.

Nandini's main advantage is her youth; it carries the promise that things might get better. But every long-term job seeker has asked herself the same question, especially when real jobs—with regular hours, benefits, and all of the once-standard trappings of professional life— seem like phantoms: what if things *don't* get better? What if we've hit some trough and, even if the latest quarterly growth rate has ticked up a percentage point or two, the promises of stable work and middle-class opportunity are being downsized and outsourced to distant continents? Those are the fears of young people who can't find work: that they're entering a game that's already rigged and whose rewards are ever-decreasing.

As part of her search, Nandini has looked at various internships, but she went that route in college. And besides, many internships come without pay, despite the practice being ruled illegal. "There's just no way I can work for free, and everyone wants you to work for free," she said.

When I first met Nandini, she had been working through Task-Rabbit, an online marketplace where people offer to do menial jobs (handyman work, cleaning, cooking, dog walking), for about five months. She discovered TaskRabbit after posting a Facebook status update in which she despaired about her employment situation. A friend responded, introducing her to the vast and murky world of digital labor markets. Nandini was intrigued but also found that some didn't seem wholly aboveboard.

"Most of them were kind of shady," she said. "They want you to go to a Costco in Forest Hills [in Queens] and check if they have x number of x product and then report it, and you get like five cents." Even when the pay for such tasks was a few dollars, she couldn't justify doing it, not when it would cost her at least that much in time and money to go to the far-flung areas of the city where these jobs tend to be.

She also wasn't too pleased that these sorts of labor markets involved little contact with other human beings. "They were all apps," she said, "and here I was trying to live a life less attached to my phone or digital screens of any kind."

Still, Nandini didn't have much of a choice, so she dove in and committed herself to TaskRabbit with the same alacrity she had shown toward her job search. She had always liked cleaning, finding it a great stress reliever, so she sought out those jobs, along with food service, cooking, office organization, event hosting, and similar work. She found that many of the people she worked for were gracious, but some seemed to have no idea how to handle a stranger working in their home. One well-to-do NYU student, who hired Nandini to clean the West Village apartment she owned, sat on the floor staring at Nandini as she worked.

Other challenges emerged. Because Nandini was bidding for jobs, she found, as many other TaskRabbit workers have, that she had to continually lower her rates in order to compete. That, combined with the 20 percent cut that TaskRabbit takes on every gig, made it difficult to earn much more than minimum wage—a particularly tough prospect in an expensive city such as New York.

She was also disappointed that, despite the labor being piecemeal and casual, it tended to take a comparatively long time—often, three to four days—for the payment to appear in her checking account. "This is how I'm making my living. I cannot wait," she said. "Whatever I'm using money for, I'm using it *today*. The rent was due yesterday."

She likes some of the work, particularly when she feels like she can help people improve their homes. "I don't control much about my life, because I can't find a full-time job," she said. "But I can control how organized I can help someone be."

Beyond that, the perks of this kind of work are few. Occasionally a client will offer Nandini a glass of wine after she's cooked and served a big meal. She rarely gets tips and thinks that many employers don't even realize that TaskRabbit takes a fifth of each transaction. She's never quit in the middle of the job, even if some haven't been what was promised in the listing. (She's scrupulous about reviewing her employers promptly and positively.) And though she doesn't assemble IKEA desks or move furniture, the work, particularly the hours cleaning apartments that may have not seen a scrub in months, can exhaust her. One day, she and another woman were hired to label six hundred bottles of eco-friendly cleaning solution in a decrepit building in Chinatown. The long hours standing and performing repetitive motions were grueling, but there was a small upside in the form of having another person to work with:

"I was so glad that I could talk to someone." She was paid $11 per hour, meaning she took home about $9.

One of the hardest jobs came from a woman living in California. She wanted someone to go sit with her elderly mother in Manhattan and speak to her. That was it—just spend some time with an old lady who doesn't receive many visitors. The mother has Parkinson's and dementia and can hardly speak or open her eyes. The woman reminded Nandini of her own grandmother, who suffered from similar conditions and died two months after Nandini and her family immigrated to the United States. As a health aide hovered in the background, Nandini spoke to the woman, trying to establish some connection. While they sat together, the woman's head kept listing to the side and her hand, grasping for some human contact, rested on Nandini's skirt. "I didn't know what to do," Nandini said. "I could only speak to her." After three hours, Nandini left. She found the nearest bench, sat down, and cried. "Maybe this is the kind of task that someone should be trained for," she said. The daughter would like to hire Nandini again, but she's reluctant to go back.

Nandini's parents know what she does for work, but she doesn't dare tell her grandparents, who still live in India. She imagines what they'd say: "I didn't let you go to America with my granddaughter so she could wash dishes for rich people." Her own family had a maid in India, where they were solidly middle class (though they would probably live less comfortably in the new, booming Indian economy).

Because she has few other choices, she continues working through TaskRabbit and applying for jobs, knowing that each résumé sent out is unlikely to be read by a human being. Entering strangers' homes—some palatial residences on the Upper East Side, others chic SoHo apartments—she's encountered her share of indignities: ungrateful employers, rooms covered with smelly clothes and trash, windowsills grimed with cigarette ash that require hours of scouring. Life is hectic now, and she has to scramble from job to job, perhaps cleaning an apartment for a few hours one afternoon before running to a charity fund-raiser, where she'd check in guests. "I can't remember the last time I saw a friend because I work seven days a week," she said. On weekends, she picks up shifts at a bookstore. Somehow, she's making ends meet.

What might be most draining is the daily race to find good tasks before the competition does. "You can bid on every single TaskRabbit job and never hear back," she said. The most she might hear is a one-sentence apology telling her that the task went to someone else. Other tasks offered less than minimum wage or would be frustratingly vague, not listing the time required or explaining the work involved. Still more are downright bizarre: requests for girls to pose as escorts and accompany men to a party; invitations to follow someone's girlfriend; demands for people of certain ethnicities.

A couple of weeks before we met, TaskRabbit had suddenly changed its bidding system, completely altering its mechanics and leaving some less experienced TaskRabbit members in the lurch. Under this new system, contractors such as Nandini no longer bid for jobs—a process which, while it pushed down wages, at least gave her some control and allowed her to seek out work on her own initiative. Now, workers list their skills and their rates, and they wait for people to bring work to them. Since this change, Nandini has seen her opportunities plummet. "I don't have any work," she said.

Although she's accumulated nothing but good reviews, Nandini thinks that being a level 10 on TaskRabbit holds her back. "There are people who are like a level 36," she said. These hierarchies, and the various rewards and gamified elements that come with them, matter. People who are higher level, who have completed more tasks, show up better in search results and are seen as more desirable. Without the ability to search and bid for tasks, she has no idea what's causing her not to receive offers—if her rates are too high, if her relatively low ranking means she doesn't appear in search results. It's tough to find out. "I might never work again," she said.

Nandini's experience isn't unique. She's one of millions of people

coping with a broken economy and taking whatever she can get. TaskRabbit has filled in the gap, but it hasn't done much to improve her lot. Instead,
it is part of a broader trend in which digital technologies are being used to
recast labor into something unrecognizable, a form of toil where subsistence
is the best to which one can aspire. To understand the implications of this
shift, we have to first look at the digital marketplaces, such as TaskRabbit,
in which people are forced to jockey for even the most desultory rewards.

- - - - - - - - - - - - - - - - -

ONLINE LABOR MARKETS

Reputation management, influence metrics, and professional networking sites have spawned a troubling shadow industry: online labor markets catering to fast, disposable work. Also called fractional
work, crowdsourcing, micro-work, and immaterial labor, these are jobs
that can be done remotely, with little to no instruction—completing
surveys, tagging photos, transcribing interviews. Occasionally a job
requires someone to go out into the physical world to confirm that
a restaurant is still open or to photograph a store display so that the
multinational company paying for it knows that it (and thousands
of other displays like it, scattered around the country or the world)
is set up properly. Usually, a would-be worker signs up, enters some
information, and allows the site to connect to her social-media profiles in order to confirm her identity. Jobs then start to come down
the pipe—some services allow workers to bid for jobs, with the lowest
price usually winning out—and the worker goes off and performs the
task for a few cents or a few dollars. Amazon's Mechanical Turk, with
its optimistically named "Human Intelligence Tasks," offers some of
the most menial work: copying receipts, drawing triangles, mimicking
facial expressions, clicking on random URLs—all of it presented with
as little contextual/explanatory information as possible. A Mechanical
Turk worker might be kept busy, he might even be mildly entertained,
but he would be lucky to earn minimum wage.

Online labor markets share a lot in common with social networks such as Twitter, Facebook, and LinkedIn. The laborer is trading less on his or her expertise and experience than on his identity and supposed trustworthiness. (This isn't to impugn the workers but rather to note that online laborers are forced to compete in a system which, much like Klout, judges their reliability based on proprietary algorithms whose criteria are often unknown.) Like traditional social networks, the owners of the labor markets don't create or contribute much. Instead, they create platforms that facilitate work by others, just as Facebook provides a platform for us to create content that financially benefits Facebook. They make a little bit of money on each transaction, sometimes by taxing both the worker and employer, meaning that the market increases in value only as it scales and that value accrues to the owners of the market. And like a social network, online labor markets can claim that they're in the business of connecting people. Social networks reduce the inefficiency of sharing; online labor markets reduce the inefficiency of hiring temporary workers. They try to make as little friction as possible between an employer and a worker. They come together in a quick exchange and part just as informally.

These labor markets practice a kind of internalized offshoring. With an increasingly unstable labor market—part of what's been described as the sluggish global economy's "jobless recovery"—companies can focus on retaining and fairly compensating highly skilled (and highly sought after) employees, such as engineers, lawyers, programmers, doctors, and scientists. Meanwhile, less complicated work can either be farmed out to low-wage freelance and temporary workers or subdivided into smaller and smaller units of work, which are then widely distributed through a cloud-based labor market. The result is an extreme form of Taylorism: in bull times, workers have more tiny tasks than they can say yes to, but they acquire no skills, they learn nothing about the product or service to which they are contributing, they have no contact with other workers, and they have no chance to advance or unionize. They simply do the task offered to them—for a very low fee—and move on as quickly as possible. Imagine a factory in which each employee wears blinders and can only see the thing immediately in front of him on the conveyor belt. An algorithm acts as the overseer, and it doesn't miss a thing. The software facilitating this transaction acts as the ultimate mediator: the employee and the employer never have to deal with each other directly. Payment can be unreliable and is wholly contingent on the employer accepting the laborer's product; if the former doesn't like what he receives, he can simply reject it and not pay the worker for his time, nor allow him to appeal or revise his work.

Online labor markets are often charged with being digital sweat-shops or, as one *New York Times* article put it, "a cloud-computing cross between Facebook and a hiring boss stopping his pickup truck in front of hungry day workers." Their defenders counter that these markets provide useful, on-demand work for the unemployed, stay-at-home parents, people looking to make a little extra cash, and others who don't belong to the conventional workforce. A 2010 survey found that most Mechanical Turk workers are in the United States and India, and that while about 70 percent have full-time jobs, they also tend to have low incomes. Like viral media, there may be a smattering of success stories, people who rise above the herd and manage to master the system, but the most productive online laborers usually manage to make only a few dollars per hour. Premise—a company specializing in what it calls "hyperdata," using rapidly collected photos of goods, particularly in grocery stores, to predict market conditions—pays its photographers 8 to 10 cents per snapshot. Premise then passes its analysis on to hedge funds and major conglomerates, who pay four- or five-figure monthly subscription fees.

Some industry leaders argue that this kind of labor opens up new possibilities of work that otherwise wouldn't exist without distributed micro-work and smartphone-connected laborers. A common example proposes that Coca-Cola, say, wouldn't bother going into thousands of grocery stores around the country to make sure that its special displays are properly arranged. It'd be too costly or difficult to organize in a timely fashion. But it could pay Elance or Gigwalk to field an army of laborers to do just that *and* have each laborer snap a photo and send it back, geotagged and time-stamped, to the Coca-Cola mother ship. And it'd only have to pay each employee $15 or so.

All of this may be true, but this line of reasoning is often presented as justification for a system that's otherwise exploitative. Just because this work is the only option for some people does not mean that we should celebrate it. Such arguments also tend to overlook other options. Perhaps Coca-Cola wouldn't hire people to check on its store displays individually, but it might hire regional marketing professionals who are responsible for working with stores in a particular area, thereby developing long-term relationships with these stores and creating meaningful, ongoing work for Coca-Cola employees.

Or take the example of Twitter and trending topics. The social network uses Mechanical Turk to hire workers to quickly respond to high-volume queries and trending topics that aren't machine-readable. This isn't necessarily skilled work, but people have to do it, since they can think creatively and grasp nuance, humor, and sarcasm that are inscrutable to an automated algorithm. (In many important ways, humans are still smarter thinkers than computers.) Twitter is taking advantage of cheap labor to propel the illusion that its trending topics are algorithmically determined in real time. In this respect, it has much in common with the original Mechanical Turk, a nineteenth-century chess-playing robot which, in fact, concealed a chessmaster in a wood cabinet underneath the board. Twitter's fleet of Mechanical Turk workers are accomplishing much the same sleight of hand, providing the illusion of seamless, automated competency, concealing the fact that company is relying on cheap, remote labor. As Ayhan Aytes describes it, "In both cases, the performance of the workers who animate the artifice is obscured by the spectacle of the machine." Amazon calls this "artificial artificial intelligence."

In an added dose of perverse irony, Twitter's Mechanical Turk workers are, in all likelihood, helping to train the machine-learning algorithms that Twitter hopes will eventually replace them. As with a number of other menial tasks common to the micro-work field, the human is there to complete the smallest unit of work that can't be done by a computer. But the data gleaned from these human workers will be used to inform the future generation of automated systems that will replace human workers or shunt them to even smaller bits of work. Nano-work will replace micro-work, until humans are engineered out of the system entirely, capturing all value generated as profits for the platform owners and the few programmers kept on as overseers.

Similar conditions describe the lives of content moderators for Facebook, Google, and other services. Drafted through online labor markets, these workers, most of whom live in the developing world, go through some preliminary training before they begin their work as moderators. Then, they sit and parse through feeds of images, triaging according to the company's policies—blocking porn immediately, but perhaps flagging a photo of a fight for further review. Pay is meager: a few cents per image or a dollar per hour, sometimes with opportunities to make a couple more bucks.* It's brutal work, numbing, boring, and rife with imagery of gore, bestiality, abuse, and violent pornography. As one moderator, or content reviewer as they're also called, told the journalist Adrian Chen: "Think like that there is a sewer channel and all of the mess/dirt/waste/shit of the world flow towards you and you have to clean it." The job of these unseen laborers is to deal with all of the horrible stuff people upload to social networks and prevent it from ever being seen. They are the ones who make sure that Facebook is a clean and well-lighted place.

Tech companies prefer not to advertise this aspect of their operations, and nondisclosure agreements prevent many of their moderators and the firms that hire them from talking. But without these workers, Facebook would be a comparative mess, and advertisers would want

---

* While much of this work is outsourced and highly distributed, a handful of U.S.-based firms have offices of content moderators who work nine-to-five and receive fast-food wages.

no part of it. By outsourcing most of this work to distributed networks of cheaply paid laborers, Facebook is able to keep its margins higher, while likely developing automated systems that will one day make these moderators superfluous. In the meantime, these laborers remain essential, even as they operate in the platform's shadows. And if social networking is defined by the machine state—that desensitized fugue we end up in when we keep scrolling through the feed, like a glassy-eyed gambler at the slots—then content moderation is the machine state at its most dehumanizing. There the feed becomes something malevolent. Contra the soothing humanitarian rhetoric of Facebook, here everything awful about human instinct is highlighted. While we enjoy the pleasures of connection, these workers are undergoing experiences that often leave them depressed, traumatized, and angry.

For this nascent market, dispensing with human labor may prove more lucrative than the current arrangement. Strangely, it may also prove more humane. One could imagine a movement forming in which labor rights advocates say that micro-work is so unsustainable and dehumanizing that it *must* be automated. Add RFID chips to all packaged food and grocery products and you can track their movement through supply chains and stores without human assistance. Perhaps companies can partner with stores to help utilize their surveillance systems to monitor the placement of goods. Firm up sentiment analysis, trending-topic algorithms, and optical-character-recognition scanning so that humans aren't forced to do such drudgery. To save content moderators from their on-the-job stress, we have to put them out of work again.

Just as Facebook or Pinterest retains control of your data, online labor markets keep workers wedded to the platform. You can't take your profile elsewhere, unless two labor markets form a partnership or decide to create an open protocol that other markets can take advantage of. (Even if there were some sort of digital CV for this industry, the lack of qualifications needed for this work, and the treatment of workers as disposable and interchangeable, would render it mostly meaningless.)

You accrue reviews, ratings, and other labor metrics, along with an account balance, according to that platform. If you have problems with a job or some personal emergency comes up, you can send a message to your temporary boss asking for more time, but given that these markets encourage employers to see laborers as easily replaced, you're unlikely to get much sympathy. Speed, profligacy, and rock-bottom rates are the name of the game. There is no room for contingencies. If you can't keep up, you'll be canned in favor of someone who will—and your ratings will go down, making it harder to get more work.

Even if you think you're doing a good job, that might not be enough. Gigwalk, perfectly in tune with the ideology of social, brags about the amount of data it collects on its workers. It keeps track of how long it takes employees to respond to messages on their smartphones. "If you are taking more than thirty minutes to respond back to a message from an employer, the likelihood of you being successful on gigs on Gigwalk drops off significantly," the company's cofounder said. Heaven help the Gigwalk worker who goes to a movie! The company here is relying on the philosophy of Big Data, which doesn't concern itself with *why* two things are related, only if they are. So if statistical analysis shows that responding to Gigwalk within an hour is correlated to a 97 percent chance of completing a job (as Gigwalk's CEO claims), then the company will update its algorithms accordingly to privilege those workers. Workers may be oblivious to the criteria on which they're being judged or may wish to try other jobs, but it's difficult to negotiate with a mostly automated system.

You can see where this is going. As governments continue to practice austerity, making lifetime employment and pension benefits a thing of the past, American corporations, despite a booming stock market and record amounts of cash on hand, follow suit. Stable employment, benefits, and retirement funds become anachronistic perks of a pre-digital workforce. Companies begin to think in terms of short-term spending, rather than long-term investment, as borrowing and hiring slow. More and more of us are forced to be contingent laborers, freelancers, or "permalancers" always on the lookout for more work, always advertising ourselves through social-media and public networks, knowing— with a sense of generalized suspicion—that our public utterances on social media may influence our future job prospects. Risk assessment algorithms may already be parsing our social-media profiles, pooling information to be used in a future background check. Forced to work constantly to pay off household debt or school loans, we don't have the time to learn the skills that would, we are told, allow us to succeed in the knowledge economy.* Large corporations start to realize that they can not only build on existing outsourcing—which has seen human resources, IT, customer service, and a range of other support staff shunted overseas—but also practice an ad hoc outsourcing at home, summoning pliable, cheap workers whenever they're needed. Managers get plaudits for being technologically progressive and nimble, cutting budgets. Stock markets reward companies operating on high margins, so more employees are fired from already profitable companies. Power accrues to software engineers, executives, high-level managers, and those controlling the algorithms and the networks; these men (and they are mostly men) are plied with spectacular working conditions and stock options to keep them happy and supportive of the status quo. Workers, in turn, have more mobility and a semblance of greater control over our working lives, but is it worth it when we can't afford health insurance or don't know how much the next gig might pay, or when it might come? When we can't unionize and lobby for better pay and working conditions? When work orders come through a smartphone and we know that if we don't respond immediately, we might not get such an opportunity again? When we can't even talk to another human being about the task at hand and we must work constantly just to make minimum wage?

Well, if you share these concerns, don't worry: Silicon Valley is

---

* Asked by a reporter to describe his politics, Mark Zuckerberg hedged before saying that he was "pro–knowledge economy."

already thinking ahead. And their plan involves no less than doubling down on social ideology and online labor, turning everything, and everyone, into something to be bought or sold. Welcome to the sharing economy.

BAIT AND SWITCH

The sharing economy combines all the elements of online labor, reputation, and social media's marketing-of-the-self to help make one's entire life purely transactional. Playing off the notion that a gig-based economy, filled with mobile freelancers, is inherently liberating, partisans of the sharing economy preach flexibility and personal empowerment. The sharing economy offers services that let you turn the core elements of your life—housing, transportation, physical labor, expensively earned skills—into rentable commodities. As with online labor, the emphasis is on reducing perceived market inefficiencies—regulation, middlemen, storefronts, the petty mundanities of paying fixed prices for goods and services. In the sharing economy, everybody is an entrepreneur and everything is negotiable. It is "people as businesses," as Airbnb's CEO declared triumphantly. Nothing should ever go unused: a parked car is inefficient, since someone else could be paying to use it. This arrangement means that trust is also fungible, an asset that must always be managed, tended to, bought and sold. Your name and personal brand and online presence are always subservient to this nebulous sense of trust. Don't have a weird Facebook profile photo, because that might not play well with TaskRabbit's vetting system or your potential interlocutor on Cookening. Make sure you always give your employers good reviews, because they might retaliate if you pan them. If you run into trouble, as did Franklin Leonard with a racist property owner on Airbnb, then perhaps you just need to recalibrate your public identity.

The sharing economy includes some online labor outlets, such as

TaskRabbit, in which independent contractors perform menial tasks, such as fetching groceries or assembling furniture, for small fees. Companies such as Lyft, Uber, and Sidecar provide taxi-type services, but they almost never call themselves taxi or transportation companies. This is because the transportation industry is highly regulated, something that Uber would like to disrupt. Government, with its pernicious regulatory apparatus, is simply making the market inefficient and costing consumers and businesspeople in both cash and intimacy with one another. (For a time, Travis Kalanick, Uber's founder, used a cropped cover of Ayn Rand's *Atlas Shrugged* for his Twitter avatar before replacing it with a drawing of Alexander Hamilton's face. Promoting individual economic liberty is presented as part of the company's mandate.)

In the case of Lyft, potential drivers have to apply for work through their Facebook accounts. Once a driver is approved, he receives a fluffy pink mustache to affix to the front of his car. Customers call for a Lyft car through a smartphone app, choosing a driver by looking at his name, his rating, and photos of him and his car. Once the customer gets inside, the driver is supposed to offer his passenger a fist bump. For a while, Lyft passengers didn't pay a fare—since that would have made Lyft seem too much like a taxi company, exposing them to certain municipal regulations—and instead were asked to offer a suggested donation. Later, Lyft instituted minimum fares and a "Prime Time amount," its euphemism for ratcheting up fares during busy hours. After each ride, customers rate the driver, and drivers who have average ratings below 4.5 stars will be fired. In what's supposed to represent a quid pro quo, customers get rated as well, meaning that you might not receive future rides if you accrue a bad score.

Usually the sharing economy is just a way for companies to outsource labor and risk to individual consumers, who take on part of the traditional responsibilities of a company while getting a fraction of the services. The company acts essentially as a search engine or aggregator, bringing the parties together, contributing little, bearing the least amount of risk of anyone involved, and pocketing a nice fee. Uber, for instance, takes about 20 percent of each fare from UberX, its popular, low-budget offering, along with a $1 safety fee. As with online labor markets, the app serves as the ultimate mediator. No one ever has to meet, which is by design. As one TaskRabbit worker remarked: "That's part of the strategy of TaskRabbit—to keep us apart from one another. We can't message each other on the Web site. The only way you get to meet another TaskRabbit is if you post a task, and I think they do this to keep us apart because they don't want us fixing the process. They don't want us unionizing."

Nor are sharing economy workers ever truly employed in the sense that most people are used to. When a group of Uber drivers assembled outside the company's headquarters to protest their firing, the company's general manager said that the drivers weren't employees and that, when they were fired, it simply amounted to deactivating the drivers' accounts. The given reason? Low ratings from passengers. This insouciance is built into Uber, which calls itself a software company, or alternatively, a transportation network company, rather than a taxi company. (Sidecar identifies as a peer-to-peer ride-sharing service.) Uber is also known for flouting local laws by setting up business in a new city without speaking to officials responsible for managing the transport sector.

There's a great deal of unacknowledged work involved in the sharing economy. Drivers have to keep their cars clean and insured, with no help from the company nominally employing them. Zipcar customers have to clean the cars after using them. They also may have to get extra liability insurance and can be charged for dents that are discovered after they drop off a rental car (that dent might've been caused by someone bumping into the car while it was parked in the rental lot, but in this mostly automated system, there are no Zipcar employees to watch over parked cars). TaskRabbits have to keep their online profiles spotless, so as to earn people's trust with small tasks, and many workers report agreeing to a task only to find that the listing played down the amount of labor involved. Tasks sometimes end up being more dangerous than advertised, and TaskRabbit doesn't offer insurance or other protections for a contract worker who falls off a ladder or gets sick on the job. Property owners using Airbnb have to clean their homes, keep essential supplies stocked, and communicate with renters. The property owner must make sure his or her insurance will cover all eventualities, as Airbnb offers a $1 million "host guarantee" that it claims is not a substitute for traditional insurance.

Members of the sharing economy also must worry about the legal gray area in which they're doing their work. Airbnb and Uber have both resisted negotiating with municipal authorities. New York City officials have declared the former illegal, saying that it amounts to an unlicensed hotel operation. Zoning restrictions and leasing terms may also prevent someone from renting a room or residence on an informal basis. And a few high-volume landlords have used Airbnb to rent dozens of properties illegally, while others have used the service to run prostitution rings or host orgies. The result can be fewer rentals available on the market or individual renters being evicted for unlawfully subletting their apartments.

Airbnb does offer some benefits. "Hosts," as property owners are called, can make extra money or better afford a home by renting out an unused room. Renters get to know locals by dealing with them directly and sometimes can stay in areas where a hotel would be too expensive. Alternatively, a renter might find himself in a neighborhood that he wouldn't otherwise visit.

But there are real costs to Airbnb, too, and not just in the potential awkwardness of staying in someone's home. Hotel taxes go unpaid, depriving cities of needed revenue and penalizing law-abiding hotels, which provide necessary low-wage jobs. Apartment complexes become filled with tourists and rotating guests who are potentially disruptive. Slumlords and other unscrupulous operators are encouraged to subdivide properties into smaller rentals from which they extract hefty fees. Rent-controlled apartments can be leveraged to extract illegal rents from unsuspecting subletters. That has become an acute problem in San Francisco, Airbnb's home city, where housing prices have risen exorbitantly due to the influx of moneyed young start-up employees. In the process, otherwise suitable long-term rental units become unavailable, turned over to tourists and short-term corporate rentals, which in turn drives up prices for locals. In the summer of 2013, evictions in San Francisco were at their highest in eleven years, due in part to the latitude the city gives landlords in evicting residents of rent-controlled apartments and the tempting payday of short-term, under-the-table rentals.

Among the most pernicious aspects of the sharing economy is the way it presents itself as a populist operation, a loose community coming together to engage in mutually beneficial, informal economic exchanges. In reality, it is anything but these things. What is shared most among participants in the sharing economy is risk—risk that the platform owners displace onto workers and customers. But the industry's self-mythologizing masks this reality. Peers, an industry lobbying group that calls itself a "grassroots organization to support the sharing economy movement," acknowledged after its launch that its "mission-aligned independent donors" included "investors and executives of sharing economy start-ups." That's not much of a surprise, considering that its work directly benefits the companies comprising the sharing economy, as well as their billionaire backers. The petitions, potluck dinners, and advocacy measures that Peers leads may help to make the sharing economy more mainstream, but above all, these efforts help to create bigger profits for the owners of the platforms powering the industry. The dozens of corporate logos for companies such as Elance and Chegg, a textbook rental service, plastered on the Peers site do little to change the notion that this organization ultimately answers to them. Airbnb users with whom I spoke reported getting e-mails from Peers without signing up for the organization's mailing list. These e-mails tended to arrive before key legal hearings or following spurts of critical media coverage.

"It's like the United Nations at every kitchen table. It's very powerful," said Airbnb CEO Brian Chesky, explaining that his company was part of a "revolution." Douglas Atkin, Airbnb's head of community, who announced the formation of Peers at a London conference, has called it a "movement" comprising "huge numbers of people, with a shared identity." Collectively, Atkin said, they were ready "to fight for their collective interests against unfair and unreasonable obstacles." It all sounds, as Tom Slee, author of *No One Makes You Shop at Wal-Mart*, remarked, like an astroturf operation, perhaps funded, if not by Airbnb itself, then by the venture-capital firms who have sunk hundreds of millions of dollars into Airbnb and similar companies. From its unveiling, Peers had the hallmarks of a well-heeled, established operation, with most of its staff hailing from MoveOn and Organizing for America, progressive political organizations that also tout their tech savvy. In a fortuitous twist, Atkin is also a cofounder of the supposedly independent Peers, and he's used his position at Airbnb to push people to sign Peers's petitions. Other tech leaders have hands in both pots: eBay founder Pierre Omidyar's foundation is a major Peers donor, after having converted the nonprofit CouchSurfing into a for-profit company, in turn precipitating a member revolt. These relationships show that Peers isn't a socially conscious nonprofit but a group using the social-justice appeal and tax benefits of nonprofit status to advocate for the industry's benefit. In April 2013, as New York State officials stepped up their investigation of Airbnb, Peers e-mailed Airbnb users asking them to sign up for times to visit offices of members of the State Senate and Assembly. "We'll provide everything for your Office Drop," the notice read, "including having a Peers leader meet you outside the office for a quick run down." In other words, Peers was telling Airbnb users when and where to show up and coaching them on what to say, but this is supposed to be a grassroots organization fighting, as it claims, "entrenched interests."

Spend enough time in the sharing economy, whether as a worker or consumer, and you begin to see the benefits of the legacy businesses these upstarts claim they are about to replace. All those supposed market inefficiencies and regulations and sterile traditions come to seem like useful, practical measures that protect both parties and provide a degree of standardization and reliability. Sure, hotels can seem impersonal, but there are actually some benefits built into that environment. When you walk into a hotel, whether a chain or a small bed-and-breakfast, you basically know what you're getting. The hotel will have trained employees, it'll pay taxes, it'll be registered with the city and regulated accordingly. You walk up to the front desk, give your name and credit card, get the room key. You learn about what the hotel offers, what you'll have to pay, when you should check out. If you don't like something, you can call the front desk or go talk to someone. Meanwhile, the hotel is insured against damages and other complications. The hotel knows how to bring in customers, fulfill their needs, and then clean up after them and bring in more. Unlike an Airbnb host, the hotel doesn't have to worry about scrutinizing each customer for reliability ("You are responsible for determining the identity and suitability of others who you contact via the Site, Application and Services," according to Airbnb's terms of service). Prices vary, but they're far from ad hoc. All of this may sound rather banal, but this kind of system, whatever it may lack in spontaneity or color, creates an environment in which a customer and hotel owner/employee know what to expect from each other. The mediating elements of regulation and formality actually help to improve the exchange.

Similarly, a taxi driver belongs to a company to which he has certain obligations and which owes him things in return. Getting in the taxi, you know exactly how you're going to pay and that the driver has gone through training—rather than being self-certified or only cursorily vetted by Uber—and is insured in case an accident occurs. If a crime happens, there's a camera in the car to capture it. The driver has a garage that he returns the car to, where it's probably shared by other drivers. He may belong to a union which, while it takes a chunk of his wages, also protects his rights and helps ensure his future employment.

If he had a dispute with his bosses, union officials and lawyers would work on his behalf. A local taxi commission keeps his industry on a tight leash, but it also limits the number of licensed cabs, making it so that drivers have more assurance that they'll have enough work and won't face competition from unlicensed drivers skirting corners on safety. This arrangement controls the number of taxis on the road while, in some cities, driving up the value of a taxi license. That can make for a difficult situation for drivers, who have to work to pay off the cost of their car or license, and for riders, who may complain of an insufficient number of taxis, but it's a more transparent system than that touted by Uber, in which everything is controlled by the company. In a traditional taxi system, a city taxi commission helps look out for the interests of customers, drivers, and owners alike.

There are also protections in place for customers. If you live in New York City, like I do, a taxi that picks you up in Manhattan is legally obliged to take you wherever you want to go, including to one of the other boroughs (with the meter running the whole time, of course). Unless you are violent, disruptive, or otherwise problematic, the driver can't refuse you a ride based on your skin color or some star rating you've accumulated. You can also expect to pay a standard fare, unlike with Uber and Lyft, which are known to institute surge pricing to leverage high demand.

Uber claims that surge pricing represents a market-based solution and offers a fair price based on availability. Except that Uber controls the market. They decide how many cars are on the road—the company has been caught asking drivers to stay off the road in order to drive up rates. At the same time, Uber has presented itself as a populist operation with a low threshold for entry. They've even established financing programs to help potential drivers get cheaper deals on cars from General Motors and Toyota. As the number of drivers (some of whom used Uber's financing program to go into debt for a shot at driving for the company) increases, Uber has more of an incentive to limit supply by keeping drivers off the road and firing those who are perceived as a liability or difficult to control. After it breaks the monopoly of taxi cartels, Uber will be free to raise prices in order to soak captive customers for greater profits.[*] When Uber dominates the terms of the market so thoroughly, how can they promise to be a fair operator? As long as an opaque algorithm dictates who gets to drive, how much people have to pay, and whether they're even allowed to do business with each other, then both driver and customer will be at a disadvantage.

With a yellow cab, you don't have any of these concerns. And once the ride is over, you pay the fee, and the two of you separate. There's no mutual assessment based on an all-too-brief encounter. You don't fist-bump or become friends or follow each other on a social network. Perhaps if you really enjoyed the ride, you'll get the driver's card and call him again for future rides.

These measures help to put all parties involved on a more equal footing. It also allows one to refuse to do business with the other without suffering great consequences—unlike, for example, the TaskRabbit worker who was almost fired from the service after complaining about a misleading job listing involving piles of laundry covered in cat diarrhea. If I approached a laundry service with such a task, they'd either laugh at me or demand a hefty price to do the work, and understandably so.

Some of these companies have helped to spur establishment players toward needed reforms. Taxi services, often seen as resisting innovation, have begun adopting smartphone apps and e-hailing systems to keep up with Uber and Lyft. Airbnb has shown that many people are interested in renting out their homes, despite laws that prevent doing so, and that cities may have to work to accommodate this need. But in the spirit of disruption, these companies tend to show up in a new city promising to lead a revolution, only to be forced—by court order, political pressure, or the realization that existing regulations do some

---

[*] Amazon practices a similar long-term plan, using deep discounts to push out competitors and attract loyal customers before eventually raising prices.

good and are unlikely to be overturned—to start negotiating with political leaders and abide by local laws. In short, the sharing economy is professionalizing, its members becoming more like the businesses they hoped to unseat. Uber, which used to only contract with independent drivers or taxi services, has expanded its training offerings and taken out supplementary insurance policies for drivers. But the company has shown an overall reticence to take responsibility for its drivers, particularly after horrific stories emerged about a driver sexually assaulting a passenger and another hitting and killing a child in a crosswalk. Overcoming its initial resistance, Airbnb has asked New York City officials to let its hosts pay hotel taxes (though, notably, the company isn't offering to pay these taxes on its own). Even with these reforms, the platform owners still wring most of the value out of the system and offer few protections to their users and laborers.

In their current forms, the sharing economy and online labor are supposed to offer transitional or temporary opportunities to people displaced by an unstable economy, but the reality is that they've become a way of life for many people who lack steady employment. (Similarly, fast food, which was once considered a seasonal job for teenagers, has now become a career for many people with few other options, despite the industry's poverty-level wages.) It's difficult to conclude that these forms of work would be growing in popularity if the overall economy weren't afflicted with widespread joblessness, stagnating wages, diminished worker rights, and an eroded social safety net. This is the environment that the sharing economy and labor markets are hoping to leverage, while perpetuating all of its worst qualities. Because rather than opening up new kinds of work, eliminating market inefficiencies, or bringing neighbors together, these platforms keep workers in a state of perpetual insecurity, making them seek lower and lower payments from remote, capricious employers who see them as disposable, one step above the automated machines that may one day replace them. Those who have assets to rent out—or share, in the industry's intentionally misleading jargon—often face unintended costs, additional labor, and the specter of illegality. There are no rights here for workers, no opportunities for advancement, no ability to unionize or ally in mutual interest—unless that interest happens to be helping Peers overturn government regulations that help provide workers' rights, ensure safe practices, and make companies pay taxes to the municipalities in which they do business. Airbnb's Douglas Atkin claimed that "we literally stand on the brink of a new, better kind of economic system, that delivers social as well as economic benefits." But the sharing economy isn't some fairer, communitarian system of barter and exchange. It more accurately resembles an extreme form of capitalism in which everyone is an entrepreneur but no one is employed. Everyone works, but they're forced to make a beggar's bargain, laboring for experiences and for occasional cash, rather than for supporting a life or a family. People sell themselves and their basic possessions for below–minimum wage payments. Life becomes a pawnshop, to which workers are always in hock. And TaskRabbit and its ilk stand behind the counter, collecting its fees all the while.

For consumers, most of these problems are invisible. That is by design. You're not supposed to know that the trending topics on Twitter were sifted through by a few destitute people making pennies. You're not supposed to realize that Facebook can process the billions of photos, links, and shareable items that pass through its network each day only because it recruits armies of content moderators through digital labor markets. Or that these moderators spend hours numbly scrolling through grisly photos that people around the world are trying to upload to the network. Uber's selling point is convenience: press a button on your phone and a car will arrive in minutes, maybe seconds, to take you anywhere you want to go. As long as that's what happens, what do consumers have to complain about? Now joined by a host of start-up delivery services, ride-sharing companies are in the business of taking whomever or whatever from point A to point B with minimal fuss or waiting time. That this self-indulgent convenience ultimately comes at the expense of others is easily brushed off or shrouded in the magical promise that anything you want can be produced immediately. This is the Amazon ethic—customer first, costs slashed to the bone, goods delivered as soon as possible through cut-rate independent contractors—to the max. That's why Amazon, so adept at hiding the exploitative conditions in its sweltering warehouses, has committed to ferrying goods by speedy drones and patented a method to ship customers items before they've even ordered them. Waiting for anything is now treated as a market inefficiency. We want what we want and we want it now. And we won't tolerate the hassles that come with dealing with other human beings—apps only, please. If some people have to suffer or laws have to be broken to make this situation possible, so be it. That's the nature of disruption, of an ideology that serves the rich while claiming to liberate the poor and middle class.

- - - - - - - - - - - - - - - - -
ESCAPING THE SHARING ECONOMY

As for Nandini Balial, she's found that, while the sharing economy may help connect her with temporary work, it doesn't provide a genuine solution to her problems. Her most fruitful work has come from people who have allowed her to take their relationship *off* of TaskRabbit, thereby saving her the 20 percent fee that the platform skims from each transaction. And with TaskRabbit's rejiggered system shifting the balance of power even more toward employers, it makes sense that Nandini and her fellow workers would try to use the platform only to set up initial introductions. On the platform, she must wait for employers to contact her; off of it, she retains more of her wages and can pitch employers directly.

Nandini now spends a lot of her time working for a twenty-nine-year-old man whom she met through TaskRabbit. This man—let's call him Joe—works in finance, so he has a lot of disposable income but not much free time. Nandini contacted him—back when workers could still seek out jobs on their own—and signed on to do some housecleaning and laundry.

"I think he was slightly discomfited by the fact that I wasn't that much younger than him," she said. "I've seen that with other Task-Rabbit posters. They'd much prefer that I be a middle-aged woman who has children"—someone who might be thought of as traditionally fitting into this kind of work.

After quickly deciding that TaskRabbit was an unnecessary middleman, Joe and Nandini began devising more tasks for her to do. Now Nandini essentially runs Joe's life. Laundry, dry cleaning, shoe repair, grocery shopping, organizing his apartment—she does it all. She hangs up his clothes, makes his bed, and has even done research to help fill out some Excel spreadsheets he was working on for his job. He gave her a key to his apartment, so now she comes and goes when needed. "I've kind of forgotten what he looks like," she said, "because I'm like his house elf: I come, I work, and I disappear."

Eventually, Joe asked Nandini whether she knew about Tinder, the dating app. She said she did. Joe had recently broken up with his girlfriend and wanted Nandini to start managing his account. He said that he would do some of the swiping—the initial yes/no choice of whether one was interested in a person. He told her to just tell women the basics about him (while pretending to be him), but she knew that wouldn't work. She decided to have fun with it, to approach the potential dates how she would want to be approached. Soon, she had a few matches. "It was going so well that I was alarmed," she said. Tinder connects through Facebook—meaning that Joe had to give Nandini his log-in information for that as well—and lists which interests you have in common with potential matches. Nandini noticed that Joe and one woman had *Game of Thrones* in common, so Nandini, pretending to be Joe, began talking about it. "Lannisters or Starks?" the woman asked, referring to the two rival families from the fantasy series. Nandini frantically texted Joe, asking him which he preferred. "What are you talking about?" he replied. He knew nothing about the show or the books; he had just liked it on impulse. Nandini's Cyrano de Bergerac performance was descending into farce. She was creating—and performing—a personality for him where none existed.

Nandini knew that Joe was logging in to check on her progress. One day, she hopped onto Tinder only to find that he had "unmatched" several women with whom Nandini had been talking. "Oh no, I'm not into them," he told her. She pushed back, telling him, "They were really cool, they liked you, they were cute." Despite the absurdity of the situation, she felt invested in the process. His response was typically glib: "Yeah, but I'm just not into that." In the end, all she could do was bill him for her time. At least, she told me, Joe gave her a raise when he saw that she was doing well on Tinder, and he promised her a bonus for each date she successfully booked. Joe managed to go on one date and compensated Nandini accordingly. His review of the evening: "Eh."

Despite the inconsistent work, the relatively low pay, the flippancy of his attitude, his seeming inability to attend the basic responsibilities of his life outside of work, Nandini has some sympathy for Joe. She knows him better than perhaps even his close friends do. She sees all the save-the-dates and wedding invitations that arrive in his mail. This workaholic finance drone seems lonely and incapable of taking care of himself. But there are limits to her understanding. "If there's one thing about your life that maybe you should do," she said, "it maybe should be this."

I was reminded of her comment about the sickly old woman, who apparently has another daughter in the New York area—one who doesn't visit. From these experiences, Nandini has learned something about TaskRabbit, the sharing economy, and how the people who turn to it see this army of contingent laborers. These labor markets are, in her view, "how the one percent get their work done." It's a way for people with means to offload their chores and responsibilities to those who are desperate enough, such as Nandini, to work for low pay and with few rights or protections.

Still, Nandini hasn't given up on her plan to work in film or television, but she's also willing to do whatever she can to find meaningful, regular work—to be an employee and not just a perpetual freelancer. "I will work for you seven days a week," she says, pleadingly. "I just need you to pay me."

In her eyes, moving isn't much of an option. It's too expensive—in most other cities, she'd have to buy a car—and there won't necessarily be better opportunities elsewhere. Certainly not, she thinks, back in Fort Worth. Her friends have all left; her brother decamped for Austin, first for the University of Texas and then a job with Accenture. Only her mother remains in their adopted home, and now the family makes plans to reunite once or twice a year in a city they can all get to. (They don't have much connection to Chicago, but it's proved geographically convenient.) But Nandini remains determined to establish herself in New York. There are practical reasons to stay, but there is also, she admits, some admixture of hope, pride, and longing. "I would rather clean houses to the end of my days than go back to Texas," she said.

Nandini is a big fan of 1950s and 1960s American television, which she watched while a student in NYU's film department. "It's just so well written," she said. "Doesn't exist anymore." After spending so many hours watching these vintage shows, she's developed what she calls her "secret aspiration." It is, she explained, "that at some point, if I proved myself, I would run into someone in a position of authority. Preferably in television. And he'd say"—and here she adopted a playful, Cary Grant–in–*His Girl Friday* accent—"'I like your chops, kid. You're hired.'"

"And I would do it. It doesn't apparently happen. It does not happen. You begin to feel a certain amount of shame to ask your friends" for help. The lifestyle, such as it is, exacts a toll. She continued: "I've been doing that for sixteen months now, and a consequence, you don't see your friends as much, and you begin to kind of live off Netflix and your credit card and TaskRabbit.

"That's the American dream, right?" she said, in a tone of tender irony. "You just swipe. You figure out how to pay for it later."

Digital Serfdom; or, We All Work for Facebook

> There is no competition in television except among competitors trying to sell the attention of their audiences for profit. As a result, while television is supposed to be "free," it has in fact become the creature, the servant, and indeed the prostitute, of merchandising.
>
> —Walter Lippmann, 1959

Exemplified by Peers, the cyber-libertarian movement has found its leaders: tech corporations. Companies are more important than countries, Mark Zuckerberg has said. Facebook is not just a Web site or even a digital community. It's a new kind of nation-state, one that promises to help connect the two-thirds of the world's population who don't have Internet access. This mission is important because connectivity is, in Zuckerberg's words, "a human right." Former Facebook employee and venture capitalist Chamath Palihapitiya agrees, telling an interviewer: "Companies are transcending power now. We are becoming the eminent vehicles for change and influence, and capital structures that matter." (Palihapitiya is also a supporter of FWD.us, Zuckerberg's political lobbying organization.) Others, such as libertarian PayPal cofounder Peter Thiel, would like to do away with government entirely; he has donated nearly $2 million to the Seasteading Institute, which hopes to build floating cities unencumbered by the trappings of government. Balaji Srinivasan, an entrepreneur and founder of a genetic-testing start-up, has called for "Silicon Valley's ultimate exit" from "the paper belt," his term for traditional sovereign governments. A number of other tech leaders agree, with some even calling for the breakup of California into several smaller states. Google cofounder Larry Page—in a speech in which he said that computers should have already eliminated world hunger—declared, "In tech, we should have some safe places where we can try out some new things and figure out what is the effect on society and what is the effect on people, without having to deploy them in the whole world."

Full of fantasy and wishful thinking, with more self-regard than self-criticism, these dreams of Silicon Valley secession are largely fatuous, but they're also revealing of the ways in which the industry operates and imagines its place in the world. These companies aren't just providing services; they're changing the world, connecting people with essential human needs, and replacing or bypassing many of the functions traditionally expected from governments. In some instances, as Page suggests, regulation may have to be thrown out entirely, so that companies such as Google can have a free hand to test their products without consequence. They want to be polities unto themselves, free of what they consider the inefficiencies (what we often call regulations, consumer protections, and the messy but necessary bureaucratic procedures of government) that define life in the rest of the country. They've already begun to beat a retreat from their surrounding communities. Facebook has offered subsidies to employees who work close to the office and has since built an apartment complex for Facebook employees. Through these and other measures, including numerous on-campus services (haircuts, laundry, free meals, gyms, massages) that normally would require someone to interact with non-company employees, social-media companies have succeeded in privatizing swathes of California's Bay Area, making it into a series of company towns.

These are steps toward the kind of digital agora sought by adherents of Richard Barbrook and Andy Cameron's Californian Ideology. It's a vision that's been cultivated over decades and that has a very real effect on the types of products and digital environments these companies create, from the personal chauffeurs of Uber to the parlous world of online influence. Here users are guests of the benevolent technocratic elite, the digital overlords creating our brave new future and allowing us to visit and enjoy the view for the price of privacy and personal data.

In an influential manifesto titled "A Declaration of the Independence of Cyberspace," John Perry Barlow, the Grateful Dead lyricist-turned-cyber-libertarian activist, told governments that they weren't welcome in the new online world. This was 1996, the time of the Microsoft-Netscape browser war, the early years of the tech bubble, and the rise of search engines, which helped us unlock the plenitude of the World Wide Web. In this heady period, it might not have seemed out of place—indeed, to many cyber-utopians, it still doesn't—for Barlow to declare to the world's governments, "Cyberspace does not lie within your borders. Do not think that you can build it, as though it were a public construction project."

In reality, the Internet was a public construction project, a government-funded initiative, in its earliest days, when the Department of Defense launched ARPANET, a communications network between research labs at four U.S. universities. And government involvement in cyberspace has persisted, particularly with the anointing of the Internet as a new battlefield, complete with cyber-weapons, teams of government-directed hackers, and mass surveillance. Cisco and Amazon are major providers of database and cloud-computing services to the CIA and other members of the intelligence community. The U.S. government remains one of Silicon Valley's most lucrative and reliable clients, and tech companies and telecoms are paid partners in U.S. government surveillance.

Even though Barlow's argument has been debunked over the years, it persists in varying forms, especially in the notion that cyberspace is somehow separate from real life, meaning that it's a place of free-dom and possibility where normal rules don't apply. (It's this very attitude that allows self-appointed scolds to say that you shouldn't ex-pect privacy online, the subtext being that you, average citizen, don't deserve it.) Certainly the Internet offers tremendous possibilities for communication and collaboration, but it's also very much tied to our quotidian existences: we use it for work, shopping, entertainment, intellectual exploration; governments use it to communicate with and spy on their citizens; companies use it to hawk their goods, spy on their customers, and do business with one another. The Internet is not some rarified place but a set of digital technologies bound together by fiber-optic cables, server farms, ISPs, and software. It is as much a physical creation, subject to the vicissitudes of nature and the laws of various governments, as it is a digital realm. It exists in the world, with real people, companies, governments, and NGOs managing it.

And yet, the Barlovian idea of the Internet as something apart from the world endures, and sometimes for good, or at least understandable, reasons. Trolls can justify their behavior by saying that the Internet is different or that it's all just for kicks; it doesn't mean anything, and they wouldn't act this way in real life. Writers, artists, programmers, and others find possibilities for expression and creation in a digital world that seems open and unregulated. (There's a touch of manifest destiny in this, the Internet as global frontier to be settled by the most enterprising among us.) Western democracies, particularly the United States, tout the Internet as a new place for free speech and dissent in the developing world, while eliding the ways in which they help many of these governments repress their citizens. Similarly, the quest for Internet freedom abroad provides the U.S. government with a useful cover for its surveillance practices at home.

The notion of an independent Internet, free from terrestrial con-cerns, has allowed companies to insert themselves as the sovereign authorities of cyberspace. It is they, not politicians or activists, who are expected to adjudicate the future of the Internet and to lobby for or against reforms such as net neutrality. They are the handmaidens to our digital emancipation. We give them tremendous amounts of personal data, allow them to identify us wherever we go online, and they in turn provide us a digital space that feels more like a global community, loaded with meaningful interpersonal connections, than any we've seen before.

This isn't so misguided. Twitter has the right to form its own policies for its community; it can police speech however it chooses. And as I argue throughout this book, companies such as Facebook and Google have tremendous power in shaping the Internet and our experiences online. Where that leaves us, however, is with an Internet geography resembling a clutch of corporate fiefdoms, atop each of which is a chief executive ruling his subjects with the hand of a benevolent dictator, claiming to act in their best interests. At the same time, these companies give the impression that they can be appealed to, that its users are part of a participatory democracy. They can determine the future of the community and its management, perhaps through one of Facebook's occasional exercises in allowing users to vote on site policies. These ventures, such as Facebook's Site Governance project (discontinued soon after a vote was held in November 2012), have the feeling of a rigged election. There are all the trappings of democracy—appeals to participation, citizenship, explanations of the issues—but the end result is that little changes. The site remains mostly the same; its inner workings are, if anything, more opaque after the vote. We all know that it will be the decisions of Facebook executives, with an eye toward company shareholders, that drive future changes to the site and that most changes will be designed to surface more user data. We might voice support for a privacy policy or two, but this is all window dressing. Our digital lives remain defined by two sclerotic governments: that of the country we live in and the social networks where we house our digital selves. For these reasons, when Airbnb executives invoke egalitarian notions of sharing or economic cooperation, one suspects that they are using populist rhetoric for more self-serving ends.

There's one other idea from Barlow that endures, and it's a dangerous one. "In our world," he claimed, "whatever the human mind may create can be reproduced and distributed infinitely at no cost." While the Internet often has the appearance of being free, of communications at no cost, there is a price. It's just not one always exacted by putting in your credit card or paying a bill (although Barlow, like many cyber-libertarians after him, doesn't acknowledge that Internet *access* does cost money, as do computers, smartphones, routers, etc). Few things are really free on the Internet. When you pass around a video of a college student who catches fire while trying to twerk, a range of actors are making money: your Internet service provider; the video site hosting it and its ad partners; the social networks monitoring your sharing habits; the television show that commissioned the video and designed it, like so much viral material, to look like a candid amateur video. The only person not making money in this arrangement is you, and you did some promotional work by sharing/publishing the video.

"If you're not paying for the product, you are the product," according to one popular digital-age axiom. It's a bit glib, but there's truth in it. The spread of so many free services and so much free content has led to the idea that the Internet is as much a place of economic freedom as it is one of personal freedom. But because we pay for so little that we do online, we're forced to pay in other ways: being tracked, submitting to video ads before we can watch TV shows, having our personal data bought and sold, liking brand pages so that we can access exclusive content or giveaways. The irony of the Internet's bounty of free media is not only the disruption of the creative industries—journalism, music, television, film—that fuel it, but also that the more we consume, the more we in turn are consumed by the networks that serve us content. We become beholden to them, laborers on the Facebook farm, where our content can be seen by hundreds or thousands, but it is *owned*, along with our digital identities, by the Facebook mother ship. We're not just the product, we're also making the product. It's for this reason that some observers have come to think of our relationship to social media as something like feudalism. They call it "digital serfdom."

- - - - - - - - - - - - - - - -

HOW WE ALL BECAME CONTENT FARMERS

Do you know that when you buy an e-book from Amazon, you don't actually own it? Instead, you purchase something called a content license. This license is essentially the right to possess and read a copy of the e-book, but only in the ways in which Amazon specifies in the agreement—the kind of digital contracts that we agree to all the time but never actually bother to examine carefully. So, if Amazon decides that you need to upgrade your Kindle software before being allowed to read more books, they can make you do that. If you've done something it doesn't like with the e-book or Amazon says that they sold it to you in error, they can pull all copies of it, without warning, from your devices. That's what happened in 2009, when, in a moment of irony that was almost too good to believe, Amazon discovered that a company had sold digital copies of George Orwell's *1984* and *Animal Farm* through Amazon, despite not having the proper rights. Amazon responded by deleting the books from users' Kindles and refunding them, though not without raising an outcry—besides losing the book, customers lost any annotations they had made—causing the company to promise never to take such an action again.

The Amazon incident is a useful reminder of how provisional the concept of ownership is on many cloud-based platforms.* If you were to buy a defective or unauthorized item at a brick-and-mortar retailer, you'd still have it with you. It'd be up to you to decide what to do with the item, because you bought it in a legal transaction; you'd own it now. Sears wouldn't burst into your home and repossess a knockoff

---

* It also shows how platform owners, such as Amazon, heavily subsidize or give away products and services with the intention of mining our data for profit and getting us hooked on the platform.

tire iron that somehow slipped into their supply chain. They couldn't prohibit you from using the tire iron because you don't have the most up-to-date instructions for it. And if you wanted to resell the tire iron, that'd be your right, too.

But that's not what happens when we live our digital lives on other people's platforms. We become subject to the baroque language of their user agreements, governed by the company's hidden bureaucracy. For better and worse, we are linked to these platforms, just as a Kindle owner is linked to Amazon. To put it another way: our online identities, like our digital media, become cloud-based. We access them whenever we want to, but we don't quite own or possess them.

It may seem like we have some semblance of free expression, but in fact it's rather limited. Your Facebook page may be your space, your representation of your personality, but you don't have the ability to replace it with a blank page with only your name, or a string of animated GIFs of cats playing, or some weird puzzle game you've coded. You can't change your page's accent color—that blue, which was chosen because Mark Zuckerberg is red-green color-blind, is there to stay. You *can* fill out the forms provided: favorite books, movies, educational history, relationship status. Maybe spice it up with some quotes from a depressed nineteenth-century philosopher succumbing to syphilis. (I don't know—it's your profile.) Add some photos and game apps, and that's about the extent of your control over the appearance and functionality of your Facebook page.

This is the case across the social web, with social networks asserting varying degrees of control over our content, from you own it (Tumblr) to we own it and may use it in advertising (Facebook/Instagram/Google+). The arrangement is built on ease of use, and the apparent difficulty of opting out. If you wanted to have a personal Web site, why would you bother securing your own domain name, coding a site (or hiring someone to do it), updating it yourself, and all of that other work when Facebook, Tumblr, Twitter, About.me, or a host of others will do it for you and connect you with numerous like-minded people?

The social web's weak ties of association simply can't be replicated by having a stand-alone site (though some technologies, such as RSS, have tried to give users the ability to be networked and autonomous). Claim a plot on Facebook's land and you benefit from the exposure and sense of inclusiveness, of being in constant conversation with your neighbors. No more depending on word of mouth, search results, or exchanging links with others. The platform owner is here to watch over us and make sure everything remains in order. Tumblr even promises, as founder David Karp said, that no one will get hurt.

It's mostly a losing bargain. We live on the promise that we will somehow earn enough—status, visibility, followers—to transcend our conditions. We keep posting, writing, sharing, offering labor to appease a distant overlord. Maybe if we rise in status enough, we'll feel like we'll have more control over our digital environments. We won't feel compelled so often to post just to appear visible to the network. We'll chuck those social-media accounts for a personal blog on a domain we own, and we'll make our fans, followers, and friends come to *us*. It's much like the viral dream: a few people will become masters of the network, popular, verified users who start making money, swag, or some other tangible benefit from all the work they've put in. They'll get that blue check mark and join the networked elite. Meanwhile, most of us will continue working, staring up at the favored few with a mix of envy and admiration, wondering how our Klout scores or follower counts measure up. This mind-set extends to the sharing economy, where everyone is cast as a micro-entrepreneur; a handyman on TaskRabbit hopes to accrue enough clients to shed the platform and set up his own shop.

But the idea of being unconnected has become a bit intolerable, hasn't it? You can think of anything and publish it immediately, which is perhaps the most remarkable communications shift of the last decade or two. It seems strange to write something and *not* post it somewhere. (This attitude, amplified, is the Facebook belief that deleting an unposted status update is self-censorship. Turn it up a couple notches and you get the hysterical absolutism of Dave Eggers's novel *The Circle*, in which a tech company mantra declares, "Privacy is theft.") For a writer, in particular, this is a lurching shift, as suddenly I am in some sense in the same business as everyone on a social network, and yet anything not immediately published, anything worried over—an article or this book, perhaps—can seem immediately aged. It seems slow, out of step. After all, I can log onto Twitter and instantly have an audience. On Facebook, there are my friends, including many acquaintances, near strangers, and people I once met but no longer can place, an amorphous social status that lends itself to a strangely immersive style of voyeuristic, but-who-are-you-really profile browsing. This kind of habit causes one to give up some things, not least the qualities of introspection, deliberation, and length, but everyone is *there*. And that's a powerful draw. We'll still have books, sure, but probably a few won't be written because their would-be authors sate this hunger for raw connection with social media and its immediate feedback loops.

If you do any sort of creative work, you are often told that it's not enough to simply write or paint or produce music. You have to publicize it and connect with like-minded people. You have to cultivate an audience. You must become a brand. For many of us, the promise of Twitter is that we will achieve this status and also that we will somehow become successful enough to no longer need it anymore. That is an ever-retreating horizon, like saying you just want to be rich enough not to have to worry about money. If the goal of social media is to become successful or famous enough not to need it anymore, to be able to determine one's own path in the world, how will you know when you're done? Or is it to sustain this sense of connectivity, of being wanted and needed by a network of people, each notification a reminder of your importance? No matter that it seems to swallow more of your time lately; time and energy have been *invested*.

A key part of digital serfdom is that our data—photos, updates, posts, likes—and connections help tie us to the platform. Since our data only circulates within the platform (and to the platform's partners), but we don't really control its circulation, we're stuck there. And because we know that our data is permanently held on the platform, never to be deleted, we feel an added compulsion to stay and manage it. We add to our photo collections, showing that we continue to experience vivid, interesting lives; we work to accrue micro-affirmations and build our follower counts; we make ourselves more popular both in the eyes of our social graph and the platform itself (algorithms, along with bots, are the most reliable audience). Or maybe we just lurk, because that awareness of others feels like its own kind of visibility.

The constant tending to one's profile, privacy settings, and alerts also keeps us chained to the platform. But without performing more digital labor, more public labor, our digital identities lapse into irrelevance. We perform digital labor because it allows us to feel as if we're enhancing our digital identity, as if each post or profile photo adds some needed complexity simply by being new. Here the recency effect is quite powerful: we need the most up-to-date version of ourselves exposed to the world.

The other key method of keeping us on the platform lies in growing it, both in its reach and in its capabilities. Redesigns and policy refreshes always make news, allowing a platform to create media and user interest with changes either cosmetic or designed to help users create more data and be more visible to one another, search engines, and advertising partners. Potential competitors, particularly the messaging apps that tend to be more privacy-friendly and have become quite popular in recent years, don't appear to pose much of a long-term threat, not as long as Facebook and Google, the great monoliths of the social web, remain flush with cash. Insurgent companies can be bought up, shut down, its talent folded into the corporate parent, or, as with Instagram and WhatsApp, left to operate with a degree of autonomy.

Another key innovation in social networks in recent years has been the development of embeddable media and widgets. We now can embed news stories, links, photos, songs, and many other scraps of media into the posts we share. Often we can then consume them within the

Web site or mobile app, making it so that we don't have to leave to enjoy other media and that the platform owner can add to our data profiles. Everything we do is filtered through the systems of the platform mediating these experiences. In some ways, this reflects a return to the walled garden of late–1990s Internet service providers such as Prodigy and AOL. It's just more of an open-door policy; everyone is welcome in the garden, assuming they play by the platform owner's rules—and let them reap the profits.

The most important part of digital labor, of keeping us complicit in the process, is that it doesn't seem like labor. As Trebor Scholz says, "This digital labor is much akin to those less visible, unsung forms of traditional women's labor such as child care, housework, and surrogacy." As with online labor markets, the digital labor of social media is highly mediated, disguised. It's made to look like play or a normal part of Web browsing. For example, CAPTCHA tests—those forms that require you to read a blurry sequence of words/letters/numbers/shapes and enter them to prove you are a human being—often double as ways of improving optical character recognition (OCR) programs. These words were scanned from books, newspaper archives, or other media, but existing OCR software can't read them. Like a Mechanical Turk worker, you provide the final bit of cognitive labor, deciphering the word for the computer. That word then goes back to whichever company paid the CAPTCHA service to help digitize their material. Andrew Ross calls these and similar methods "the micro-division of labor into puzzles, stints, chores, and bits." One of the largest of these services, reCAPTCHA, is owned by Google.

This kind of disguised labor has percolated through many areas of digital life. An Android app called Twitch takes over your phone's lock screen and, rather than having you enter an unlock code or pattern, asks you to answer a quick question or rate photos. Similar apps might pay a user a few cents to structure some data or confirm a statement on a *Wikipedia* page. The notion is that, because the action takes about as long as to punch in an unlock code and because we often pull out our phones when we're bored or idle, then we also wouldn't mind doing a little bit of work. An appeal to a larger goal is usually made—answer a question about your personal health to aid some distant researcher, or help contribute to a weather report. By calling this practice crowd-sourcing and claiming it's for the greater good, the managers of this highly distributed workforce can collect lots of lucrative, useful data without expending many resources of their own.

Media companies have picked up on this opportunity. Duolingo, a free language tutorial service, uses its students as labor. The phrases they are asked to translate in the course of their practice come from clients such as *BuzzFeed* and CNN. In fact, *BuzzFeed*'s expansion into various non–English language markets was predicated on using Duolingo users—and paying Duolingo—to translate its famous listi-cles into foreign languages. CNN articles have been republished in several languages through this process, too. Each article is chopped up into many pieces, but each piece is shown to more than one student, a redundancy that's supposed to improve the quality of the transla-tion. The reasoning behind this is beguiling: these people are getting free language education, and they're already doing work by perform-ing translations as tests. Why not make it mean something? Why have that work go to waste? It's a compelling way of turning the issue around without providing an answer of one's own.

For one thing, it involves these students giving away their expertise—their in-demand native languages—for free. The other is that by partnering with Duolingo, media organizations are getting labor at a vastly below-market rate, and they're also undercutting de-mand for professional translators and journalists who do (or would do) this work for a living. Crowdsourced translations are translations that won't be done by a fairly compensated expert.

CNN and *BuzzFeed* are rich, for-profit media companies and could afford to pay for this kind of work. These are the tasks for which over-seas bureaus and foreign stringers exist. The news coverage surrounding the announcements of these partnerships was revealing. *BuzzFeed*, in particular, was praised for what the *Wall Street Journal* called its "brazen, nutty growth plan." The article's writer, Farhad Manjoo, cooed that "when you click on a *BuzzFeed* link, you're subsidizing language education for millions of people around the world." In some circuitous way, this is true, but the corporations involved are extracting more than their share of the profit. There's a strange moral calculus in play here, where a company's ability to use software and a distributed labor system to reap gains is seen as creative and radical. The media outlet's expansion is seen as a worthy end in and of itself, and the manner in which it's achieved matters only insofar as it shows the power of clever engineering.

Other companies justify their use of free digital labor by appealing to a sense of altruism, brand loyalty, or, frequently in the case of translations, a vague lofty internationalism. Theirs might be a more direct relationship, not mediated and highly distributed through a game or app, but they embody similar labor relations. The Smithsonian Institution, despite its ample funding, has crowdsourced translations from what it called "digital volunteers." TED—the expensive, extremely popular ideas conference—uses unpaid translators for subtitling its widely watched videos. These translators are credited and some are even invited to the conference, but it strikes a discordant note that a well-heeled organization, one that caters to elites who often claim to have panaceas for the developing world, would use unpaid labor. Facebook used volunteers to translate its menus and documentation into other languages, helping to accelerate its early international expansion. When Google launched its Chrome Web browser, it asked a number of artists to contribute free work that would be used to decorate the browser. The company promised exposure—the consistent refrain now offered to creative types, as if by constantly deferring compensation they can somehow, someday land a big payday, or perhaps just a job. When this might be, no one can quite say, but it's similar to the philosophy that social media's exercises in self-branding and self-promotion will eventually pay off for enterprising users. (Many white-collar workers, especially recent college graduates, know another form of this treatment: unpaid internships, which cater to those who can afford to work for free.)

For years, AOL relied on its Community Leader Program, comprising thousands of remote volunteers charged with moderating message boards and chat rooms, enforcing the terms of service, serving as chat room hosts, or writing content. The staffers were unpaid but received modest discounts on their AOL memberships. The program, at its most expansive, involved 30,000 community leaders who saved the company an estimated $7 million per month. By 1999, the volunteer corps was down to 15,000 people. That's when seven of these leaders, tired of their treatment by AOL and beginning to see themselves as unpaid workers rather than volunteers, went to the Department of Labor to ask if AOL owed them back wages. When the DoL didn't intervene, the community leaders filed a lawsuit. Others went on strike and were fired in retaliation. In an effort to reassert control, AOL tried to reorganize the program, reducing volunteers' responsibilities, but it soon fell apart. The lawsuit became a class-action case; 2,000 members of the program signed on. The case settled in 2009, years after the Community Leader Program was shut down, for a reported $15 million. A couple of years later, AOL bought the *Huffington Post*, another new-media company built partly on the backs of volunteers, none of whom saw any of the $315 million purchase price.

As these examples illustrate, the knowledge economy relies on extracting maximum information and data from users at minimum cost. It is true, as Tiziana Terranova says, that "the Internet has been always and simultaneously a gift economy *and* an advanced capitalist economy." Some things we're comfortable giving away for free, others we're not, and the decision on what is exploitative may differ between well-meaning individuals. But as Terranova goes on to explain, we shouldn't "mistake this coexistence for a benign, unproblematic equivalence." Or, as another scholar describes it, these companies are engaged in a "variable scale of labor relations," and some of them can be considered wrong, if not illegal.

We see this in the fantastic wages and lifestyles afforded to programmers, and in the record income inequality and urban stratification to which these companies have contributed, as huge sums accrue to a small class of elites while legacy businesses are disrupted out of existence. Because once you achieve scale as a platform owner, it's enormously lucrative. Craigslist, whose free classified ad model decimated the newspaper classifieds industry, is privately held, but its revenue per employee is estimated to be more than $3 million, tops in Silicon Valley. Some thirty-odd people work for the company. Thirteen worked for Instagram when it was bought, and about eighteen months after its $1 billion sale, members of the tech press were saying that Facebook got a bargain, even though the photo-sharing service hadn't made a cent before the deal. (When you've got the user base in place, advertising and revenue are considered a formality.) Facebook snapped up WhatsApp for about $19 billion—a huge windfall for its fifty-five employees, less so for the hundreds of millions of users who made the messaging app so popular. Compared to this, AOL's $7 million per month of free labor may pale. But Facebook and its ilk have, in some sense, expanded on AOL's volunteer program only by getting *all* of its customers to do free labor. Their whole business model is built around this kind of "free." Looking more broadly, much of digital media now comprises what the Marxist theorist Toni Negri called a "social factory." In a social factory, the work and value-creation that might be traditionally found in a factory shifts to society itself. Facebook is a literal manifestation of the social factory. We do the work, by clicking, writing, posting, giving over our content, data, and attention. This work is diffused throughout our society, through our day jobs and entertainment and most basic communications. We might not even realize it's work. The writer and game designer Ian Bogost describes this form of always-on but rarely acknowledged labor as "hyperemployment": "We do tiny bits of work for Google, for Tumblr, for Twitter, all day and every day."

It's enough to make one think that platform owners don't do much at all. In digital serfdom, the digital lords appear to be little more than caretakers fattening themselves on our data production. As *Slate*'s Will Oremus describes it, "The site's users are so productive that all the employees really have to do is keep the lights on and the servers running. We do the work, and they harvest the profits." They also designed the platform, sure, but once it's up and running and users are steadily working, their job is largely focused on maintenance and maximizing productivity in the form of more user data and better partnerships with advertisers, data brokers, and large media organizations. They also buy up the occasional would-be competitor. Users are paid through visibility and status—metrics, micro-affirmations, a shot at viral fame—and in the end, our culture, relationships, and experiences are mined for profit.

There is still an exchange going on here—free data for a free service. What makes it more unsettling, though, is how our work as data producers for someone else comes to be seen as the normal order of business for the Internet. It's tough to imagine a situation beyond this one. It's become practically unthinkable to pay for an online service, with the exception of Netflix or the occasional app. Surveillance, tracking, obtrusive advertising sold at rock-bottom rates, and virtually unlimited, free content (much of it low-grade and designed to pander to SEO and sharing best practices) are the complementary aspects of this arrangement. So too are limitations in how we can use these services, whether it's strict restrictions on APIs, identity construction, speech, or ownership of our own creative expression. The entire platform becomes dedicated to producing conditions optimal for advertisers, which can only be at odds with the best interests of the users themselves. The relationship is far from passive, nor is the platform owner a neutral intermediary. The platform and advertisers are complicit, essentially asking each other, "What can we do to make them click?" The result is a centrifugal process of tracking and profile building, all in search of some vaguely defined relevancy—ads can always be more "relevant," these companies say—that really equates to higher click-through rates. Absent a focus group, no one approaches you in public and asks whether a billboard was meaningful to you and, if not, to submit more personal information so that the next billboard you see *might* be more convincing. Why should we accept the same practice online, in secret, done to us thousands of times every day?[*]

---

[*] For all of the exploitation, manipulation, and erosion of privacy that comes with the current slate of social-media companies, they are built upon an even more abusive industry: the massive manufacturers in China, such as Foxconn, which produce the computers, smartphones, and networking equipment powering Silicon Valley. Despite a great deal of reporting on the subject and some superficial improvements, the conditions at these factories remain horrific, with hundreds of thousands of workers packed into crowded dormitories, forced to work on their feet all day, exposed to dangerous chemicals or repetitive stress injuries. With low pay and little freedom of movement, we rarely hear their stories, except when news of a spate of employee suicides reaches the outside world. What's more is that this labor—far more dehumanizing and manipulative than any aspect of digital serfdom or the sharing economy—is closely connected to the extraction of rare metals such as coltan, a so-called conflict mineral mined in slave-like conditions in the Congo. The Amazon jungle, specifically the area near the borders of Brazil, Colombia, and Venezuela, has also emerged as a key source for coltan mining. There smugglers dominate the trade, often forcing dispossessed minority groups, as well as children, to work in dangerous, illegal mines.

While this situation might seem distant from the swift trade in personal data practiced by Google and Facebook, it would be irresponsible not to make these connections. Similarly, when technology companies such as Apple practice planned obsolescence, churning out newer and slightly better versions of their devices once or twice a year, they—and their enablers in the gadget-hungry tech press—are encouraging the perpetuation of these abusive labor practices. They are also ensuring the proliferation of even more e-waste—piles of electronic gadgets and junk, laden with toxic chemicals and metals, that accumulate in dumps in West Africa, India, and China, where scavengers dig through them for scraps to sell. Stuck on the consumerist treadmill of perpetual upgrades, consumers are complicit in this process. Our guilt may not be tantamount to that of a Foxconn or Apple executive, and there is some difficulty in opting out of these technologies, given the powerful marketing forces behind them and their increasing importance in work and social life, but we do bear some responsibility, even as we make our small contributions far down the production chain. Silicon Valley depends on a pyramid of labor relations ranging from the highly moneyed and pampered (executives, programmers, managers) to users contributing free digital labor in return for services to miners and factory laborers in the developing world. Women also make less than their male counterparts throughout the industry. A single company may be taking part in all of these varied labor relations simultaneously. Amazon's Mechanical Turk provides a stirring exam-

THE FLÂNEUR FANTASY

Maybe it's not all so dire, and we have more sovereignty in the social web than pessimistic commentators like me tend to think. I've explored feudalism and factory work as metaphors for digital life. Others have taken a sunnier view, explaining that, at our best, we're actually all "cyber-flâneurs" in the digital street. Flânerie traces to nineteenth-century Paris, when bohemian life flourished in the form of grand arcades, cafés, artisans, shops, a lively literary scene, and an abundance of leisure time. Flâneurs were the ultimate aesthetes—highly cultured, social beings who browsed and conversed and treated urban life as something to be consumed and experienced through the senses. The original window shoppers, flâneurs wandered through the crowded arcades, seeing better versions of themselves reflected back at them in tailors' windows. There is a touch of conformity about the way flânerie is idealized now, but it was intended to be an individualist lifestyle, akin to dandyism and other vaguely rebellious socio-artistic movements. Although it was highly stylized, there was something casual and almost careless about flânerie; you were meant to take your time and not have a particular goal. Serendipity was the order of the day. A variety of factors led to the dissolution of flânerie as a way of life—the renovations of Paris's boulevards, the introduction of the automobile, World War I, and the advent of advertising, mass production, and consumerism as fundamental parts of urban life. Perhaps the final death knell was in 1940, when the German literary critic Walter Benjamin, then working on an immense but unfinished manuscript about Paris's bygone arcades, committed suicide while fleeing the Nazis. An era and one of its most devoted chroniclers had disappeared for good.

---

ple of mindless drudgery and low pay, but the company is also known for horrific warehouse conditions, including bringing in ambulances rather than air conditioning on scorching days. And it is, naturally, a principal conveyor of products from the vast factories of China's industrial belt.

Flânerie had a resurgence in the 1990s, when the Internet began to take on many of the characteristics of late-nineteenth-century Paris. At the time, the Internet's boulevards were open and chaotic, not yet corralled into the highly structured environment of Facebook, the robotically indexed archives of Google, or the closed systems of a thousand different apps. We often went online not to find an answer to a question or to purchase something, but to explore. It was a strange, uncharted space full of amateurs and craftsmen. We tinkered with GeoCities pages and ended up in chat rooms not necessarily because we wanted to have a particular conversation but because the very experience was novel, even intoxicating. We passed through the Internet mostly anonymous, unmonitored, and at our own pace. E-mail hadn't yet become burdensome work; many of us actually enjoyed it.

This isn't the state of affairs now, though for some the cyber-flâneur dream endures. The Internet's arcades have been reimagined in Facebook, Twitter, Tumblr, and Pinterest (that ultimate site of browsing and consumption)—never mind that these cultural offerings are often served up to us by algorithms or marketers. I have my doubts that flânerie was ever the apotheosis of individuality that its proponents claim; it was, after all, based much on fashion, social standing, cultivating a certain appearance and intellectual affect, and impressing other members of the bourgeoisie. Flânerie wasn't open to everyone. On that score, at least, the social web has democratized cyber-flânerie by making it seemingly available to all. We can all go online and stare at the conveyor belt of digital culture, liking, rating, buying, sharing memes, every click an advertisement for our taste. Serendipity may even return—it's now a popular idea in tech circles, something that Silicon Valley innovators promise they can engineer into our digital lives, years after they did away with it by creating such structured, filtered online worlds.

The cyber-flâneur can't browse for the sake of browsing because he can never be alone. It's not only Facebook's principle of frictionless sharing, its surveillance-based business model, and the bizarre, creepy idea promulgated by Zuckerberg and his compatriots that *everything* we do is better with friends. We have also done this to ourselves, by giving in to the Facebook Eye, allowing our online experiences to be meaningful and important in direct relation to how they're shared. The cyber-flâneur can't exist in a society that's adopted sharing as a cultural ideal, as a form of personhood. He can't depend on other people to validate his experiences or to rate them so that he may increase his sense of self-esteem. Cyber-flânerie depends on a kind of uncaring aloneness that's incompatible with the idea that sharing life's novelties completes them or that aesthetics are only as stable as the number of affirmations they attract. It also relies on a kind of stasis, or at least a deliberative quality, that's lost in the real-time social web, where the newest thing is axiomatically the most relevant and exciting.

We better resemble tourists than flâneurs. We come upon something—an article, a viral video, a birthday notification—and we dispatch it with the confidence of an experienced traveler. It's the hardened tourist who says that he's "done France" or "done Thailand" and doesn't feel the need to go back. He's been there, ingested the place, wrung some photos and Day-Glo memories out of it. The trip has been—fitting for a society besotted with the language of psychology and business alike—*processed*. Digital life is little different, especially with the premium placed on being able to travel and explore distant places from one's laptop. We go places—we read, we post that article, we share that image—and we almost never return to them. For a minute, we might think that that *Onion* video is one of the funniest things we've ever seen, but soon it's gone, and we move along to the next item.

The same features that have chained us to platforms—surveillance, tracking, advertising, the inability to take our data (i.e., our digital selves) with us—have killed off the cyber-flâneur. These elements make us feel like leaving the platform would be a great loss. And yet we aim to experience the online world with a bohemian sort of joy— Look at this cat photo! Check out this inspiring TED talk! Did you read this incredible story?—but end up turning our consumption of

pop culture into work for others. How do we reconcile this tension between consuming the world as we want to and knowing that every act
of enjoyment translates to a micro-payment in the pocket of Google,
Twitter, Facebook, or some faceless advertising network?

Some commentators have tried to tackle this problem by considering social media in terms of "prosumption," in which users are simultaneously both producers and consumers. Over the years, being
a consumer has required an increasing amount of self-service—from
pumping our own gas to dialing phone numbers (which initially left
some people baffled, since they were used to dealing with operators,
who were eventually put out of work). In a book of the same title,
George Ritzer defined what he called the "McDonaldization of society," in which consumers are asked or expected to do some work
themselves. He points to anything from busing one's tray at a fast-food
restaurant to scanning your own groceries at the checkout counter or
using electronic kiosks to check in at an airport—essentially, anything
that might once have been done by a paid employee but now falls under the self-service label. These practices help corporations to increase
efficiency and keep down costs by outsourcing labor to the customers
themselves. Sometimes the consumer's labor is acknowledged, as with
IKEA, which allows users to select their own (often quite heavy and
cumbersome) items from the warehouse, cart them to the register, and
then assemble them at home. Besides decreasing prices, this kind of
work allows the customer to feel like he is engaging in some quasi-DIY
craftsmanship, even though he's assembling a mass-produced piece of
furniture from pictographic instructions designed to appeal to a global
audience.

The passing down of labor to consumers follows a trend visible in
anything from the masculine urban independence of Home Depot and
Lowe's (motto: "Let's Build Something Together") to the homespun,
crafting spirit of Etsy, which promises to monetize previously unrec-
ompensed hobbies. This societal shift has been long in the making. In
1981, the Roman Catholic priest and philosopher Ivan Illich coined

the term "shadow work" to define work that's both passed down to consumers or traditional work that's rarely acknowledged as such. By this definition, Illich would include self-service gas stations but also domestic housework. This latter idea is particularly important, as cleaning, child care, cooking, and many other domestic tasks that are labor-intensive and time-consuming have long been diminished as not "real" work. That, in turn, helps to hollow support for feminist movements, the rights of homemakers, and a strong social safety net which, for example, would provide child-care services that would allow women to put aside their domestic labor and enter the paid workforce.

The sharing economy is loaded with shadow work, which you might discover when you learn you're required to clean out your Zipcar but not the (similarly priced) rental from Hertz or Enterprise. Shadow work also includes many tasks—driving, paying for gas, maintaining equipment, buying office supplies to make up for a budget shortfall—that hit poor people hardest. Shadow work can easily turn a living wage into something below subsistence level. It's the very hidden nature of shadow work that makes it even more problematic. Because it is unacknowledged, there's often little incentive to discuss it and to weigh its various costs. Or, more perniciously, shadow work is expected to be just the cost of entry, a burden that someone is expected to bear. Say you receive a decent job paying $35,000—enough to support yourself and pay rent and your student-loan bills, though not much else. But the job requires you to drive forty miles each way and doesn't provide a transportation subsidy. You also have to provide your own computer; pay a hefty health insurance premium; and instead of a human-resources manager to keep track of sick and vacation days, you have to enter that information into a database yourself, part of a company-wide system that requires employees to clock in and out each day. All of this is shadow labor, and it takes a toll on workers, reducing their autonomy and ability to make a good living.

To be sure, there are forms of shadow work that are beneficial. Sometimes we want to do things ourselves. Shadow work can also give us new skills and cause us to feel like we have a stake in something. Despite the unpaid labor, this trade-off can be worth it. The rise of literacy meant that people began to write their own letters, and scribes and paid letter-writers began to disappear. But eventually, a new class of scribes—secretaries, typists, office administrators, clerks, copyists, notary publics—appeared, providing a meaningful stream of paid work and expertise. Yet the subsequent advent of digital communications meant that these support staff soon were made to seem redundant. Administrative work was outsourced to third-party firms, including many overseas. Typing, and then e-mail, came to be seen as an essential skill—if not always an acknowledged form of labor—to the extent that office workers now, on average, spend 28 percent of their time on e-mail. As digital communication increasingly fragments among various platforms—phone, e-mail, social media, texts—this kind of burden is likely to increase, which is why a number of corporations have begun to look at ways of limiting communications or banning e-mail entirely. Still, these practices are usually examined with an eye toward maximizing efficiency and company profits, not ensuring that workers have sufficient protections and are compensated for all of their work.

It's in this kind of context that we should look at the labor we put into social media. Twitter is work, Facebook is work. Words are being written, content produced and shared, ads sold against it. A welter of data, some of it structured by us, is produced, and this has value. Yes, this work is often voluntary. You put in what you want, and if you don't like that Facebook is profiting off of your relationships and communication with friends and your very identity, then you can quit. But the flip side of network effects—of a network rising in value and utility as more people join it—is that there can be a real social cost to opting out. And professional cost, too. Engagement in social media and other digital products has become required for many white-collar jobs, representing another way in which the work/life divide is broken down. The digital work of producing for Tumblr or Pinterest, then, becomes part of the work for producing for one's day job. And as our online profiles are connected to our real names, complete with biographies listing our affiliations, we become the always-on, public representatives of our employers, our schools, and our communities. When your school begins monitoring your social-media feeds or your employer asks you to insert a disclaimer in your Twitter biography, your social-media output is now linked to your status with that institution. You become potentially liable toward them for everything you do online, but with few of the benefits of the association. If you screw up and make an offensive joke online, you might be fired. If you comport yourself as a good digital citizen, it will go unnoticed, you will receive no annual bonus. At the same time, your ability to conduct yourself online as you choose will be subtly circumscribed.

Prosumption, then, is just a happy gloss on digital serfdom. It pays some lip service to the shadow work that underlies the social web, but it ultimately treats it as more empowering than it truly is.

- - - - - - - - - - - - - - - - - -

FROM DIGITAL SERFDOM TO NETWORKED INDIVIDUALISM

Digital serfdom is the dark side of a mostly laudable dream: that we can use networked technologies to allow people from varying backgrounds to connect easily, and to publish as often as they like. But intoxicated with the ease and reach of social media, we didn't bother to ask some important questions. What are the costs of these free services? What does it mean to grant Internet companies so much power over our expression, our identity, our manifold digital selves? Do we want the standards for digital life dictated from Palo Alto and Mountain View? Does the convenience of Facebook make up for the lack of ownership we have over our data? Is an information economy built on widespread surveillance just? Are we astute enough to separate the benefits of networked cultures from the crude business models of the current social-media giants?

At the moment, there's no combination of technologies and policies to allow people to claim ownership of their scrap of the Web's real estate and still be closely networked to others. Nor does there seem to be any desire for such an arrangement. The possibility that this might be needed hasn't entered the public consciousness.

What might the reconsidered social web look like, though? It could be a nonprofit social network explicitly devoted to maximizing user rights, ownership of data, and public conversation. It could be a government-owned Twitter, as the writer Navneet Alang has argued, although conservative reticence to government involvement in commerce and liberal/libertarian opposition to government surveillance would likely handicap such a project. Moreover, it would require seeing social media as a utility, which is to say as something practically essential and in the public interest—veering close to the kind of cyber-utopian posture that has little served users. The media theorist Danah Boyd has compared Facebook to a utility, while the writer Benjamin Kunkel has said that social media's public value, along with its tendency to concentrate the profits from the labor of millions into the hands of a small coterie of tech moguls and investors, means that social media should be socialized. All of these are valid arguments and areas of inquiry, though I think they are in some sense premature. While many of the elements of social media—sharing, swift communication and publication, an ease of transmission, the shifting of once private communication to a quasi-public space—were present in earlier communication platforms, social media is still new. Facebook and Twitter might be gone in ten years, to be replaced by whatever other platform emerges. Or, as seems likely in the case of Google, they might become more deeply insinuated in our lives, especially as Google's social layer and its forays into wearable computing and physical-world tracking and advertising seem destined to turn all of reality into its own proprietary, augmented reality.

Given Silicon Valley's cult of disruption, it's likely that future innovations in digital communication and broadcasting will be seen as just as revolutionary as the advent of social media. The tech industry is expert in nothing if not its own self-mythologizing. This image of perpetual upheaval, of boom-and-bust as both cyclical and salutary, doesn't help the industry in checking its practices or thinking long-term. Take the lust for personal data: with every competitor seemingly grabbing as much of it as they can, sometimes without even knowing how it might be useful in the future, social-media firms feel compelled to do the same. There's little time, or market benefit, to stop and question these practices, to take a sober look at how to develop sustainable, respectful, long-term relationships with consumers and their personal information. Instead, Internet companies by and large seem content to use network effects to their advantage, growing as quickly as possible and capturing so much data (including the material we willingly upload and the social graphs we develop) that it begins to seem too costly in time and energy to move elsewhere. This attitude only serves to trap users, making them more servile toward platform owners.

Government regulation would help to bring some social-media companies' practices back in line and to return a modicum of power to users. While often a step behind the corporations they police, the EU and European governments have a strong record of protecting consumers' rights. There are some gaps in this record. Ireland's own status as a corporate tax shelter has made it a kind of low-rent economic beachhead for American technology firms. Yet the Continent has shown an admirable appetite for policing Internet firms' activities. One proposed EU data protection law calls for fines of up to 5 percent of a company's global revenue, or a maximum of 100 million euros, for violating the law's provisions, which require companies to boost protections for data turned over to government authorities, give users more control over how their data is used, and allow people to request data to be permanently deleted. Company violations of privacy policies would also earn steep fines.

Consumers could start opting out, supporting social networking sites, such as App.net and the ill-fated Diaspora, which privilege user rights, even though the laws of network effects mean that often these networks aren't as expansive as we'd like. The friends, colleagues, and public figures we'd like to follow still aren't there. Users could also make a very simple, but radical, demand: charge us for these services. Many of the structural problems associated with social media come from the fact that we don't pay in cash, so we are required to pay in other ways. As we've seen, there is potentially no limit to the collection and trade in user data; and as that data circulates, it can be used against us in unforeseen ways. By combining regulation of the data trade with a simple agreement to allow users to pay for advertising-free social networking, consumers could have much more power over their networked presence, along with a sense of security and well-being that might make them even more enthusiastic users of these services. The social-networking giants would fear losing some customers, but they would gain in reputation and prestige—in some ways akin to HBO, for which cable customers pay a premium in order to see high-quality programming without advertising. Following a similar path, social-media companies would also be able to develop reliable revenue channels that aren't based on tricking users into viewing or clicking on ads or giving up more personal information. There's honor in this, in having a fair, above-board exchange with consumers, free of the air of hucksterism. It all seems rather antique—imagine, paying for something you use on the Internet!—but it would help to restore balance to an information economy that has done more than any non-defense-related industry to make ubiquitous surveillance the law of the land. It could also help prevent future tech bubbles from emerging and decrease the income inequality so endemic to Silicon Valley, where sky-high valuations and massive venture capital investment seem wholly divorced from the ability of companies to make a profit.* Paying for digital services would also reduce the fetishistic attitude toward scale. No longer would com-

---

* Municipalities would also need to stop giving tax breaks to tech companies that threaten to move elsewhere. And corporate tax rates and taxes for the wealthy should be increased to subsidize the social safety net and pay for shared services, such as public transportation.

panies have to focus on expansion at any cost, although that would still, of course, be a prime goal, as it is for most capitalistic concerns. But companies would be freer to expand at a more modest pace, allowing them to be responsive to their customers. We might see more small and midsize Internet companies emerge, beyond the app developers that fork over a percentage to platform owners and mostly hope to get big enough to be bought out by an industry giant. Larger media companies such as The New York Times have already seen the benefits of charging digital customers to read articles beyond their paywalls. The result hasn't been a return to the profits of a couple of decades ago, but it has diversified revenue streams, made the company less reliant on advertising, and helped to inculcate a needed change in the culture: that quality digital goods and services, no matter how intangible or easily obtained, are worth paying for.

Paying for something gives us the right to demand things in return. "You're getting a service for free" is too useful a rejoinder for the industry and its defenders. Under this broad justification, almost any sort of exploitation can be whitewashed.

The Myth of Privacy

At the same time that these systems will bring a greatly increased flow of information and services into the home, they will also carry a stream of information out of the home about the preferences and behavior of its occupants.

—"Teletext and Videotex in the United States," a 1982 report commissioned by the National Science Foundation

It was the sort of junk mail sent out millions of times per year, this one finding its way to a house in Antioch, Illinois, a Chicago suburb. The average American receives forty-one pounds of junk mail annually, some of it lobbed in a blind fusillade from big retailers hoping to reel in anyone and everyone, others carefully targeted through mailing lists, loyalty cards, surveys, demographic data, and other ingredients of the data mining mill. The letter from OfficeMax belonged to the second category—its recipient was chosen for specific reasons—but something had gone wrong. The envelope was addressed to Mike Seay, but below his name the words "Daughter Killed in a Car Crash" had been printed. OfficeMax blamed an unnamed third-party data broker for mistakenly printing the information on the envelope. A distraught Seay, speaking to a reporter, raised some important questions: "Why would they have that type of information? Why would they need that?"

There is no obvious reason why someone would need to know that information, but it turns out that any piece of personal information is potentially valuable. Relevance is only an algorithm—or a sales pitch—away. Data brokers are supposed to be the unseen cogs in the surveillance economy, collecting vast amounts of information on hundreds of millions of people and analyzing it for patterns and likely outcomes. They are devoted to extracting value from the raw information of our lives for their own gain; they sell this information to stores, insurers, banks, tech companies, HR departments, and basically anyone who comes calling. Government agencies are part of the trade as well, with the DMV selling to data brokers and the TSA passing on information to debt collectors.

The letter that Mike Seay received was bizarre, grotesque. It seems indefensible, if not also counterproductive, to use such information to inform a sales pitch. And yet, the kind of data collection and processing that eventually resulted in the letter being brought to Seay's door are standard in the surveillance-driven advertising economy, the default business model of today's technology industry. Beyond being an isolated act, Seay's experience shows how the meaning of privacy has been radically reconfigured. Even as we rightly worry over how digital media is altering social conceptions of privacy, a quieter, more dramatic, and more insidious revolution is taking place, one that has roots in the sort of information collection and consumer profiling that corporations have been performing for decades. Retailers, credit card companies, social networks, data brokers, and advertisers have been turning consumer privacy into the ultimate commodity. The consequence of this transformation is that we are more exposed, less in control of our public identities, than ever.

HOW WE'VE GOTTEN PRIVACY WRONG

Technological elites like to argue that privacy is dead or somehow superfluous. Kevin Kelly, one of the founders of *Wired* magazine, cleverly summarized the situation: "Privacy is mostly an illusion, but you'll have as much of it as you want to pay for." His remark sounds like sophistry, but he's largely, if unfortunately, correct. Our popular conception of privacy is rooted in some fundamental misunderstandings. There's an inherent deception in the privacy claims made by the companies collecting and managing our data. Facebook may tout its granular privacy controls, but those controls, besides being difficult to navigate, apply to how other *users* see your information—not what Facebook sees. Facebook's all-seeing eye misses nothing, including the messages and photographs you've chosen to delete; you or your friends may no longer see this deleted content, but Facebook still does. (Nor do we always know exactly what information Facebook is collecting.) The effect is akin to locking the barn door after the horse is gone, while still believing the horse to be inside.

Privacy is, above all, the currency we draw on to pay for a range of free Internet services, most notably social networks. We offer Google and Facebook information about ourselves, while simultaneously being assured that our data is being used responsibly and that we have a wide range of privacy controls. It's more accurate, then, to say that privacy is submitting to market pressure, becoming increasingly commoditized. Your privacy has been taken, chopped up into packets of data, and circulated through commercial transactions beyond your view.

A now-famous paper produced by Bain & Company and published in 2011 by the World Economic Forum argued that personal data should be considered a "new asset class." "Personal data will be the new 'oil,'" the report's authors claimed. "It will emerge as a new asset class touching all aspects of society." This kind of hyperbolic talk reflects the wild promises of high-powered consultants determined to earn their (surely enormous) fees, and it brushes away some important details. For one thing, oil, unlike data, isn't potentially infinite. But the report does reflect a very real appetite for personal data that can be collected in unprecedented quantities, aggregated, and mined for patterns—and profit, too. That this profit opportunity often comes into direct conflict with traditional notions of privacy, ownership of one's personal information, and a sense of the self as something more than a collection of data points does not mean that the anti-privacy crowd is guaranteed to win. It does, however, mean that a great deal of money and corporate and political power is now being brought to bear on figuring out how to monetize personal information as never before. Bain and the WEF are not alone in this kind of thinking. Carlos Dominguez, a senior vice president at Cisco, has written about the rise of a new "trust economy," in which privacy is monetized but companies that act more virtuously are rewarded.

It might be, as Kevin Kelly theorized, that privacy will be something for which you have to pay. But it won't be called that—it'll be described as more personalized service or a way to get special offers. AT&T has offered some customers of its U-verse with GigaPower Internet service two choices. A "standard" plan is $99 per month; a "premier" plan is $29 cheaper—which should be a red flag. What comes with the "premier" plan? It allows AT&T to "use your individual Web browsing information, like the search terms you enter and the Web pages you visit, to tailor ads and offers to your interests." You let AT&T surveil your Web browsing, and the price of your Internet access—already inflated owing to a lack of competition—gets slashed. It's the obverse of the plan I proposed in the last chapter, in which we might be allowed to pay to escape data collection and advertising. Instead, AT&T essentially offers to pay you to submit to *more* data collection.

We get offered this kind of bad bargain more than we might think. A car insurance company that offers teenagers discounted insurance in exchange for good grades is asking for that student to give up some of his privacy. Still, that might be an exchange that the student, or his parents, is willing to make. Health insurance companies might offer customers a discount for submitting biometric or genomic data. The historical tendency is for what is at first seen as optional to later become expected, and then required.

For those who, in past generations, may have worried about the privacy implications of a Social Security card—which was meant to give one access to government pension benefits and now is used to get anything from health insurance to a cell phone—or who thought that the advent of computerized financial records might give too much power to corporations, this is a nightmare indeed. But privacy has always been personally defined while also tending to undergo broader shifts depending on societal and cultural mores, as well as the introduction of new technologies. It's fairer to say that, in recent years, privacy has been reconfigured by the combination of ubiquitous, mobile computing and a surveillance society that extends from homeland security to advertising to the mutual surveillance (and over-sharing) of social media. Privacy's entry into the marketplace as a commodity—or as an impediment to the harnessing of personal data, that valuable new commodity—reflects a pattern native to neoliberalism. In this system, economic freedom is equated with individual freedom, leaving little room for ethics and other nonmarket values. The problem of corporate power is ignored, while government power—so long as it stands in the way of businesses doing whatever they want—is vilified. It's either evil or a bureaucratic impediment to innovation. In short, for many corporations, your privacy is an economic challenge, not a legal or moral one. It's what stands in the way of social networks, market research firms, and other companies greasing their profits. Whether by rhetoric, legal challenge, deceit, or simply clever programming, they'd like to erase privacy as we know it. Encouraging transparency, disclosure, and social networking is just a means to an end. It allows them to, in the name of progress, encourage you to become complicit in forking over your data. It also offers a useful excuse for what they've been doing all along, in secret.

If privacy is to survive in the social-media age, in which human relationships and expressions are now being pushed through the data-mining and value-extracting machines of capitalism as never before, then its champions will have to argue forcefully for its importance. To do that, we will have to reaffirm that privacy is a long-standing legal right, that it's still important in an age of constant connection, and, more important, that it should be individually defined. No one should be able to impose his definition of privacy on you.

We also will have to reclaim privacy from the crowd that says it no longer exists—that it's just a myth—and from nostalgists who tend to view it mythically, as some prelapsarian state that can be recovered. We can't return to a pre-digital world of living off the grid, which never existed in the first place, but we can better understand the relationship between this flood of personal data and its uneasy relationship with privacy.

- - - - - - - - - - - - - - - - - -
WHAT EXACTLY IS PRIVACY?

Privacy remains a difficult subject in part because there's no standard conception of what it is. In some respects, it's personally defined—we all have varying ideas of what we would like others to know about us, and how they should be allowed to learn that information. In *Privacy in Context*, Helen Nissenbaum writes of the "conceptual murkiness" surrounding privacy and says that it can be "a claim, a right, an interest, a value, a preference, or merely a state of existence."

Privacy isn't about hiding information; it's about having the right to choose when and how you reveal information. As Danah Boyd has said, "Privacy isn't the opposite of being in public." We're all, to some degree, in public now. Rather, "privacy is the ability to control a social situation." To that end, people require a sense of agency, including both social and technical skills (as well as the legal rights and computer code underpinning them). This doesn't mean that privacy is incompatible with information sharing. As Nissenbaum explains, "What people care about most is not simply *restricting* the flow of information but ensuring that it flows *appropriately*." Data tends to flow between networks and crop up in unexpected or unwanted places; part of privacy is being able to control that flow. So privacy, then, is a mix of living in public and private and having personal control over the two. It's also about feeling that this control will be respected by other actors, from your best friend to the bank or government handling your sensitive information.

In recent years, a sort of consensus has built among tech industry leaders that privacy is changing. The argument usually goes—and is presented as the enlightened conclusion of the progressive tech set—that with the advent of social media, we are all sharing more, we are exposing more of our lives, social norms are evolving, and privacy is less important. Somehow, we've become more tolerant of one another, forgiving personal failings and foibles that technology, in part, plays a role in revealing. Disclosure has, per Zuckerberg and Sheryl Sandberg, become synonymous with authenticity. In the spirit of Kevin Kelly, privacy may still exist as an *idea*, but it is an increasingly outmoded and impractical notion. Privacy, they say, retards innovation and the formation of relationships based on the new ethic of radical transparency.

The problem with this line of argument is its narrow-mindedness. It is both ahistorical and reflective of a milieu whose social standards, if they are even uniform, may not represent other areas of the country, much less the world. Nor do they account much for individual preference or the kind of agency for which Danah Boyd argues. Surely we would like to live in a world where we are not judged too harshly for leaked private messages or an indiscreet photo that accidentally goes viral, and perhaps we are getting there, but this softening of stigma and judgment hasn't become the natural course of affairs. And privacy is not just about protecting individuals from embarrassment but also about keeping essential information out of public view: addresses, phone numbers, bank and health records, Social Security numbers. It is, in many cases, about protecting information which, if revealed, could produce real harm and distress. This is all the more important for disadvantaged populations, such as poor and oppressed minorities, who may not have the means to secure their own privacy and whose needs may be overlooked by corporations and governments. "Privacy has a social value," Daniel J. Solove says. "Even when it protects the individual, it does so for the sake of society." This is why, in regard to surveillance and privacy, the "what do I have to hide?" excuse is so fallacious. You may feel you have nothing to hide—though to test such a prospect, I'd ask you to let me sift through all of your e-mails and personal records—but privacy is a social bond. We have a civic duty to protect it for others.

One of many such stories can be found in a October 2012 *Wall Street Journal* story about two closeted University of Texas students who were outed to their families when they were added to a Facebook group for the Queer Chorus, a campus choir they had joined. The students, along with the group's moderator, didn't realize that the group's privacy settings were such that the students' Facebook friends could see that they had joined the group. The unwanted outing led to upheaval in the personal lives of the outed students, who had come from highly conservative, religious families. One student slipped into depression and was forced to cut off contact with her father, who had become verbally abusive upon learning that his daughter was gay. Besides being a troubling example of the unintended consequences of Facebook's confusing and misleading privacy features, the story illustrates the primacy of having personal control when it comes to disclosing sensitive information. It also shows that, even if one wants to live in a world of openness and transparency, individuals must be granted the autonomy (and attendant protections) to come to this position on their own.

By arguing that we live in an age of evolving privacy standards—or in which they no longer exist—Facebook can provide ideological cover for its own privacy violations. The claim that we shouldn't expect privacy online can in turn justify future privacy violations. It need not be true or intellectually coherent as an argument; it just needs to be repeated enough by powerful actors who don't have to answer to anyone. Similarly, the growing use by police of license plate scanners and cell phone tracking systems, along with widespread government surveillance of communications networks (often done in league with telecom and tech firms), has given this argument additional ballast. Given that the offline and online worlds are increasingly intermingled, an affront to privacy in one sphere may affect the possibility of retaining privacy in another. Why, for example, should Google worry about the surveillance and privacy implications of Google Glass when every day we live and work in cities covered in CCTV cameras? When then New York Police Commissioner Ray Kelly said that privacy was now "off the table" and that he was "a major proponent of [CCTV] cameras," it became harder to blame tech companies for taking a dismissive attitude toward privacy—or for cooperating with law enforcement and intelligence agencies when they came calling.

By employing the language of progress or saying that consumers simply were not ready for the privacy changes that they were implementing, companies can deflect privacy controversies. As a result, we see companies adopting an incrementalist approach, whereby privacy is gradually stripped away from users; Facebook's privacy settings over the years, for instance, have only gotten more byzantine, more frequently requiring users to opt in and offering weaker controls against the spread of information to friends and the public. The steady trend is toward surfacing and exposing more user data. Occasionally, Facebook has had to backpedal, but the message is always that while consumers are not yet ready for these changes, they soon will be. In November 2007, the company launched the Beacon advertising platform, which automatically posted on Facebook information about users' activities on 44 third-party sites. An immediate backlash caused the company to pull the product. But some elements of Beacon, such as the syncing of Spotify with Facebook, have since returned, so that the songs a user listens to on Spotify are now posted in Facebook's activity feed. And Facebook didn't lose much, not when its core business comprises a data-collection operation far bigger than anything that Beacon might have offered.

Mark Zuckerberg once described a theory that has since ossified into something called Zuckerberg's Law. According to his formulation, we share two times more information each year. The law is less one of social behavior than of corporate planning. What better way to make Zuckerberg's Law a reality than by creating an environment in which nearly every action users take on the network is shared somewhere, whether with one's friends or with Facebook itself? This is the true nature of how privacy has been hollowed out in recent years, as everyday life has become synonymous with data production and mass corporate and governmental surveillance. But to shore up our defenses, we must first understand where the ideas underpinning privacy originated, and how they can be employed to protect individuals' rights today.

PRIVACY'S HISTORICAL ROOTS

Ever since Samuel Warren and Louis Brandeis published "The Right to Privacy," their foundational 1890 *Harvard Law Review* article, privacy and technology have been closely linked. "Recent inventions and business methods," they said, meant that measures should be taken to preserve "the right to be alone." Warren and Brandeis worried that "instantaneous photographs and newspaper enterprise have invaded the sacred precincts of private and domestic life; and numerous mechanical devices threaten to make good the prediction that 'what is whispered in the closet shall be proclaimed from the house-tops.'"

Give the rise of an aggressive, muckraking press; photography; the rapid communication made possible first by the telegraph and, later, the telephone and other new technologies; the two legal scholars had some legitimate concerns. When Warren and Brandeis warned that "to occupy the indolent, column upon column is filled with idle gossip," they probably couldn't foresee how gossip would become an industry unto itself, practically a currency of exchange in the celebrity and entertainment economies, in which stars, and those aspiring to be stars, deliberately trade in their privacy for fame. Brandeis, who wrote much of the article, was instead thinking of his own bourgeois social circle, which had suddenly found itself the object of prurient interest as journalists began reporting on society weddings and social functions. He probably wasn't much concerned about the average man or woman on the street. Still, the authors' initial worry about "intrusion upon the domestic circle" has some relevance for today, when the home is no longer an inviolable space but rather prone to being recorded, Instagrammed, caught on Google Maps's Street View, or mentioned in a Facebook status. The network has entered the home, producing its own set of privacy complications and cultural mores that must be negotiated.

There are a few other key features of "The Right to Privacy" that make it useful for how we think about privacy today. Warren and Brandeis argued that the law as it already existed offered privacy protections; it's just that they had to be articulated and affirmed. This is a useful point when debating those who say that privacy is somehow unfounded in the law. At the same time, it's clear that the two scholars understood that privacy may have to adapt in the face of technological change. For example, they say that, in addition to addressing privacy concerns over the press and photography, the law should protect citizens against "the possessor of any other modern device for rewording or reproducing scenes or sounds." Writing a year before Thomas Edison patented his Kinetographic Camera, Warren and Brandeis knew that privacy depended both on the behavior of individuals and on yet-to-be-created technologies used to capture reality and "reword" it—re-creating it in pictures, text, and sound, with all the attendant possibilities for distortion and misinformation. Previously, assaults on privacy had resulted from "violating a contract or a special confidence;" now, Warren and Brandeis worried, "modern devices afford abundant opportunities for the perpetration of such wrongs without any participation by the injured party." In some sense, they anticipated problems that would later arise with stalking, paparazzi, and surveillance, especially with their concern about photographs taken "surreptitiously."

But they probably couldn't imagine how common and easy to use these tools would become, and how quickly one's likeness could be spread through the world, whether you wanted it to be or not.

- - - - - - - - - - - - - - - - - -

THE OLD BARRIERS COME DOWN

Although they didn't have a term for it at the time, Brandeis and Warren were contending with what some theorists call "context collapse." Technology and a growing cultural appetite for news and prurient gossip were eroding barriers between certain social settings. Traditionally private events and personal details were now being exposed to a wider public—in some cases, for the good. Today, the digitization of all manner of information, relationships, and cultural production and consumption means that it is increasingly difficult to keep these phenomena separate. Social media has perhaps done more than any recent technology to further this condition, as we now use the same platforms to interact with friends, lovers, family members, bosses, strangers, and people for whom we might have no fixed label at all. These people frequently see the same updates and profile information about you, even though your relationships with them might differ wildly. What does it mean that you now share the same photos and updates with your spouse and your boss, with your hair stylist and a tour guide you hired for a day in France?

With such social barriers eroding, privacy, then, is not confined to one app or social network or even to the online world. Information bleeds between networks, particularly as universal log-ins allow our Facebook or Twitter identity to *become* our identity, portable between sites and services. Privacy becomes networked, as each context may have local standards, controls, and practices, but the information itself—the stuff which privacy dictates we have control over—slops over the sides of the vessel, spreading and multiplying. This can happen in many ways. People share things in other environments: a private message is read aloud; a tweet is posted on Facebook; a screenshot or photograph is taken of a supposedly ephemeral Snapchat. Requests to keep information private go ignored. A friend tags you in a photo without thinking about whether you want to be identified in it. Some people don't know better—they don't think how their sharing of some information may violate someone's privacy, particularly as sharing becomes its own kind of sociocultural value.

Danah Boyd has argued that this new model should be thought of as "networked privacy." Under these conditions, privacy and security wax and wane. Privacy also becomes a burden—a delicate, shifting state that must be perpetually monitored, maintained. It's work, as anyone who's pored over friends' photos the morning after a party, looking for those unflattering shots that might lead to an awkward request to take them down, has found out. Facebook, prompted by Google+'s Circles feature, has made some effort to allow users to divide their friends into different lists, the better to target who can see one's updates. Still, this is only a modicum of control—these people are still lumped together as your Facebook "friends," after all; there's no distinction on that level. And social networks tend to treat privacy as an all-or-nothing affair. Users can accept a site's privacy controls, or they can refuse—by leaving. Rarer is the ability to address privacy concerns as they arise or to adjust one's privacy settings on the fly.

Social networks have also paid little attention to the fact that "privacy has always been interdependent," as Lior Jacob Strahilevitz called it in his essay on "collective privacy." This oversight is particularly ironic given social media's goal of bringing people closer together and highlighting commonalities. Tagging photos on Facebook is one obvious example of the collective nature of privacy. A user might be tagged in an update or a photo with which she doesn't want to be associated. If she hasn't selected the option that allows her to approve tagging requests, then her privacy may be compromised as soon as the update appears. Securing this kind of privacy requires both consciousness on the part of the network's programmers and those involved in creating the Facebook post. An individual should be allowed to decide whether she wants to be associated with this photograph, but the person uploading the photo should also make sure that those pictured want to appear online, tagged or not. These interdependencies can be difficult to negotiate and situational: a couple getting married may ask guests not to share updates about the event, for fear of offending those who aren't invited; a dinner party's host may say that he doesn't mind posts to Instagram, only for one of the guests to hide his face when the iPhone is pointed his way. Of course, with these same contextual barriers being eroded, and with practically any human feeling or endeavor now trackable and shareable, the sense of privacy as interdependent may be lost. Or it may be simply reconfigured: if anything is shareable, if we are intermittently transparent to one another but always transparent to ad networks and intelligence agencies, then perhaps our collective privacy is none at all.

- - - - - - - - - - - - - - - - - -

FACEBOOK AND THE NEW NORM

No social-media firm has been as explicit about its desire to overturn popular notions of privacy as Facebook. In January 2010, during an on-stage interview at an industry awards event, Facebook's Mark Zuckerberg said that privacy could no longer be counted as a "social norm." Many companies would hesitate, he said, to institute privacy changes for 350 million users (the site's user base at the time). "We decided that these would be the social norms now," Zuckerberg crowed, "and we just went for it."

These "norms" have changed often. At a May 26, 2010 event, Zuckerberg promised to make privacy controls "simpler." The site's controls have hardly gotten easier to manage, and it's not uncommon for a user to institute a certain level of privacy only to return months later and find out that that option is no longer available. I once foraged through the hedgerow of Facebook's privacy controls and selected an option so that none of my Facebook friends could see photos in which I was tagged. Now, not only can all of my friends see these photos, but I can also no longer find a setting within Facebook's privacy controls to hide my photos. But, as Joseph Turow notes, Facebook's privacy settings "are irrelevant when it comes to advertisers. In offering the data anonymously, Facebook claims the right to use even aspects of profiles that members have chosen not to make public."

Through these and other measures, Facebook's great achievement has been to repeatedly chip away at the edifice of privacy and ensure that each move—each removal of a privacy control, each introduction of a new feature that exposes more user information—is eventually accepted. We are all frogs in the Facebook pot, slowly being brought to a boil. In the view of writer and digital activist Cory Doctorow, "Facebook trains you to undervalue your privacy." As Doctorow indicates, this practice of undervaluing privacy is not so much a side effect as a core value. It's essential to Facebook's business model that its users feel less and less attachment to their privacy so that they can share more, churning out ever more data. Facebook's premium on frictionless sharing means that sharing should be natural and easy between users, but this lack of friction also applies to the process of Facebook's own information gathering.

The amount of data that Facebook collects on its users is enormous and would be the envy of any intelligence agency, if they didn't have access to it already. In 2011, Max Schrems, an Austrian law student, requested and received a copy of his data file from Facebook; it was 1,222 pages and contained information that Schrems hadn't intended to turn over to the social network, such as the geographic coordinates of where he logged in from, people he had unfriended, and other data he had deleted. Schrems went on to file a complaint with the office of the Irish Data Protection Commissioner (Facebook's European headquarters are in Ireland), claiming that they had violated various European privacy and data-protection laws. His action spurred the Irish government to audit Facebook's practices of data collection and retention and recommend a number of changes. Facebook also claimed that it would introduce a site-wide policy of deleting some user data that was more than a year old. But absent vigorous activist campaigns such as Schrems's, such promises are rarely followed up on by independent auditing.

Facebook has made similar promises about the data it gathers through its ubiquitous Like buttons. The company's official justification for its use of the Like button as a tracking mechanism goes as follows: "We record some of this information for a limited amount of time to help show you a personalized experience on that site and to improve our products." The information collected in this manner is deleted or anonymized after ninety days. Facebook also says that this browsing information is not sold to third parties.

Do you trust them? Will it always be this way, or might Facebook, when its stock price starts flagging, decide to start retaining data longer or to sell it to some of the market-research firms that would love to get their hands on it?

Writer and technologist Anil Dash has written that the company is "advocating for a pretty radical social change to be inflicted on half a billion people without those people's engagement, and often, effectively, without their consent." (Dash was writing before Facebook membership surged past the 1 billion threshold.) These privacy policies, he warned, represent an ideology of radical transparency that can have unintended consequences from some users: "Facebook is philosophically run by people who are extremists about information sharing. Though I choose to talk about my politics, or my identity, or my medical history or my personal relationships, I can do so primarily because I have the privilege to do so thanks to my social standing, wealth, and the arbitrary fact of being born in the United States."

By putting small pieces of Facebook across the Web, the social network has essentially arrogated itself the right to watch and catalog us wherever we go. While we retain some ability to limit what we show to other people on Facebook, we have few ways to limit what Facebook itself learns about us. And once that information ends up in the black box of Facebook's data centers, we have no idea how it might be used.

- - - - - - - - - - - - - - - - - -

INTERNET TRACKING ENTERS THE PHYSICAL WORLD

In an age of fabulously cheap digital storage and data-as-a-commodity, there is little reason for social networks to stem their customer surveillance. Regulatory responses have been remarkably lenient: Google's $7 million fine for using its Street View cars to indiscriminately suck up financial and password information from unsecured home WiFi networks represented a rounding error for the company. (Previously, the FCC fined Google just $25,000 for obstructing its investigation into Street View.) And occasional flare-ups of user backlash have proved fleeting in the face of powerful, useful, and free services. With more than one billion active accounts, Facebook has built up a formidable network effect, in which the cost of opting out, for many users, is too high. Facebook and its peers have also seen little pushback from spreading their tracking mechanisms well beyond their own networks; they've become an integral part of the social web. In this way, Facebook can extend its logic of persistent user surveillance to the Internet writ large. A user may leave Facebook.com, but he remains under Facebook's careful watch. We become conditioned to Facebook's prying eyes, in the same way we became conditioned to Gmail reading our e-mail so as us to provide us contextual ads. (Imagine the backlash if the U.S. Postal Service started opening every letter, reading its contents, and inserting a contextually relevant advertisement. On the other hand, they do already scan the address information—the metadata—of every letter, cataloguing it for America's intelligence services.)

Some browsers, such as Google Chrome and Mozilla's Firefox, have installed Do Not Track features, which are supposed to stymie the ability of advertisers, targeters, and ad networks to track browsing habits. But users must activate this capability in the browser's settings, and as of March 2013, only 11.4 percent of desktop Firefox users had activated Do Not Track. There's an even bigger flaw in this system, though: Web sites are under no obligation to respect these requests, and in fact, most don't. This Do Not Track capability was instituted after members of the advertising and tech industries, government officials, and privacy advocates came together to support the mechanism, which was a central part of the Consumer Privacy Bill of Rights that the Obama administration presented in February 2012. Nine months later, CNN said that "the entire plan is on life support"—a victim of faltering negotiations between privacy advocates, who claimed that Web giants such as Yahoo and AOL weren't negotiating in good faith, and the companies themselves, who said that the other side expected too much. In a meeting with the W3C, the international consortium that helps devise standards for the Web, a vice president of the Direct Marketing Association reportedly "proposed that Do Not Track signals should actually permit data collection for advertising purposes, the very thing the mechanisms were designed to control." The Association of National Advertisers then published an open letter to Microsoft CEO Steve Ballmer, criticizing his company for automatically enabling Do Not Track on its Internet Explorer 10 browser (which at the time hadn't even yet been released). Even the most cursory privacy measures, it seemed, would be vigorously contested.

But the effort was doomed from the beginning. The Privacy Bill of Rights called for corporations to sign up voluntarily, with enforcement entrusted to the congenitally toothless FTC. The bill itself was mostly a set of vague recommendations, along with some general principles, such as "Consumers have a right to secure and responsible handling of personal data." There was also some dispute over the meaning of Do Not Track, with *Wired*, for example, questioning whether logging which stories readers browsed in order to serve up recommendations counted as tracking.

In the absence of concerted industry action and meaningful government regulation, some people have taken measures into their own hands. Coders have created a number of anti-tracking tools, often in the form of free browser plug-ins that users can install. Apps such as Ghostery, DoNotTrackMe, and Disconnect block the more than 2,000 "retargeters" that use cookies, ad networks, and other techniques to track your browsing history and present you related ads across the Internet. These plug-ins can also block surveillance from social widgets, including Like buttons. Some browser makers have also stepped up, with Safari and Firefox automatically blocking cookies from third-party sites that the user hasn't visited. But some of these blocking apps aren't quite what they claim to be. Evidon, the company that makes Ghostery, takes some of the data it collects from Ghostery users—there are eight million of them—and sells it to advertisers.

Anti-tracking and -targeting measures may provide short-term solutions for users seeking some modicum of privacy or anonymity while browsing the Internet, but they do little to overturn the industry's status quo, which remains single-mindedly focused on knowing more about user activities than the users themselves.

Critics claim that advertising is essential to the digital economy. It's what makes so many Web sites and services free. This may be true, but consumers have no responsibility to help support a broken, if widely used, business model. Whatever implied social contract existed between advertisers and users has been torn up by the industry. Never before has so much information been collected, so much commercial surveillance performed, on such a broad cross-section of consumers, with all of it digitized and freely traded among data brokers. As the current ardor for Big Data shows, information harvesting can be an essentially endless process, with the only limits being technological. As Helen Nissenbaum writes, "This faith in information, envisioned as an asset of enormous value, creates a virtually unquenchable thirst that can only be slaked by more information, fueling information-seeking behaviors of great ingenuity backed by determined and tenacious hoarding of its lodes. Inevitably, as our awareness of this landscape grows, so grows a sense of privacy under assault."

We console ourselves with bromides about how no one should expect privacy online—as if this is an unchangeable situation to which we should simply be resigned. At least, we're reminded, it's not like this out in the physical world. There, we are constantly bombarded with advertising that targets us but on a far more general, and less personalized, level—a Gucci ad in GQ caters to the magazine's readers' presumed interest in luxury goods; a billboard for the new "Iron Man" movie broadcasts more widely, hoping to interest any and all passersby (Hollywood blockbusters are mass products on the largest possible scale; nearly anyone is a potential customer). But that practice is changing. With privacy being increasingly leaky, contexts breaking down, offline and online networks intermingled, social-media accounts linked across devices and platforms, and advertising networks and companies such as Facebook buying up reams of consumer data, new possibilities in targeting are opening up, particularly with the addition of facial recognition software. We are moving toward a world in which the same kinds of technologies that track you online are now tracking your movements and behaviors in the physical world. And often, it's the same companies involved, insinuating their tracking and collection technologies into all aspects of your life.

- - - - - - - - - - - - - - - - - -

TARGETING INDIVIDUALS

If you were walking down Oxford Street in west London in early 2012, you may have seen a bus-stop billboard featuring a 40-second video ad for a campaign to educate girls in the developing world. That is, if you were a woman. If you were a man (or recognized as one by the advertisement's built-in facial-recognition system), you would have instead been shown a shorter clip, encouraging you to visit the Web site of Plan UK, the organization behind the ad. The shorter ad was envisioned as a way of turning the tables on men, who usually have more choices and opportunities than women; only women, in this case, were allowed to see the full message. In Japan, NEC has produced digital billboards that also recognize a subject's gender and market different products accordingly. An American company, Immersive, has done the same, touting its "software that turns any camera into an intelligent sensor." As these pieces of software have improved, they have also been able to gauge a customer's relative age and their reactions to the ad (how long they look at it, for example). The resulting analytics may be employed to further hone the campaigns. And the next step, of course, is to individualize the targeting process, so that the software will be able to match your face to a photo from one of your social-media profiles. The ad, then, wouldn't have to stop in the mall; it could follow you onto your cell phone or appear next time you log into Facebook. It might appear as you drive by a digital billboard and then continue the conversation on a TV screen in an office elevator. We would be told that these ads are simply the most relevant to us, that they are finely targeted to further our engagement or are responding to the interest we've shown in the product in the past. The ads that we're used to following us around the Internet would follow us throughout our world. This persistent targeting should be called what it really is: surveillance, stalking, harassment, visual pollution.

There are some new technologies that allow for limited targeting of public advertisements without violating user privacy or imparting a sense of being surveilled. Plan UK's ad comes close, but its use of facial recognition might be troubling to some. Consider another clever but more respectful use of targeting. In May 2013, the ANAR Foundation, a Spanish organization that aids children facing abuse, put up an advertisement featuring a large photo of a child's face and some words against child abuse. The advertisement, which was on street-level displays, such as bus stops, made use of Lenticular printing, which allows viewers to see different images from different angles. (Sometimes used for large movie posters, Lenticular printing can impart a sense of movement.) In this case, ANAR calculated the average height of a 10-year-old child and produced another image—with bruises on the child's face and a message addressing children directly, along with a hotline number—that only people of that height could see. An adult, looking down from a higher angle, would see a basic message against child abuse. A small child, presuming he or she could read, would see a more targeted, urgent message. There are ways in which this technology might be misused or put toward more vulgar commercial ends, but in this case, it was an ingenious approach to spreading information in the public interest and to directing it toward those who need it most.

As wearable computing and the proliferation of digital displays, interfaces, sensors, and cameras bring down the wall between offline and online, the possibilities for tracking, targeting, and data collection increase immensely. Social networks and the logic they represent— persistent surveillance of users, industrial-level data collection—are becoming integrated into our surroundings. Some designers speak fancifully, but not improbably, about a future in which anything is potentially a display, where digital interfaces seamlessly appear and disappear, as needed, in the objects around us.

It's difficult to keep track of all the various programs and initiatives that tech companies have under way to monitor our activities. But here are some that will give you an understanding of the scope of the effort. We know that Facebook's Like buttons—similar to Twitter's social widgets—allow it to learn a great deal about your Internet activity, your life, your relationships, your personal history. The company has even worked on tracking where users move their cursors onscreen— for example, to see if they hover over certain ads but then decide not to click. Facebook's partnerships with the large data brokers Acxiom, Epsilon, and Datalogix allow it to know what you buy in retail stores, since these firms gather data about frequent-buyer cards, such as the ones you may use at CVS and the grocery store.

Google Now sifts through your smartphone data, calendar, e-mail, GPS, search, and many other sources of information to find out about your daily activities and keep you up-to-date with tips, directions, re- minders, and advertisements. Google has tested using location information to detect when smartphone users enter retail stores in order to see if online searches (and ad impressions) lead to in-store visits. Depending on your device and software configuration—whether you've opted into your phone's location services, have Google apps on your iPhone, or (the easiest method) have an Android phone—Google may be able to monitor your location almost constantly. These sorts of efforts allow Google to tell advertisers that its mobile ads work and to serve you ads when they think you're most susceptible to them.

Stores, in turn, are working on tracking their customers like never before. Nordstrom, Bloomingdale's, American Apparel, Verizon, and other major retailers have utilized software that picks up on smartphones' Bluetooth and WiFi signals in order to monitor how customers move through stores. The data can be useful in assessing store design, how long customers spend in certain areas, the paths they take, and so on. Stores can also attach unique identifiers to each phone and track customers over repeated visits. The information can be sold to brands who want to know how consumers interact with displays. A clothing company might see that some female customers are repeatedly stopping in front of their new luxury line, lingering, and then leaving, leading them to believe that their prices are too high.

Other stores have used cameras and facial-recognition programs to gauge customers' moods and responses to different stimuli. An Italian company sells mannequins, called EyeSee, which contain cameras equipped with facial-recognition software that can recognize customers by approximate age, gender, and ethnicity. NEC has developed facial-recognition software that allows stores to recognize VIP customers when they enter. An app called Facedeals invites businesses to install facial-recognition cameras, which recognize customers based on their Facebook photos and then sends deals and coupons to their phones. The app, which also checks customers into the location, must be authorized by customers. The deals offered are based on the customers' "like histories," so "personalized deals can now be delivered to your smartphone from all participating locations—all you have to do is show your face." SceneTap also places facial-recognition cameras in bars and uses them to track the ages and gender ratio of the patrons—data which is then viewable on maps in the SceneTap app and on its Web site. "These apps are bridgeheads, or perhaps trojan horses, for more powerful (and probably more intrusive) services to come," the technologist Alessandro Acquisti told *Ars Technica* after SceneTap's launch.

Google has received a patent for "pay-per-gaze advertising," physical advertisements with embedded sensors that can tell when customers are looking at them. It's perhaps the most literal example of the attention economy. Under such a system—which, Google's patent notes, can include "billboards, magazines, newspapers, and other forms of conventional print media"—advertisers wouldn't pay Google just based on number of impressions or clicks. They'd also pay Google every time you look at the ad. Your gaze becomes a metric of value, making it almost impossible to walk down the street and not be caught up in an economic exchange between two companies. Google's sensors are also supposed to be able to pick up on pupil dilation and other emotional cues, providing Google and its partners with information on how ads affect people on an instinctive level.

The same patent mentions Google's work on overlaying ads on Google Glass, perhaps the signal product reflecting Google's vision of providing each customer an intensely personalized experience. From search to ads to shopping to an automated digital personal assistant, Google is promising to shape your experience of the world around what it and its commercial partners believe you need to see.

Google has also introduced customized maps that are different for each individual user. Two designers who worked on the project claim that "the more context it has about you, the more useful it can be"—a constant refrain for supporters of these kinds of technologies. These maps are likely to be different each time they appear, meaning that no map is the same twice or the same for two different people. This might seem like an exciting use of user data, but it also raises some problems.

What happens when two people, perhaps two students working on an assignment, are separately looking at the same area, and Google decides to show them two different versions of it? Will my map be loaded with the places advertisers want me to see? Will Google decide that I don't need to see poor regions of my city, because they think I don't go there enough anyway, and my demographic data indicates I'm middle class? What other kinds of selection biases might occur? If Google thinks I haven't been to a park in a while—maybe it's winter or maybe I choose to leave my smartphone at home when I go out for a jog—will it stop showing me parks and other public spaces? Google already guides the routes I take when traveling. Perhaps, they'll decide to start directing me past restaurants that advertise with them and locations featuring billboards with their pay-per-gaze technology. As I pass these restaurants, I might receive ads or coupons in my Gmail inbox offering me a discount. Along the way, my entire urban experience potentially comes under the influence of Google.

The maps example also raises some of the same problems we run into when thinking about sorting algorithms. Knowing that each action can influence various overseeing algorithms, do we adjust our behavior accordingly? Do we do things just so that the algorithms monitoring us won't go off track, so that it'll still "like" the things we like? You might already do a form of this—say, give a thumbs-up to a song on Pandora because you want to hear more of that type (assuming you trust Pandora's system to usefully recognize a particular song type). You may start paying with cash at bars, worrying that, if your health insurer has access to credit card data, it might think those five beers were only for you, rather than you and your friends. You might rate art films, but not the new Marvel movie, so that you appear more cultured to the dating site offering you discounts and the Facebook app that lists what films you've recently seen.

From this point of view, these webs of tracking, advertising, and surveillance technologies are profoundly coercive—as surveillance tends to be. For better and worse, ads landscape our environments.

They influence the culture and invite us to view ourselves in certain ways. Someone who's constantly receiving advertisements for weight-loss drugs, gyms, and plus-size clothing will be receiving different messages than his neighbor who's shown ads for beach vacations, fancy watches, and the latest French bistro. With these systems working in concert, people living side by side, even in the same home and sharing the same devices, might be served far different information and guided to different stores and opportunities. (Some companies already claim to be able to recognize different users who spend time on the same gadgets.)

The tracking industry works with advertisers and social-media companies to leverage users' insecurities. In October 2013, a marketing firm called PHD released a report about when women feel least attractive, encouraging advertisers to target women with "quick beauty rescues" on Sundays and Mondays and social opportunities (including the expensive accessories and clothing society says they require) later in the week. The study even broke down by the hour when women are more likely to feel unattractive, with 5 a.m. to 7 a.m. being the peak, or trough, as the case may be. The ad network MediaBrix has developed a method for targeting gamers in what it calls "breakthrough moments" in mobile games. You might be struggling to crack a level on a strategy game on your phone, but when you finally do, an advertisement will pop up to congratulate you. MediaBrix also has products that introduce sponsored digital rewards (Congrats on beating that level! Here's a trophy from Gillette) or that allow gamers stuck at some point to choose to watch a video in order to overcome the obstacle. These methods are less instinctually repugnant than PHD's report on women's insecurities, but MediaBrix is operating on a similar principle: target users when they're at their most vulnerable.

Microsoft, too, has a patent for targeting users based on emotional states. The patent discusses examining what users are looking at, their perceived reactions, and serving up ads accordingly. Advertisers would also receive more control over their audience: "Advertisers provide targeting data that includes the desired emotional states of users it intends to target," the patent reads. Samsung, whose Galaxy smartphones can already be controlled with eye movements, has been refining facial recognition technology. One of its patents would organize and regulate users' interactions based on their emotional states. The site *IPWatchdog* commented, "It appears that Samsung is seeking to improve social-media communications by limiting interactions between members who provoke negative emotions, or increasing interactions between members who instill positive emotions." Besides being able to push certain types of users into desired exchanges—which would be helpful for regulating unwanted speech and for keeping users happy and chained to the platform—this technology could be used to provoke certain emotional responses in users and target them accordingly.

These systems are far more manipulative than any market research done in the past because advertisers now have the ability to reach us at virtually any time. They also know far more about us than their predecessors ever did, while making us complicit in the process by encouraging check-ins, structuring data, location services, and other data production/sharing that is, we are told, designed to improve a service. A growing crop of biometric tools—sleep measurement apps, fitness monitors, the thumbprint reader introduced on Apple's iPhone 5S, the gene-sequencing service 23andme.com—means that corporations are set to know us at the physical, even genomic level. ("Your DNA will be your data," says one particularly creepy HSBC ad spotted at JFK airport.) They may even anticipate health problems before we realize we have them. Read your fitness tracker's terms of service agreement. Are they required to notify you if they detect a health problem? Do they reserve the right to sell your personal information to health insurers? An FTC study found that thirteen health and fitness apps sold data to seventy-six other companies, sometimes including users' names and e-mail addresses. Within weeks of being acquired by Facebook, Moves, a fitness app, announced that it would begin sharing data with its new parent company. The change would allow Facebook, should it desire, to follow their users into the physical world, gaining information both about their movements and their physical health. It's easier to push ads at someone if you know when he's hungry, tired, or ill.

Although these new devices are in turn creating new methods of surveillance, the cookie remains a useful and widely used tracking technology. But soon it might be obsolete. Apple, Microsoft, Google, Verizon, and other companies are in an arms race to improve tracking technology. This is why Do Not Track buttons and industry-government negotiations over such rights are meaningless. Future tracking technologies will be integrated at the hardware level, making them harder to disable with software, while your face will serve as another kind of cookie, to be measured and parsed by CCTV and facial-recognition systems.

The MAC address—a unique identifier, similar to a serial number, stored in each cell phone's hardware—has already been deployed as a tracking mechanism. In August 2013, the city of London was forced to take twelve recycling bins off the streets after it was reported that the bins were tracking the movements of passersby by noting their MAC addresses. Renew, the company behind the bins, had been soliciting local businesses, offering them ad-targeting information about people walking through the area. On their busiest day, the bins identified 106,629 people, each of them, on average, more than eight times. The bins also had Internet-connected screens to show ads. *Quartz*, the business site that broke the story, noted that the company that developed the tracking "orbs" in the bins described its technology as "a cookie for the real world." While the bins were a commercial project, they were installed before the 2012 Olympics, when the UK government was instituting extraordinary security measures, including putting surface-to-air missiles on top of apartment buildings. If given access to these bins (or if access were obtained through other means), British intelligence services could have a real-time map of people, and their data-rich devices, walking through the area. Seen this way, these spying bins are little different than the license-plate readers on police cars and build-ings throughout the United States. They are designed to identify and track people moving through public space—a type of mass surveil-lance that should be considered anathema.

From locations to moods to the latest research on when women claim to feel insecure—all of this data is crunched in the service of parting individuals with their money, or in getting them to do the equivalent micro-labor, such as clicking on an ad, causing a tiny pay-ment to pass between advertiser and ad network owner. Do this bil-lions of times and you can be the next Google.

Any advantage helps, even small psychological cues. Sociologists and behavioral economists call these "nudges": subtle reminders or ges-tures that can help people make better decisions in their self-interest. This might be useful at times; perhaps you want your fitness monitor to light up with an alert or to contact your doctor if certain vital signs deviate from expected patterns. In the hands of marketers, nudges might tell your mapping app to steer you past a particular billboard or excitedly congratulate you when you finally, finally beat that level in Candy Crush. Or maybe credit card offers will start appearing when marketers think you're most impulsive or when they know that your checking account is looking low. As the process is perfected, the con-sumer would be relegated, Mark Andrejevic explains, "to the role of feedback mechanism in an accelerating cycle of production and con-sumption."

The underlying irony here is that consumers produce the informa-tion which, through this constant feedback system, helps steer their behaviors. This is done by our browsing, social networking, publishing posts, and other forms of data production, but also sometimes more deliberately—liking brand pages, structuring data, doing check-ins, requesting coupons and special offers. In perhaps the most direct ex-ample of this participation, Acxiom, one of the industry's largest data brokers, unveiled a Web site called AboutTheData.com, which was supposed to allow the company to get ahead of bad press generated by industry practices and various governmental investigations. The site, which gives people the ability to opt out of tracking by Acxiom, also was intended to show that the company wasn't afraid to operate more transparently. But About The Data's most diabolically savvy feature is that consumers can "correct" errors in their profiles—in other words, they can improve and structure their personal data. This might, as Acxiom claims, give consumers a better experience, but it is also clearly to the company's advantage, allowing them to make consumers complicit in filling out their data profiles. And while the site lets users suppress some data, it also doesn't show them everything that the company has on them.

Joseph Turow, the author of *The Daily You: How the New Advertising Industry Is Defining Your Identity and Your Worth*, fears possibilities of social discrimination. In an interview, he explained that "it's simply the idea that increasingly companies will use data about us in order to make decisions about how important we are, and some people will win and some people will not." This might be, he added, how the world has always worked, but that doesn't mean that it *must* be that way, nor that we should delegate this power to machines. "What we have here is a winner/loser scenario that takes place algorithmically," he said, "basically through the tracking of people and the using of predictive methods to figure out who's important and who isn't, on definitions of people—they don't even know it's going on. But companies are defining us, constructing us and making decisions about our importance without even our having any clue that this is taking place in any serious way."

How might it look to be at the losing end of one of these decisions? It might not just be the depressing ads you receive or, compared to a less prosperous neighbor, being overcharged for items on a shopping site. You might be denied disability insurance because the insurer looked at your social-media profile and decided that you didn't look depressed. That's what happened to Nathalie Blanchard, a Quebec woman whose insurance benefits were revoked after she posted pho-

tos of herself at a birthday party and at the beach—excursions which
her doctor recommended in order to help battle her depression. You
could be denied a loan because a bank thinks that your small number
of Facebook friends means that your life is unstable or that you are
unreliable. That's how the financial services company Lenddo determines credit worthiness. High schools and universities—many already
monitoring current and potential students, whether for purposes of
discipline or admission—may decide to start using predictive analysis
to determine which students may become violent. The city of Chicago
used a similar system to make lists of people likely to commit or be
victimized by violent crimes and then tasked police officers and social
workers to target these individuals.

"People will worry how they relate to one another and to machines and may even change their behavior because they want to
be treated better," Turow said, before adding that because we don't
know the parameters of the algorithms judging us, we're "constantly
guessing."

There's social damage done by these practices. They create a sys-
tematic disrespect for people's privacy. They privilege certain types
of people over others. They make everyday people worry about what
kind of information—including biometric data, that most personal
and revealing kind of information—is being collected on them. And
we may not even know when it's being used. For instance, Facebook's
tag suggestions for photos draw on facial recognition. You may have
disabled Facebook's facial recognition feature, but it could still have
a faceprint of you that it could use toward its own ends. Google scans
photos uploaded to Picasa and Google+ to detect child pornography
and report violators to law enforcement agencies. That's an understandable use of this technology, but once private companies get in
the business of seeking out criminal activity for law enforcement, it's
worth asking how far these policies go. If a terrorist attack or some
other crisis were to occur, would Google give government agencies
broader access to its stores of user photos in order to help identify

suspects? And how many innocent people might be caught up in this new dragnet?

- - - - - - - - - - - - - - - - - -

RECONSTRUCTING PRIVACY

With all of these challenges arrayed before us, the future of privacy will require contending with the surveillance economy, both reining in its excesses and plotting out viable alternatives. These practices are unlikely to disappear anytime soon, but they must be regulated and managed, as well as opened up to public scrutiny. All too often, the very companies that preach a culture of openness and radical transparency only advance this attitude to the extent that it serves their interests. Practically speaking, it is Internet users who remain radically transparent, to one another and, to the greatest extent, to the companies that collect their data. What data is collected, the algorithms that sort data and decide which ads to show users, how long certain data is preserved, the extent of industry cooperation with governments—these areas are rarely considered open. Transparency is a rhetorical ploy designed to serve those who want to retain this kind of power.

Individuals should take some responsibility to better inform themselves about who is collecting what data and what's being done with it. Many consumers fail to even examine privacy settings on the apps and social networks they use. According to *Consumer Reports*, about 13 million of approximately 150 million U.S. Facebook users "said they had never set, or didn't know about, Facebook's privacy tools." And they often have an antiquated idea of what privacy is—one that is monolithic and fails to take into account the ways in which privacy is now both networked and context-dependent. They also don't consider how the wholesale dispersal of personal information may yield unintended, and unwanted, consequences down the road.

Even so, the greater onus in securing privacy certainly lies with the companies collecting this data and on the authorities entrusted with monitoring them. Consumers remain on the defensive, with well-meaning activist organizations such as the Electronic Frontier Foundation (EFF) fighting a perpetual rear-guard action. Social networks' privacy settings often act as a red herring, a way of obscuring the vast data collection efforts being undertaken by network administrators and the many ancillary companies that have found ways to profit off of corporations' limitless appetite for personal information. A more responsible Facebook or Google would not simply offer better or straightforward privacy settings; they would decline to collect this data in the first place, much less share or trade it with so many outside partners. They would encrypt all data they transmit, encourage the adoption of other security measures, and use the court system to fight government data requests. They would also diversify their business practices so that their financial prospects were not contingent upon surveilling users and manipulating them into passivity with promises of better advertisements for prescription drugs or cheap electronics.

Decentralized or peer-to-peer social networks, such as Diaspora and App.net, have failed to take off, but the attention these companies have garnered indicates a nascent consumer interest in better data policies, particularly for placing an emphasis on user ownership of data—which ultimately means user ownership over their own public identities. While it would be foolish to attribute Twitter's success to its relatively strong record of fighting for user rights, particularly when dealing with government requests for user data, these practices have earned it some deserved praise, including from the EFF. Twitter's behavior may help set new standards for the industry and prove commercially useful as the notion of personal data and privacy as commodities increasingly (if unfortunately) trickles down to consumers. Microsoft, which like Twitter publishes transparency reports about what it shares with the U.S. government, based a high-profile ad campaign around the concept of being "Scroogled"—a portmanteau meant to highlight what it sees as Google's privacy lapses, such as "Google's policy of supplying the name, e-mail address, and neighborhood of users who purchase apps on Google Play." This ad campaign has come under some criticism. Microsoft has long been cast as a stodgy old man, if not an outright villain, of the tech scene. The company certainly engages in some data collection practices comparable to Google's—like most big tech companies, it's deeply reliant on the Web's surveillance infrastructure—and the subsequent revelations of its participation in the NSA's PRISM program have done nothing to burnish Redmond's image as a privacy defender. The ad campaign also has the unfortunate consequence of presenting privacy as a commodity, on the same commercial plane as, say, how much storage space the company's cloud storage service offers. Although appealing to market dynamics may be important for the privacy debate, a purely economic approach to privacy has tended to favor large corporate actors while leaving users and regulators sidelined. Should there be a populist backlash to social networks' privacy policies, it should be grounded, at least in part, on moral arguments—if only because, in an age of networked privacy and contextual collapse, privacy transcends traditional boundaries. It is not merely a problem of one company's practices, or even of the whole digital world, but of our entire surveillance-saturated society. If governmental surveillance can be opposed on legal *and* moral grounds, shouldn't Google or Facebook's surveillance of its users submit to the same reasoning? Were the argument left purely to the market, those with the deeper pockets are likely to win, as they have been for too many years now.

Big Data and the Informational Appetite

You are the sum total of your data.
No man escapes that.

—Don DeLillo, *White Noise*

To understand the depth of our privacy problem, we have to look at the ideological, economic, and cultural roles that data collection and data mining have assumed in recent years. There is now so much data produced on our behalves—about a terabyte per capita per year—that a major industry has arisen, one that fits familiar patterns of techno-utopian thinking. Big Data, as this emerging field is called, promises to take the incredible amount of data collected—browsing histories, sensor information from smartphones, GPS coordinates, social-media activity, purchasing information, medical reports—and turn it into useful insights. Big Data has found supporters in health care, insurance, scientific research, education, energy, and intelligence. While some commentators have argued that the utopian possibilities of Big Data are overblown, others offer more dire outlooks: "the surveillance possibilities of the technology," according to the director of the Human Dynamics Lab at MIT, "could leave George Orwell in the dust." At its most far-reaching, Big Data promises predictions about the behaviors of individuals and population groups, as well as to forecast anything from traffic to weather conditions to street crime. The faith in Big Data—which is really just a trendy, catchall term for various types of bulk data collection and analytics—has led social-media companies to think they can know and predict our behaviors, that they can, as Eric Schmidt says, know us better than we do ourselves.

It's the enthusiasm for Big Data, along with the attendant idea that one's analytical capacities increase with the amount of data collected, which causes Gen. Keith B. Alexander, then the director of the NSA, to justify widespread surveillance by saying, "You need the haystack to find the needle." In Alexander's eyes, Americans' phone records—all of them—form that haystack. For Silicon Valley firms, the haystack is potentially the full array of a user's life—whatever can be tracked. On second consideration, "whatever can be tracked" applies to the U.S. intelligence community as well. Witness its since-abandoned Total Information Awareness project (now replaced by a smattering of connected programs and partnerships with tech firms, telecoms, and foreign intel agencies); the Mastering the Internet project, maintained by GCHQ, NSA's close British partner; or the National Reconnaissance Office, which in December 2013 launched a spy satellite on a rocket painted with an image of a world-straddling octopus and the words, "Nothing Is Beyond Our Reach." The ODNI, the office that oversees the entire U.S. intelligence community, was so proud of this event that it even tweeted photos of the rocket and a separate one of the logo, which also served as a mission patch. It is a matter of pride to confess one's informational avarice.

No wonder, then, that the NSA and Silicon Valley have made such good partners. Both are in the data collection and targeting business, and Silicon Valley collects heaps of data which the NSA would love to have.* Silicon Valley is merely targeting consumers with ads and prompts and nudges that might get them to click or to buy something. They are bound together by common interests, philosophies, and methods.

---

* The NSA even refers to government agencies requesting data and analysis from it as "customers."

One of the main problems with Big Data is that it produces correlations but not causations. We learn that two things seem to be related—for example, that people with a specific set of personal characteristics are prone to depression or bad driving—but we don't learn *why*. This is ironic given that Big Data is the ultimate fact-producing discipline: it promises answers, actionable ones. But data itself can be messy and often must be smoothed over, interpreted, supplemented. It doesn't always lead us where we hope to go. As the researchers Danah Boyd and Kate Crawford have argued, Big Data "encourages the practice of apophenia: seeing patterns where none actually exist."

Ironically, it's the very unreliability of Big Data–style analysis that prompts ever more data collection. If you think you've detected some false patterns or aren't finding the kinds of correlations you sought, why not just collect and analyze more? If you can't process all the data you've stored—a problem that the NSA has faced—just build more data centers and hire more mathematicians and data scientists. Whether you're Facebook or the U.S. government, the money is out there to do just that.

Even the apparent presence of a pattern can lead us toward some false choices. A health insurance company may believe that people who buy six key grocery items are 30 percent more likely to develop diabetes, but does that give the insurer the right to raise this group's premiums or deny them coverage? Is your purchase history, or your poor gym attendance, as indicated by your smartphone GPS, a pre-existing condition? Should the insurer be allowed to push you toward better health by sharing this analysis with a company that sends out coupons for quinoa and kale? Must the insurer notify customers that their personal information is being used in this way? Already some health insurers have begun lowering premiums for people who use fitness monitors and let their employers and insurers collect that data. The obverse of this arrangement is that those who don't submit to this kind of surveillance are penalized. They have to pay more just to keep their basic health information private.

David Lyon, the surveillance theorist, links the rise of Big Data to the growth of risk management as a central practice for governments and business. The more data that can be collected, the thinking goes, the more that risk can be anticipated and mitigated and hedged against with complex insurance policies. Eliminating any sense of danger or unpredictability therefore becomes the most important goal, with concerns about civil liberties, privacy rights, and adverse consequences far behind.

It's almost immaterial whether Big Data, as a discipline, works. (The answer to that isn't really yes or no, anyway.) Important decisions are being made based on Big Data techniques, and Big Data, in turn, gives ballast to the tracking and data collection technologies I outlined in the last chapter. The deliberate erosion of user privacy, the push toward constant, frictionless sharing, the institution of a wide array of tracking mechanisms, the digitization of every thought, gesture, and scrap of culture—all of these measures are designed so that social networks may harvest more user data. Given the direct relationship between data collection and targeted advertising, advertising-supported social media may be incapable of guaranteeing users' privacy. When combining the supposedly far-reaching possibilities of Big Data with the urgent profit motives of being a publicly traded corporation, it's difficult for firms such as Google and Facebook to set limits for themselves. As Facebook's VP of infrastructure told GigaOm in August 2012, "Everything is interesting to us." At the time, the company was harvesting 500 terabytes per day, including 2.7 billion likes and 300 million photos. For the platform owners, digital serfdom yields an endless bounty.

Riding the Big Data wave, Facebook's data collection efforts now extend far beyond its own social network. It has solicited potential advertisers for their customers' e-mail addresses and phone numbers, so that they could, for example, show users ads for clothing they had looked at on a third-party retailer's site. In March 2013, Facebook bought data from four major market research firms, including two,

Acxiom and Datalogix, that at the time were under investigation by the Federal Trade Commission. These data brokers vacuum up information about consumers, aggregating them into various classes that allow marketers to better target them with advertising. Someone designated as a single, suburban mother who loves to exercise might get advertisements for running gear, minivans, and dating sites. The *New York Times* summed up the potential consequences of the deals: "It could mean seeing advertisements based not just on what they 'like' on Facebook, but on what they eat for breakfast, whether they buy khakis or jeans and whether they are more likely to give their wives roses or tulips on their wedding anniversary. It means that even things people don't reveal on Facebook may be discovered from their online and offline proclivities."

Through these partnerships, Facebook gains information about what you buy and when and where you buy it. You might then start seeing Facebook ads for milk before your regular shopping trip or be offered a CVS coupon before going to fulfill your prescription. In practice, these campaigns have been very fruitful for Facebook and their advertising partners, with various reports claiming improved click-through and purchase rates. While some of this information is hashed or anonymized, Facebook can still learn much about the buying habits of users similar to you. Sometimes, anonymizing is as simple as assigning a person a number instead of a name; I might not be "Jacob Silverman" in a company's eyes, but it's just as useful for them to know me as 5402831. And given that past studies have shown that large data sets can sometimes be de-anonymized, Facebook may have hopes of matching its user profiles to the ones put together by these data brokers. One study by Belgian researchers working with a data set drawn from 1.5 million cell phone owners found that 95 percent of them could be identified with the help of only four location points. With these kinds of successes in mind, Facebook could set its software to work to uncover information about its users' shopping habits, their daily commutes, when they are more likely to make purchases, and other personal proclivities. Given all of the information that users already voluntarily give Facebook, from likes to location check-ins, the possibilities for better understanding users, for filling out the data set representing their selves with even more detailed information, must seem endless. Were Facebook to make deals with brick-and-mortar retailers who are trying to apply Amazon-style customer surveillance to their stores, it could have a nearly seamless view of its users' lives, from morning to night.

It's worth noting that while these data brokers are considered distasteful by many, they are not recent entrées to the scene. It's a $150 billion industry that for decades has been working with advertisers, employers performing background checks, defense contractors, health insurers, and other industries to provide information about consumers. And while some of their practices have earned federal investigations and popular opprobrium, deals of the kind executed by Facebook are legal. As a 2007 study by scholars at the University of California, Berkeley, noted, "Most consumers are under the misimpression that a company with a 'privacy policy' is barred from selling data." Nor are many consumers aware of how much can be done with this data. Data brokers can both learn your habits and even, as one widely cited story chronicled, learn that a young woman was pregnant before her parents knew. These firms are also contributing to a recent tendency of e-commerce sites to show different prices to consumers based on where they live and how much these companies think they might be willing to pay. In a very real way, data brokers can know things about you that you don't know yourself, leading to the kind of manipulative efforts that make an unsuspecting user and his money, or his data, easily parted. Finally, their view of consumers tends to be extraordinarily callous, as Mike Seay learned when he received the mailing mistakenly printed with information about his dead daughter. Data brokers sell lists of elderly people who are interested in sweepstakes, recovering alcoholics, people believed to be struggling with personal debt, and victims of domestic abuse and chronic disease. For a time, Medbase2000, an Illinois-based company, offered a "rape sufferers list" for sale on its Web site. The cost was 7.9 cents per victim. After a Senate hearing into the practices of data brokers, a *Wall Street Journal* reporter contacted the company. The president of Medbase2000's parent company claimed that the list was "hypothetical," a test for possible products. (Why this sort of list was used as an in-house hypothetical was never explained.) The company then removed the page, along with a list of HIV/AIDS patients, from its site.

- - - - - - - - - - - - - - - - - -

THE INFORMATIONAL APPETITE

Why has Big Data so quickly become a part of industry dogma? Beyond the tremendous amounts of money being thrown at Big Data initiatives, both in research dollars and marketing efforts designed to convince enterprise clients of Big Data's efficacy, the analytics industry plays into long-held cultural notions about the value of information. Despite Americans' overall religiosity, our embrace of myth and superstition, our surprisingly enduring movements against evolution, vaccines, and climate change, we are a country infatuated with empiricism. "A widespread revolt against reason is as much a feature of our world as our faith in science and technology," as Christopher Lasch said. We emphasize facts, raw data, best practices, instruction manuals, exact directions, instant replay, all the thousand types of precise knowledge. Even our love for gossip, secrets, and conspiracy theories can be seen as a desire for more privileged, inside types of information—a truer, more rarified knowledge. And when this knowledge can come to us through a machine—never mind that it's a computer program designed by very fallible human beings—it can seem like truth of the highest order, computationally exact. Add to that a heavy dollop of consumerism (if it can be turned into a commodity, Americans are interested), and we're ready to ride the Big Data train.

Information is comforting; merely possessing it grounds us in an otherwise unstable, confusing world. It's a store to draw on, and we take threats to it seriously. Unlike our European brethren, we evince little tolerance for the peculiarities of genre or the full, fluid spectrum between truth and lies. We regularly kick aside cultural figures (though, rarely, politicians) who we've determined have misled us.

Our bromides about information—it's power, it wants to be free, it's a tool for liberation—also say something about our enthusiasm for it. The smartphone represents the coalescing of information into a single, personal object. Through the phone's sheer materiality, it reminds us that data is now encoded into the air around us, ready to be called upon. We live amid an atmosphere of information. It's numinous, spectral, but malleable. This sense of enchantment explains why every neoliberal dispatch from a remote African village must note the supposedly remarkable presence of cell phones. They too have access to information, that precious resource of postindustrial economies.

All of this is part of what I call the informational appetite. It's our total faith in raw data, in the ability to extract empirical certainties about life's greatest mysteries, if only one can deduce the proper connections. When the informational appetite is layered over social media, we get the messianic digital humanitarianism of Mark Zuckerberg. Connectivity becomes a human right; Facebook, we are told, can help stop terrorism and promote peace. (Check out Facebook.com/peace to see this hopeless naïveté in action.) More data disclosure leads to a more authentic self. Computerized personal assistants and ad networks and profiling algorithms are here to learn who we are through information. The pose is infantilizing: we should surrender and give them more personal data so that they can help us. At its furthest reaches, the informational appetite epitomizes the idea that we can know one another and ourselves through data alone. It becomes a metaphysic. Facebook's Graph Search, arguably the first Big Data–like tool available to a broad, nonexpert audience, shows "how readily what we 'like' gets translated into who we are."

Zuckerberg is only one exponent of what has become a folkway in the age of digital capitalism. Ever connected, perhaps fearing disconnection itself more than the fear of missing out, we live the informational appetite. We have internalized and institutionalized it by hoarding photos we'll never organize, much less look at again; by tracking ourselves relentlessly; by feeling a peculiar anxiety whenever we find ourselves without a cell phone signal. We've learned to deal with information overload by denying its existence or adopting it as a sociocultural value, sprinkled with a bit of the martyrdom of the Protestant work ethic. It's a badge of honor now to be too busy, always flooded with to-do items. It's a problem that comes with success, which is why we're willing to spend so much time online, engaging in, as Ian Bogost called it, hyperemployment.

There's an inherent dissonance to all this, a dialectic that becomes part of how we enact the informational appetite. We ping-pong between binge-watching television and swearing off new media for rustic retreats. We lament our overflowing in-boxes but strive for "in-box zero"—temporary mastery over tools that usually threaten to overwhelm us. We subscribe to RSS feeds so as to see every single update from our favorite sites—or from the sites we think we need to follow in order to be well-informed members of the digital commentariat—and when Google Reader is axed, we lament its loss as if a great library were burned. We maintain cascades of tabs of must-read articles, while knowing that we'll never be able to read them all. We face a nagging sense that there's always something new that should be read instead of what we're reading now, which makes it all the more important to just get through the thing in front of us. We find a quotable line to share so that we can dismiss the article from view. And when, in a moment of exhaustion, we close all the browser tabs, this gesture feels both like a small defeat and a freeing act. Soon we're back again, turning to aggregators, mailing lists, Longreads, and the essential recommendations of curators whose brains seem somehow piped into the social-media firehose. Surrounded by an abundance of content but willing to pay for little of it, we invite into our lives unceasing advertisements and like and follow brands so that they may offer us more.

In the informational appetite, we find the corollary of digital detox and its fetishistic response to the overwhelming tide of data and stimulus. Information is power, particularly in this enervated economy, so we take on as much of it as we can handle. We succumb to what Evgeny Morozov calls "the many temptations of information consumerism." This posture promises that anything is solvable—social and environmental and economic problems, sure, but more urgent, the problem of our (quantified) selves, since the appetite for information is ultimately a self-serving one. How can information make us richer? Smarter? Happier? Safer? How can we get better deals on gadgets and kitchenware? The informational appetite is the hunger for self-help in disguise.

Viral media thrives because it insists on both its newness and relevance—two weaknesses of the informational glutton. A third weakness: because it's recent and seemingly unfiltered, it must be accurate. Memes have the lifespan and cultural value of fruit flies, but they're infinite, and they satisfy our obsession with shared reference and cheap parody. We consume them with the uncaring aggression of those who blow up West Virginia mountains to get at coal underneath. They exhaust themselves quickly, messily, when the glistening viral balloon is deflated by the revelation that the ingredients of the once-tidy story don't add up. But no matter. There is always another to move onto, as well as someone (*Wikipedia*, Know Your Meme, Urban Dictionary, et al.) to catalog it.

The informational appetite is the never-ending need for more page views. It's the irresistible compulsion to pull out your phone in the middle of a conversation to confirm some point of fact, because it's intolerable not to know right now. It's the smartphone as a salve for loneliness amid the crowd. It's the "second screen" habit, in which we watch TV while playing games on our iPhone, tweeting about what we're seeing, or looking up an actor on IMDB. It's Google Glass and the whole idea of augmented reality, a second screen over your entire life. It's the phenomenon of continuous partial attention, our focus split among various inputs because to concentrate on one would reduce our bandwidth, making us less knowledgable citizens.

The informational appetite, then, is a cultural and metaphysical attitude as much as it is a business and technological ethic. But it also has three principal economic causes: the rapid decrease in the cost of data storage, the rising belief that all data is potentially useful, and the consolidation of a variety of media and communication systems into one global network, the Internet. With the ascension of Silicon Valley moguls to pop culture stardom, their philosophy has become our aspirational ideal—the key to business success, the key to self-improvement, the key to improving government and municipal services (or doing away with them entirely). There is seemingly no problem, we are told, that cannot be solved with more information and no aspect of life that cannot be digitized. As Katherine Losse noted, "To [Zuckerberg] and many of the engineers, it seemed, more data is always good, regardless of how you got it. Social graces—and privacy and psychological well-being, for that matter—are just obstacles in the way of having more information."

The CIA's chief technology officer isn't immune. "We fundamentally try to collect everything and hang on to it forever," he said at a 2013 conference sponsored by *GigaOm*, the technology Web site. He too preached the Big Data gospel, telling the crowd: "It is really very nearly within our grasp to be able to compute on all human generated information." How far gone must you be to see this as beneficial?

Compared to this kind of talk, Google's totalizing vision—"to organize the world's information and make it universally accessible and useful"—sounds like a public service, rather than a grandiose, privacy-destroying monopoly. Google's mission statement, along with its self-inoculating "Don't Be Evil" slogan, has made it acceptable for other companies to speak of world-straddling ambitions. LinkedIn's CEO describes his site thusly: "Imagine a platform that can digitally represent every opportunity in the world." Factual wants to identify every fact in the world. Whereas once we hoped for free municipal

WiFi networks, now Facebook and Cisco are providing WiFi in thousands of stores around the United States, a service free so long as you check into Facebook on your smartphone and allow Facebook to know whenever you're out shopping. "Our vision is that every business in the world who have people coming in and visiting should have Facebook WiFi," said Erick Tseng, Facebook's head of mobile products.* Given that Mark Zuckerberg has said that connectivity is a human right, does requiring patrons to log into Facebook to get free WiFi impinge on their rights, or does it merely place Facebook access on the same level of humanitarianism?

All of the world's information, every opportunity, every fact, every business on earth. Such widely shared self-regard has made it seem embarrassing to claim more modest goals for one's business. A document sent out to members of Y Combinator, the industry's most sought-after start-up incubator, instructed would-be founders: "If it doesn't augment the human condition for a huge number of people in a meaningful way, it's not worth doing."

As long as we have the informational appetite, more data will always seem axiomatic—why wouldn't one collect more, compute more? It's the same absolutism found in the mantra "information wants to be free." We won't consider whether some types of data should be harder to find or whether the creation, preservation, and circulation of some data should be subjected to a moral calculus. Nor will we be able to ignore the data sitting in front of us; it would be irresponsible not to leverage it. If you think, as the CEO of ZestFinance does, that "all data is credit data, we just don't know how to use it yet," then why would you not incorporate anything—*anything*—you can get on consumers into your credit model? As a lender, you're in the business of risk management, precisely the field which, in David Lyon's view, is so well attuned to Big Data. The question of collecting and leveraging con-

---

* When I found this quote, Tseng's Twitter avatar was a photo of himself in front of an ad on the London Tube. The ad text read: "I think I'm being watched."

sumer data, and making decisions based on it, passes from the realm of ethics to a market imperative. It doesn't matter if you're forcing total transparency on a loan recipient while keeping your algorithm and data-collection practices totally hidden from view. It's good for the bottom line. You must protect your company. That's business.

This is how Big Data becomes a religion, and why, as long as you're telling yourself and your customers that you're against "evil," you can justify building the world's largest surveillance company. You're doing it for the customers. The data serves them. It adds relevance to their informational lives, and what could be more important than that?

Ethics, for one thing. A sense of social responsibility. A willingness to accept your own fallibility and that, just as you can create some pretty amazing, world-spanning technological systems, these systems might produce some negative outcomes, too—outcomes that can't be mitigated simply by collecting more data or refining search algorithms or instituting a few privacy controls.

Matt Waite, the programmer who built the mugshots site for the *Tampa Bay Times*, only later to stumble upon some complications, introduced some useful provocations for every software engineer. "The life of your data matters," he wrote. "You have to ask yourself, Is it useful forever? Does it become harmful after a set time?"

Not all data is equal, nor are the architectures we create to contain them. They have built-in capacities—"affordances" is the sociologist's term d'art—and reflect the biases of their creators. Someone who's never been arrested, who makes $125,000 a year, and who spends his days insulated in a large, suburban corporate campus would approach programming a mugshot site much differently from someone who grew up in an inner-city ghetto and has friends behind bars. Whatever his background, Waite's experience left him cognizant of these disparities, and led him to some important lessons. "What I want you to think about, before you write a line of code, is what does it mean to put your data on the Internet?" he said. "What could happen, good and bad? What should you do to be responsible about it?"

The problems around reputation, viral fame, micro-labor, the lapsing of journalism into a page-view-fueled horse race, the intrusiveness of digital advertising and unannounced data collection, the erosion of privacy as a societal value—nearly all would be improved, though not solved, if we found a way to stem our informational appetite, accepting that our hunger for more information has social, economic, and cultural consequences. The solutions to these challenges are familiar but no more easier to implement: regulate data brokers and pass legislation guarding against data-based discrimination; audit Internet giants' data collecting practices and hand down heavy fines, meaningful ones, for unlawful data collection; give users more information about where their data goes and how it's used to inform advertising; don't give municipal tax breaks (Twitter) or special privileges at local airfields (Google) to major corporations that can afford to pay their share of taxes and fees to the cities that have provided the infrastructure that ensures their success. Encrypt everything.

Consumers need to educate themselves about these industries and think about how their data might be used to their disadvantage. But the onus shouldn't lie there. We should be savvy enough, in this stage of late capitalism, to be skeptical of any corporate power that claims to be our friend or acting in our best interests. At the same time, the rhetoric and utility of today's personal technology is seductive. One doesn't want to think that a smartphone is also a surveillance device, monitoring our every movement and communication. Consumers are busy, overwhelmed, lacking the proper education, or simply unable to reckon with the pervasiveness of these systems. When your cell carrier offers you a heavily subsidized, top-of-the-line smartphone—a subsidy that comes from locking you into a two-year contract, throughout which they'll make a bundle off of your data production—then you take it. It's hard not to.

"If you try to keep up with this stuff in order to stop people from tracking you, you'd go out of your mind," Joseph Turow said. "It's very complicated and most people have a difficult time just getting their lives in order. It's not an easy thing, and not only that, many people when they get online, they just want to do what they do and get off."

He's right, except that, when we now think we're offline, we're not. The tracking and data production continues. And so we shouldn't be too hard on one another, particularly those less steeped in this world, when they suddenly raise fears over privacy or data collection. Saying "What did you expect?" is not the response of a wizened technology user. It's a cynical displacement of blame.

It's long past time for Silicon Valley to take an ethical turn. The industry's increasing reliance on bulk data collection and targeted advertising is at odds with its rhetoric of individual liberty and innovation. Adopting other business models is both an economic and ethical imperative; perpetual surveillance of customers cannot endure, much less continue growing at such a pace.

The tech companies might have to give up something in turn, starting with their obsession with scale. Katherine Losse, the former Facebook employee, again: "The engineering ideology of Facebook itself: Scaling and growth are everything, individuals and their experiences are secondary to what is necessary to maximize the system." Scale is its own self-sustaining project. It can't be sated, because it doesn't need to be. The writer and environmentalist Edward Abbey said, "Growth for the sake of growth is the ideology of the cancer cell." The informational appetite is cancerous; it devours.

When the CEO of Factual speaks of forming a repository of all of the world's facts, including all human beings' "genetic information, what they ate, when and where they exercised," he may indeed be on the path to changing the world, as he claims to be. But for whose benefit? If the world's information is measured, collected, computed, and privatized en masse, what is being improved, beyond the bottom lines of a coterie of venture capitalists and tech industry executives? What social, economic, or political problem can be solved by this kind of computation? Will Big Data end a war? Will it convince us to convert our economy to sustainable energy sources, even if it raises gas prices?

Will it stop right-wing politicians from gutting the social safety net? Which data-driven insight will convince voters and legislators to accept gay marriage or to stop imprisoning nonviolent drug offenders? For all their talk of changing the world, technology professionals rarely get into even this basic level of specificity. Perhaps it's because "changing the world" simply means creating a massive, rich company. These are small dreams. The dreamers "haven't even reached the level of hypocrisy," as the avuncular science fiction author Bruce Sterling told the assembled faithful at SXSW Interactive, the industry's premier festival, in March 2013. "You're stuck at the level of childish naïveté."

Adopting a populist stance, some commentators, such as Jaron Lanier, say that to escape the tyranny of a data-driven society, we must expect to be paid for our data. We should put a price on it. Never mind that this is to give into the logic of Big Data. Rather than trying to dismantle or reform the system—one which, as Lanier acknowledges in his book *Who Owns the Future?*, serves the oligarchic platform owners at the expense of customers—they wish to universalize it. We should all have accounts from which we can buy and sell our data, they say. Or companies will be required to pay us market rates for using our private information. We could get personal data accounts and start entertaining bids. Perhaps we'd join Miinome, "the first member-controlled, portable human genomics marketplace," where you can sell your genomic information and receive deals from retailers based on that same information. (Again, I think back to HSBC's "Your DNA will be your data" ad, this time recognizing it not as an attempt at imparting a vaguely inspirational, futuristic message, but as news of a world already here.) That beats working with 23andMe, right? That company already sells your genetic profile to third parties—and that's just in the course of the (controversial, non-FDA-compliant) testing they provide, for which they also charge you.

Tellingly, a version of this proposal for a data marketplace appears in the World Economic Forum (WEF) paper that announced data as the new oil. Who could be more taken with this idea than the technocrats of Davos? The paper, written in collaboration with Bain & Company, described a future in which "a person's data would be equivalent to their 'money.' It would reside in an account where it would be controlled, managed, exchanged and accounted for just like personal banking services operate today."

Given that practically everything we do now produces a digital record, this model would make all of human life part of one vast, automated dataveillance system. "Think of personal data as the digital record of 'everything a person makes and does online and in the world,'" the WEF says. The pervasiveness of such a system will only increase with the continued development and adoption of the "Internet of things"—Internet-connected, sensor-rich devices, from clothing to appliances to security cameras to transportation infrastructure. No social or behavioral act would be immune from the long arms of neoliberal capitalism. Because everything would be tracked, everything you do would be part of some economic exchange, benefiting a powerful corporation far more than you. This isn't emancipation through technology. It's the subordination of life, culture, and society to the cruel demands of the market and making economic freedom unavailable to average people, because just walking down the street produces data about them.

Signs of this future have emerged. A company called Datacoup has offered people $8 per month to share their personal information, which Datacoup says it strips of identifying details before selling it on to interested companies. Through its ScreenWise program, Google has offered customers the ability to receive $5 Amazon gift cards in exchange for installing what is essentially spy software. The piece of software, a browser extension, was only available for Google's own Chrome browser, which surely didn't hurt its market share. Setting its sights on a full range of a family's data, Google also offered a more expansive version of the program. Working with the market research firm GfK, Google began sending brochures presenting the opportunity to "be part of an exciting and very important new research study." Under this program, Google would give you a new router and install monitoring software on anything with a WiFi connection, even Blu-Ray players. The more you added, the more money you'd make. But it doesn't add up to much. One recipient estimated that he could earn $945 from connecting all of his family's devices for one year. A small sum for surrendering a year of privacy, for accepting a year of total surveillance. It is, however, more than most of us are getting now for much the same treatment.

Making consumers complicit—yet far from partners—in the data trade would only increase these inequities. We would experience data-based surveillance in every aspect of our lives (a dream for intelligence agencies). Such a scheme would also be open to manipulation or to the kind of desperate micro-labor and data accumulation exemplified by Mechanical Turk and other online labor marketplaces. You'd wonder if your friend's frequent Facebook posts actually represented incidents from his life or whether he was simply trying to be a good, profitable data producer. If we were to lose our jobs, we might not go on welfare, if it's even still available; we'd accept a dole in the form of more free devices and software from big corporations to harvest our data, or more acutely personal data, at very low rates. A gray market of bots and data-generating services would appear, allowing you to pay for automated data producers or poorly paid, occasionally blocked ones working, like World of Warcraft gold miners, in crowded apartments in some Chinese industrial center.

Those already better off, better educated, or technically adept would have the most time, knowledge, and resources to leverage this system. In another paper on the data trade, the WEF (this time working with the Boston Consulting Group) spoke of the opportunity for services to develop to manage users' relations with data collectors and brokers—something on the order of financial advisors, real estate agents, and insurers. Risk management and insurance professionals should thrive in such a market, as people begin to take out reputational and data-related insurance; both fields depend as much on the perception of insecurity as the reality. Banks could financialize data and data-insurance policies, creating complex derivatives, securities, and other baroque financial instruments. The poor, and those without the ability to use the tools that'd make them good data producers, would lose out. Income inequality, already driven in part by the collapse of industrial economies and the rise of postindustrial, post-employment information economies, would increase ever more.

This situation won't be completely remedied by more aggressive regulation, consumer protections, and eliminating tax breaks. Increasing automation, fueled by this boom in data collection and mining, may lead to systemic unemployment of a kind we've never seen. Those contingent workers laboring for tech companies through Elance or Mechanical Turk will soon enough be replaced by automated systems. It's clear that, except for an elite class of managers, engineers, and executives, human labor is seen as a problem that technology can solve. In the meantime, those whose sweat this industry still relies upon find themselves submitting to exploitative conditions, whether as a Foxconn worker in Shenzhen or a Postmates courier in San Francisco. As one Uber driver complained to a reporter: "We have a real person performing a function, not a Google automatic car. We have become the functional end of the app." It might not be long before he is traded in for a self-driving car. They don't need breaks, they don't worry about safety conditions or unions, they don't complain about wages. Compared to a human being, automatic cars are perfectly efficient.

And who will employ him then? Who will be interested in someone who's spent a few years bouncing between gray-market transportation facilitation services, distributed labor markets, and other hazy digital makework? He will have no experience, no connections, and little accrued knowledge. He will have lapsed from subsistence farming in the data fields to something worse and more desultory—a superfluous machine.

Automation, then, should ensure that power, data, and money

continue to accrue in tandem. Through ubiquitous surveillance and data collection, now stretching from computers to cell phones to thermostats to cars to public spaces, a handful of large companies have successfully socialized our data production on their behalves. We need some redistribution of resources, which ultimately means a redistribution of power and authority. A universal basic income, paid for in part by taxes and fees levied on the companies making fabulous profits out of the quotidian materials of our lives, would help to reintroduce some fairness into our technologized economy. It's an idea that's had support from diverse corners—liberals and leftists often cite it as a pragmatic response to widespread inequality, while some conservatives and libertarians see it as an improvement over an imperfect welfare system. As the number of long-term unemployed, contingent, and gig workers increases, a universal basic income would restore some equity to the system. It would also make the supposed freedom of those TaskRabbit jobs actually mean something, for the laborer would know that even if the company cares little for his welfare or ability to make a living, someone else does and is providing the resources to make sure that economic precarity doesn't turn into something more dire.

These kinds of policies would help us to begin to wean ourselves off of our informational appetite. They would also foment a necessary cultural transformation, one in which our collective attitude toward our technological infrastructure would shift away from blind wonder and toward a learned skepticism. I'm optimistic that we can get there. But to accomplish this, we'll have to start supporting and promoting the few souls who are already doing something about our corrupted informational economy. Through art, code, activism, or simply a prankish disregard for convention, the rebellion against persistent surveillance, dehumanizing data collection, and a society administered by black-box algorithms has already begun.

Social-Media Rebellion

I'm Nobody! Who are you?
Are you—Nobody—too?
Then there's a pair of us!
Dont tell! they'd advertise—you know!

How dreary—to be—Somebody!
How public—like a Frog—
To tell one's name—the livelong June—
To an admiring Bog!

—Emily Dickinson

David Roberts was tired. For almost a decade, Roberts, a journalist who wrote about climate change for the environmental Web site *Grist*, had been following some version of the same routine. His day started with his wife and kids leaving the house early in the morning. By 9:00 a.m., he was on the computer and usually stayed there—a short break to walk his dog being the main exception—through the late afternoon. (He ate lunch at his desk.) Around 6:00 p.m., his wife and kids would return home, and they'd eat dinner together. After putting the kids to bed, he and his wife would take some time together, maybe watching TV until 10:30 or eleven. That's when Roberts's wife went to bed and his workday continued. He'd get back on the computer and write until one or two in the morning.

"I was getting very little sleep," Roberts told me. "I was going to sleep at two or three and getting up at like eight, which is awful, awful."

The schedule was eating into him. He was tired, physically and mentally. He didn't have the opportunity to exercise or travel. He was kicking around an idea for a novel, but his work commitments made it impossible to get started. Plus, his beat was climate change— about which you could say that not much was changing, except that everything was getting worse. And each day he had to repeat it again, spending eleven-plus hours at his computer and much of the rest of his waking hours peeking at his iPhone.

It wasn't all bad. He loved his colleagues, though he mostly communicated with them via e-mail, and he appreciated that he was free to write about the subject mostly as he wished. He could work from home and see his family, and he was writing about an important topic, perhaps the central issue of our time. He was respected as a climate change expert and had 37,000 Twitter followers, and he loved being able to turn to them for conversation, to answer questions, or just idle chatter: "It was like this really erudite, diverse cocktail party that was always going on and that I could always wander into and just get into a chat." But still the routine, the never-ending labor of the contemporary knowledge worker, got to him. After almost ten years of this, he wasn't sure how much more he could handle.

Part of that was the onslaught of information with which he had to deal. "There's so much coming in all the time," he said. "By the time I got to my computer, there's this mass of stuff waiting." He'd dispatch that in time, but this kind of informational busywork—e-mail, social media, keeping up with mailing lists and news sources, talking with his colleagues—would crop up constantly throughout the day. Roberts described it as "running in place."

Earlier in his career, he'd write seven to eight blog posts a day. That had changed in recent years. Many of those posts, which tended to be short, would become tweets—"look at this thing" messages, he called them. At the same time, he began writing far longer posts that would go up once a day or every other day. He could only manage to get this writing done late at night, after his family was asleep, along with many of his colleagues and usual correspondents. The flood of information winnowed to a gentle eddy, and he was free to write.

Roberts had no quotas about how much he had to write, tweet, or otherwise produce—a great boon for a journalist these days. But he felt some sense of guilt, or at least that murky unease which signaled that there might be more to do.

"There was never a point at which I was done," he said. "It was always there—hanging." That may not be a feature unique to journalism, he acknowledged, but it was arduous all the same.

He recalled a "constant sense of being behind and being obliged to sit down and crank something out. It felt after a while like I was just kind of going to the mines to break rock. I just wanted to have some time to sit and think."

His descriptions matched uncannily with my own feelings and those of many people with whom I've spoken—people whose day jobs essentially require them to handle information of one sort or another, and the unemployed or intermittently employed who hope to join their ranks. My situation differs from Roberts's in a few particulars, but as a freelance journalist, one who theoretically could always be working, finding new assignments, searching for stories, tracking invoices, or simply building up social capital and connections on social media, it can seem as if there is no distinction between work and not-work. There is always more information to consume, more work to be done, more pitches to write, or colleagues' work to catch up on. Especially when one's income is variable and contingent, it seems that the knowledge worker's job is never done.

More challenging is the sense of guilt and nagging anxiety that can come with putting the devices away and trying to carve out time to be alone or with friends. Like Roberts's wife, my partner works mostly conventional hours at an office. When she comes home, we want to see each other, compare notes on the day, clean up, cook, read together, watch *The X-Files*—all those shared tasks and felicities that constitute domestic life. But inevitably, work, the network, pervades the home. Sometimes, because of the peculiarities of my day—if I used the afternoon for meetings or errands or simply frittered it away on social media, e-mails, aimless reading, or some other trifle—I'm just getting started at actually putting words to paper. Naturally, many people have it worse, but it's hard all the same to build a kind of order into one's life.

This experience speaks to an important, but under-valued, shift that's taken place over the last couple of decades. The same devices we use for work are now those we use for entertainment, education, communication, calendar-keeping, and so much else. The distinctions between work and not-work are becoming muddled in part because the devices themselves—smartphones, tablets, laptops—are used for both. A cell phone game can just as easily be interrupted by an e-mail alert that must be attended to; after sorting through some messages and links, one might find that thirty or forty minutes have somehow passed, like sand through your fingers. And, as is vividly shown with digital labor markets and the digital serfdom that is the cost of being on social media, what appears to be entertainment is, in fact, a kind of work. The blurriness between the two is mostly by design, and it's one with a healthy lineage, from every HR-led happiness campaign to the tricked-out campuses of tech companies, where toys, bikes, free food, and other amenities are designed to inculcate an atmosphere of fun and abandon. (If someone has ever told you "try not to let it feel like work," it's probably because he was giving you work.)

As he approached his tenth anniversary at his job, Roberts decided to do something. He spoke with his wife, and they decided that they could afford for him to take some time off. They'd have to cut some expenses and change their lifestyle, but it could be done. Then he talked to his employer, and they agreed that he could take a year off, unpaid. He intended, and would be welcome, to return, but of course his mind might change after twelve months away from the desk. Next he wrote up a long essay about his work, his experience at *Grist*, what it felt like to be always connected, and why he had to do something about it now. He said he would be gone for a year and would use his computer a little bit—he wasn't giving up the Internet entirely—but he needed time to work on his own projects, to get in better shape before he was over the hump of middle age, and to be with his family.

Roberts had resources that some people don't. He acknowledged to me "the extraordinary privilege of all this." Most people can't afford to quit their jobs, much less take a year off with the assurance that they can return at the end of a sabbatical. Some people would inevitably call his decision self-absorbed. And he was addressing the kinds of tangled questions—Is social media somehow addictive? When should we disconnect? Should we blame our troubles on the tools, the tool-makers, ourselves, or something else entirely?—that cause sociologists and cultural critics to nudge their spectacles up their noses and start declaiming.

It wasn't a surprise, then, that Roberts's essay would land with the kind of small tremor that can shake a corner of the culture and cause a flurry of responses in the form of think pieces, social-media debates, and curt dismissals. (I encountered the essay when it was being heatedly debated by a few writers I follow on Twitter.) He had grasped onto something real and urgent, even if the terms of debate were still being puzzled out.

In recent years, a contentious discourse has appeared around the notion of the Internet as a place of addiction, the only cure for which is a detox. Cold turkey. I think there's more truth than fiction in the addiction metaphors used to describe Internet technologies; they are, after all, designed for "stickiness," to keep users around for as long as possible (and when they are gone, to get them to return). But there's a great deal that's problematic with this line of thinking, starting with the monolithic treatment of "the Internet." The term is a useful short-hand for a set of related tools, technologies, and digital spaces, and so I often turn to it, in the same way that a range of Web sites, apps, and services fall under the moniker of "social media." But the Internet has, over the years, become something more generalized and deeply embedded in our lives, far more like a utility than a discrete place or piece of equipment. (That is also what makes these tools and services so difficult to forsake.) Digital detox, then, as this release from Internet addiction is often called, fails as an idea because it treats "the Internet" as one thing with a limited set of purposes. Digital detoxers rarely talk about quitting one Web site or changing some of their practices or protesting a company's behavior; they're ready to slag it all. To them, the Internet and all of its constituent technologies are the problem, and they're happy to leave it all behind, and, just as often, to pay handsomely for help in doing so: digital detox is a lucrative field for some, with guidebooks and weekend retreats promising a return to the supposed purity of a pre-digital life.

That brings us to the second problem with digital detox: it caters to a wealthier class of Internet users, the kind of people who can afford forsaking the Internet entirely. Never mind its many uses for doing white-collar jobs, finding work, communicating with loved ones, reading articles, and watching TV shows (one wonders if, as TV and films are increasingly delivered to us as streaming media, digital detoxers will give up on these, too, out of principle). Armed with their DSM–5s, raised on metaphors of addiction, these disaffected Internet users are ready to pathologize their basic communication habits. They're sick, in a sense, though the responsible party may be themselves or a distant Internet overlord. And because they're addicted, they require a dramatic break with their old lives, which inevitably leads to quitting the Internet in the most dramatic, public way possible—blog posts, Facebook announcements, alerting their friends to their new life choices, giving away gadgets and forking over passwords to spouses. There's a rich vein of irony here, as many digital detoxers claim that they are looking to escape the over-sharing and endless daily banalities common on Facebook, Twitter, Instagram, et al. By publicly complaining about their untenable Internet lives and their plan to escape it all—and by often doing so on social media, no less—they are maintaining the very kinds of habits they seek to change. They are performing their discontent.

For journalists, quitting-the-Internet essays have become a tired genre, although many of them contain more insights than they're often granted. The tech Web site *The Verge* had an employee quit for an entire year and then write about his experience. Media theorist Douglas Rushkoff and *New Yorker* journalist Steve Coll wrote articles about quitting Facebook. It's a custom that's launched a thousand trend pieces, some of them charting detoxers' newly happy analog lives, others revealing their quick relapses or decision to set up shop on other social networks.

To those of you who have read these and similar articles, you may be equally familiar with some of the critiques I just raised. But here's where I part ways with many critics of digital detoxers, even as I find much to dislike about digital detox. There's a very real need reflected in the practice of digital detox and in writing about it, even if the latter is often a contradictory, or merely excruciatingly self-conscious, play for attention. A widely shared sense exists that our digital tools have somehow become the masters of us. It can feel as if we don't have control over our digital lives—over our data, our privacy, our self-control, our ability to step away from the screen for a while. As we become more educated about the practices of social-media firms, perhaps we'll become more sophisticated and pointed in our thinking. We'll learn, as David Roberts did, that there's nothing *wrong* with quitting Twitter, nor with airing that desire in writing, and the reflexive backlash that such a move provokes is evidence that some kind of reform is needed in the culture. Content to graze in the seemingly boundless pasture of digital life, we've become strangely intolerant of those who say they'd be happier elsewhere. We see the mark of Luddism hovering over even the most considered critiques. Only elderly artists such as Ray Bradbury ("It's distracting. It's meaningless; it's not real. It's in the air somewhere.") are granted the privilege of being house heretics. His rejectionism was considered earned by virtue of his position, but mostly it was cute and nonthreatening, the charming fury of a contrarian well into senescence.

But Roberts's essay was different than most of the detox laments—certainly more thoughtful and encompassing than the precious reports from detox retreats that appear in the *New York Times*'s Style section. And my conversation with him affirmed that while the problems he described had some roots in technology, they were not solely technological; the same went for the solutions. In his essay, he noted that he wasn't giving up Internet use entirely; there's something silly and unnecessary about that. It seeks a kind of purity that's both unhealthful and impractical. But as was his choice—and, it seemed, his luxury—he decided to change his habits, to spend a year away from the relentless pace of digital journalism, wherein every small outrage builds to something cacophonous, until the next pseudo-event intervenes. That cycle was particularly exhausting for an environmental journalist used to covering the simultaneous advancement of environmental destruction and our political leadership's paralysis in doing anything about it. Roberts's decision to pull back from this world was a reflection of his beat and the state of journalism as a profession, with its demands to be always on, always connected, and always reacting to the latest press release or micro-scoop or faux scandal to come down the pipe. That this style of journalism is deeply connected to the architectures of digital life and social media meant that it was easy to see Roberts's essay as just another entry in the digital detox canon. But for anyone who has found his or her work life overwhelming, especially in an age when digital technologies have helped to erase the barriers between work and home life, Roberts's decision was more understandable. He wasn't quitting the Internet. He was quitting, for a time, his job and the tools and lifestyle that went with it. A fellow journalist, like me, might see this as a reflection on the state of the industry. A Marxist might see it as a commentary on the structural inequalities of the so-called knowledge economy. A Twitter executive might see it as the loss of a good customer—the accounts of prolific media figures are, after all, good vectors for advertising. But it wasn't digital detox, even if it had some of the hallmarks of it.

As the Internet becomes more like a utility, more of a background element in our lives, connecting us with different technologies, people, and services, it's important to make these distinctions. It's what allows all of us to be good critical consumers, to speak specifically about Facebook's odious privacy practices rather than just "Facebook," "social media," or "the Internet." Of course, at times, we'll fall back on these familiar terms. We like metaphors and handy references. And we may even consider ourselves addicted. Working on this book, I, idly and frequently without reason, switched over to my browser and went to Twitter, Tumblr, or Facebook, looked around for a minute or ten, and then closed the tab. I did this too many times to count; surely, there's an app that could tell me, but I don't want to know. The gesture was routinized in me. I did it usually without thinking, except to berate myself later for feeling like responding to the latest outrage on Twitter was more important than my work. ("I think in tweets now," Roberts wrote. "I feel the need to comment on everything, to have a 'take,' preferably a 'smart take.'") Likely, my behavior was a combination of instinct, habit, and reflex, but I've also spent hours in bed or in line at the grocery store scrolling through my feeds not in search of anything but just because it was *there*, holding out the promise of something new that seemed, by default, more important than whatever I was doing. This promise of news from elsewhere was more important than talking to someone else next to me in line, or finally closing my eyes to try to sleep, or just spending some time daydreaming, trying to be bored. I've forgotten how to be bored.

Is that addiction? I don't know. But I do know that the issues surrounding social media are far more significant, and communal, than to be spoken of in the single pathologized language of addiction. Every time I clicked away from this book to check a social-media feed was a small failure. It was also an instance in which I searched for entertainment, connection, friendship, passion, something interesting. I did some good research there, accidentally at times. It was a moment in which whatever tracking systems that had their eyes on me were able to add a few more data points to their lode. I was fidgety and alone, perhaps, sometimes lonely, but I was also networked into the circulating systems of data, tracking, surveillance, communication, cultural exchange, identity formation, and advertising that undergird nearly all social-media platforms. Perhaps I was addicted, yet I was all of these other things too—all of which are, in my view, far more important.

This is why, when it comes to social media, it's hard to talk about opting out. Partly, it's that the form is relatively new and ever-changing; we haven't developed a shared vernacular. But it's the communitarian feeling of social media as well: if you're opting out, you're opting out of communicating with me, too. I need your broadcasts for my timeline to feel richer, or at least to feel like I have an audience. I want to know what's happening in your life; "keep me in the loop" is a literal request here. There's also an implied arrogance in digital detox: Why are you too good for this?

When our sense of ourselves depends on being seen, on being visible and circulating through the network, then when someone chooses to opt out, the whole enterprise can be called into question. Writing in the late seventies, Christopher Lasch noted that ordinary people now face "an escalating cycle of self-consciousness—a sense of the self as a performer under the constant scrutiny of friends and strangers." He might as well have been writing in 2014. That self-consciousness is already difficult enough to negotiate, particularly in a society suffused by so many forms of surveillance, peer judgments, and idealized forms sold to us through pop culture. If we depend on others for validation—for retweets, likes, and hearts to show that our broadcasts are good enough—then our very sense of ourselves is made unstable. We become even more conscious of these meta-level concerns. We seek out more material—news articles to comment upon, photos to post, friends to interact with—because it is only by engaging with this stuff that we can find a way forward. We chew through content because our consumption of something is both a way of expressing our identity and a way of feeling like we've accomplished or learned something. In Lasch's view, for "the performing self, the only reality is the identity he can construct out of materials furnished by advertising and mass culture, themes of popular film and fiction, and fragments torn from a vast range of cultural traditions." Today we are still magpies, always on the hunt, never finished with the process of building ourselves out of the memes, videos, articles, and soundbites circulating around us.

When I spoke to Roberts, he was six months into his break. "It's been great," he said. He joined a gym, lost weight, went on several trips. "It's pleasant to not have things hanging over your head," he said with some relief. But it wasn't all relaxation and self-discovery. With his wife the sole breadwinner, his family came to rely on him more for chores and support and also just simply being there, which he was happy to oblige. He rarely spent more than a couple of hours a day on the computer, most of that time taken up with e-mail and working on a novel. The smartphone retained its strong lure: "I have not cured that," he said. Words with Friends and RSS feeds were still temptations, but he wasn't posting on social media.

He also wasn't thinking about it in the same way. Since he left his job, he noticed a change in how he approached the world. If he saw something remarkable—encountering a beautiful sunset while on a hike, or laughing at a funny comment by one of his kids—a familiar thought process kicked in. "Some part in the back of my mind is like, well, I can't tell somebody about this," he said. "So is it even real? What good is it? What good is this experience if I can't tell people about it? Which is kind of perverse if you think about it."

It was that old self-consciousness, that magpie identity, reasserting itself. The Facebook Eye combined with Lasch's self-consciousness— here was a beautiful world before him and it was hard to enjoy it alone. He could pass on reports about what he'd seen to his wife, his kids, some friends later, but there was no possibility for that instant communion with a distant-but-present digital audience. Some ability to shape his sense of self, to do it publicly in the way so many of us have become accustomed to, was lost. How strange that the twin feelings of being connected and the documentary impulse have come to feel like important, even essential, aspects of daily experience. They reify what we've seen and give us the opportunity to present it for approval, for vicarious enjoyment, or, if not out of sheer habit, then for other reasons that we don't quite understand. Yet all this comes, as we've since learned, at a price.

"Being online and being on social media is a constant exercise in identity construction," Roberts said. "You're constantly defining yourself: I'm this kind of person, this is what I like, these are the jokes I make." Roberts enjoyed that at times, but it's not what he wanted to feel *all* the time. He wanted to engage with the world, but also to allow it to show itself. He wanted to be present in it, without thinking of how he could package it for a tweet. He spoke of reading Emerson and consulting with a scholar of Buddhist mysticism. "That kind of experience"—contemplative thinking—"is the opposite of what you're doing online," he said. "What you're doing online is all ego, and that's not ego. I think we're losing the cultural value of that, and we're losing our capacity for it. You're constantly thinking, *How can I use this experience as part of this identity construction I'm doing?* You never just take things in."

I understood what he meant. I had felt the same impulses when I started this project and sometimes felt overwhelmed by them. Though instead of Emerson or Buddhism, or the consolations of mindfulness, I found some philosophical ballast in Heidegger, Husserl, and other thinkers who emphasized that there is no division between us and the world. That we must let the world reveal itself to us. Otherwise, we risk making things into objects, rather than accepting them as beings of their own. Unfortunately, it so often seems that we're guilty of this. We can't just look at the world as it is—appreciating it, noticing its odd details and everyday miracles—without making our presence known and commenting upon it. I worried that my constant attention toward digital media was returning some of this dualism to my world. It wasn't solely responsible for it, but the sense that my online identity and relationships were so constructed—objects assembled out of my ego, out of impulse and vanity and desperation—left me feeling anxious and inauthentic. I felt the need for my work as a journalist, as a creator of the very kind of media people around me were always circulating, to be validated. And each tweet or Facebook post could, at my most self-conscious moments, seem like a referendum on whether I was interesting or funny enough or presenting myself in the way I hoped to be received. It was hard to decide whether to post something to shore up my sense of self or to beat a retreat and delete as much as I could.

Lasch knew all this almost five decades ago. The great sociologist wrote: "Awareness commenting on awareness creates an escalating cycle of self-consciousness that inhibits spontaneity. It intensifies the feeling of inauthenticity that rises in the first place out of resentment against the meaningless roles prescribed by modern industry. Self-created roles become as constraining as the social roles from which they are meant to provide ironic detachment. We long for the suspension of self-consciousness, of the pseudoanalytic attitude that has become second nature."

We long for the suspension of self-consciousness—yes, that is it. But there's also something painfully deliberate and artificial about this, isn't there? Why can't the people Lasch is writing about just *be*? And yet, this is exactly how our media exists today. Its tendencies and biases have only become amplified, the metrics and hierarchies made more quantified and visible. It is a series of objects—widgets—here to inform and serve and shape us, which is what makes it so existentially vexing. Where do I end and where does the mediated world begin? Can I come to terms with my self-consciousness, if not modulating it, then learning to appreciate it?

This neuroticism is also one of the characteristic feelings of modern consumerism. It's precisely these anxieties that advertising—the meal ticket of every social network of consequence—has long tried to leverage. As Roberts said to me, "As soon as you're ad-based, attention is your currency. You're not trying to improve your customers' lives. You're trying to get them to look at you as often as possible, and so you're fated to be distracted and annoying." This model recurs throughout the history of mass media, from radio to TV to the Internet. Now it's woven into the very fabric of our informational economy and our identities to boot.

Important differences lie in the medium's architecture and the peculiar culture surrounding it. But when you learn how curiosity-gap headlines manipulate readers into clicking, or that SEO pollutes search results with bad content, or that Facebook's feed is designed to be, according to Zuckerberg's vision, "like television, a compelling, satisfying flow that doesn't make demands on you," or that every action you take on the network is tracked and fed into databases that megacorporations mine for value—well, it can all seem rather absurd. The machine is rigged against our better interests.

To top it off, the wizened social-media user has seen it all before. Every joke has been uttered, every hashtag exhausted. Spend enough time with digital media, and it becomes a familiar, enervating routine. The feed scrolls on; only the time stamps change. There's a reason why we call it the news cycle—it repeats, ad infinitum. We know how it works, we know all the tropes, we know every meme (well maybe not *this* meme but one just like it), we know about the outrage cycle and how politicians and celebrities use social media to generate press coverage. We know that all of this Sturm und Drang is secondary toward the great gods of advertising, who fund everything. And yet we seem to have no capacity to change it. As Thomas de Zengotita says, "We are all method actors now," acting out our familiar roles.

We do have a few choices. You can accept this situation as it is. Perhaps you'll escape the "pseudoanalytic attitude," try to be less snarky or willfully clever, and vow to be earnest and sincere. You might decide you don't care anymore, brush off self-consciousness, and start being a little crazy on social media. You could find refuge in anonymity. You could write a book or try for some more radical critique.

Or, finally, you could quit. Blow it all up. Delete what data you can, close your accounts. Install the Tor browser. Ditch your iPhone. You don't need to go detox in the country. You won't win any great victory except to claw back a little privacy and interiority for yourself. (In a milieu so obsessed with questions of narcissism, it's fitting that by turning away we can recover some part of ourselves.) It's a Pyrrhic victory through and through. These systems are so pervasive that you're bound to be caught in the net somehow. There's loss here, no doubt, and you'd have to work for it. But it's a choice, and it might be all the better for not being a popular one.

David Roberts chose a version of this. He opted out and accepted the consequences. When I spoke to him, halfway into his sabbatical, he still planned to return to his job; he believed in the work and missed talking to his colleagues, many of whom were good friends. His rebellion would be a temporary one, but he hoped to make some changes in his habits. A new routine would be his refuge. Still, he couldn't help but worry: "When I go back to standing at my computer all day, am I going to be any more equipped to have a healthy relationship to the technology and to the work than I was before?"

- - - - - - - - - - - - - - - - -

CLOGGING THE NETWORK

There's no shame in opting out of social media, even as that prospect becomes increasingly untenable for many, but it's worth seeing what kinds of protests we can launch from *within* the network and whether these platforms can be sites of resistance and change. If we want to reform this culture, then we need to be, as one critic wrote, where "the majority of people live and communicate . . . trying to devise *conflictual* ways of using the existing networks." The digital may be novel, but rather than being a separate space, it is increasingly intertwined with our lived reality. These technologies touch all of us. We might as well challenge them together.

Some people have taken up that call. But one might question what counts as resistance here, because it comes in unexpected forms. For instance, every day, people log onto Twitter and announce, falsely, the death of somebody famous. It might be a simple message, or it might be a link to an obscure, just-off-the-presses news item, indicating that Nicolas Cage died snowboarding in Switzerland or Jeff Goldblum rode a Segway off a cliff. Some of them originate with a site called Global Associated News, which allows users to generate their own celebrity death stories, featuring methods of demise one might expect for the global rich (car crash in Australia, Jet Ski accident in Turks and Caicos, yacht sunk off of Saint-Tropez). It doesn't matter that these stories are patently untrue—Global Associated News emphasizes that it's a parody site—or that Internet users should be used to hoaxes, unlikely rumors, and other false reports. While these shared items often die on the vine, sometimes they receive a few retweets or shocked replies, because the urgency of responding wins out over the impulse to verify. Or, in rare cases, one of these reports burbles up through the blogosphere and eventually appears as a mainstream news item, if only so an overworked blogger can satisfy her daily quota by writing about a hoax that got some attention on social media. In most cases, though, they remain under the roiling surface of digital opinion, to be stumbled upon accidentally or through an intentional search. These items aren't great pranks—we're mostly used to this nonsense, and even those who are fooled are rarely impressed at being so gullible. There's no real humor here. But in their wry disregard for convention, these messages do represent a form of rebellion. They're users throwing a wrench into the network by intentionally spreading misinformation. And there's meaning to this.

As Liel Leibovitz, a media scholar at NYU who studied the phenomenon, describes it, celebrity death hoaxes on social media are "a form of labor dispute masquerading as good fun. As the apps set up to calculate the value of your Twitter account by impressions and followers—frequently referred to around the time of the IPO—suggest, users are aware of the fact that the platform's value is derived entirely from their labor, for which they're unpaid. The death hoax is a reaction to that, however unconscious: by pursuing this seemingly harmless prank, users are devaluating the service's main currency, information, by diluting it with highly visible rubbish."

Leibovitz's analysis gets at the heart of what it means to be a citizen of these networks. We are all, by virtue of signing up, members. Citizenship is something more—certainly not an ideal state, but more like an awareness of what it means to be beholden to a social-media company, with its baroque terms of service agreements and ownership of our digital identities. It's a sensitivity to class and to the political economy of digital media—its basis in surveillance, its reliance on user-supplied data, its erosion of privacy. With citizenship comes a growing sense that we are largely disenfranchised, that digital serfdom means not just producing data but producing data and content that is *true*. If we withhold information from the platform owner, or better yet, if we add disinformation to the mix and try to get it to spread through the usual viral channels, then we are acting against the platform owner's desires. This isn't violent insurgency, nor is it street protests on behalf of civil rights. The goals here may not be lofty, much less articulated. But these "willful acts of obfuscation," as Leibovitz calls them, say a lot about the standards that platforms expect from us and how we might subvert them. They're less about remaking the platform than using their own built-in capacities to illustrate their structural inequities and deficiencies. In a world where frictionless sharing is an ideal, this kind of sharing reintroduces friction; it throws off some sparks on its way through the data stream.

DEVALUE YOUR DATA

You could probably tick off a few of the typical social-media avatars: a smiling photo from vacation or a party; a snapshot of a person alongside friends or a partner; a yellowed middle-school yearbook relic; a poster for a play you're in or a political cause you support, perhaps one related to a recent catastrophe (Donate to the Red Cross, We're All Virginia Tech). What about a picture of a monkey or a solid black box? Less common, yes, but maybe not atypical—people throw up something weird as their profile photos all the time. There's something interesting going on here, though, an impulse we don't often consider. These people are deliberately trying not to be seen. Why? Maybe they don't like how they look, or they don't want their photos online, or they just don't like the idea of being present in this fixed form at all times. It could seem strange, if it weren't so common now, to have this digital self always out there, standing in for you, available to whoever has the proper permissions and wants to take a look. Why should it be there? Some young Facebook users, looking to protect their privacy and retain this kind of control, de-activate their Facebook accounts after every session, reactivating it with each subsequent log-in. A related practice, called "whitewalling," entails erasing all wall posts and messages after they've been read.

These photos of something other than one's self are the first way in which many people begin to question the network's expectations. A false avatar is a small example of defiance. It says that you don't care. You don't need to be seen or to impress anyone. And if you leave that avatar up for a long time, your friends and followers begin to pick up a feeling of prolonged absence. You are away from the network, doing more interesting things, and while you may still post, you're less present in your posts, because they're not accompanied by that photo of you. It's the picture of your favorite quarterback, or an obscure impressionist painting, or a Gila monster. You are being less than visible and transparent, because you prefer not to. Your Bigfoot avatar may appear in a Google+ or Facebook ad, but it will be worth less because your face won't be there to help sell the product. And perhaps more important, you're being a less valuable user by not providing a photo that the platform owner can use to improve its facial-recognition software.

So let's consider what else we can do to make the networks worth less. That is, worth less to the oligarchic platform owners, but not worth less to us—in terms of culture, entertainment, identity play, security. Let's start by attacking that great digital-age shibboleth: data.

First, don't structure data. Unless you want to, of course. Unless you really want to be part of that hashtag—assuming a corporation hasn't injected itself into it yet by tweeting bad jokes or buying up all related ads—or you think Facebook needs to know exactly what you're doing and who you're with. But really, there's not much reason to do this. Don't do the platform's job. Don't append your location to tweets, don't check in anywhere, don't tag brands and businesses. Disable location services, disable notifications (you'll be more relaxed as a consequence). Take a look at apps you've authorized and remove the ones you don't use or that look suspicious.

Second, start lying about yourself. Give the networks false information, particularly about where you live and what you do. Put weird characters in your name, invent a middle name. Use a pseudonym if you want, until the network's administrators flag you as one of those no-fun, inauthentic fakes. Retweet bullshit (except, it's prudent to say, during times of crisis and confusion such as the Boston Marathon bombing). Start experimenting and breaking with convention. Don't be a jerk, but have fun with it.

I'm speaking aspirationally here, as I'm probably, by all measures, a hypocrite. I abide by many social-media conventions. I punctuate properly and give hat tips and other forms of credit. I respond to most people who tweet at me, and I have an irritating habit of checking the Connect tab on Twitter and feeling a flush of anxiety when my tweets receive no reply. It's a bit wearying and dull, being polite and outraged and a curious consumer as the day's news cycle dictates. Am I speaking independently, interacting with networks in the way I choose, or am I allowing my impulses to be guided by larger forces? All of this, this image of the networked individual as buttoned-up, restrained by programming and social convention alike, is what boosts the appeal of Weird Twitter, bots, and historical tweets (@1814now and @RealTimeWWII are among my favorites). None of them is truly avant-garde, but they show us something different. They allow for a breaking open of the staid digital space and the introduction of something unexpected. That is also why Twitter users tend to go nuts when a corporate account goes off-message and starts tweeting gibberish or tries to use the anniversary of Pearl Harbor to promote canned spaghetti. The mask is lifted and suddenly a real person, perhaps one only minutes away from losing his job, is revealed. Like when a newscaster who suddenly loses track of the teleprompter and begins ad-libbing, we feel ourselves witness to a human moment. When everything feels scripted, these moments count for something.

For the uninitiated, Weird Twitter is a small subculture featuring bizarre jokes, non sequiturs, intentionally crappy graphic design, meta-commentary on the medium itself, chaotic spelling and capitalization, and a general sense that these people, if not quite insane, are doing an admirable job of mimicking insanity. Few would wave the Weird Twitter flag proudly—it's a bit uncool to do so, and the term itself is almost passé—but the label is a good shorthand for a loose collection of people who, as one friend of mine described it, perform a kind of "dadaism, with less art and more snark." These accounts are at their best where they run into "normal" Twitter, when someone trying to debate the contents of a Walmart-promoted tweet encounters someone with a meat-loaf avatar claiming that "mr wallmart" is a three-hundred-year-old man who lives in his basement. It's strange and disorienting, like seeing a person trying to give a sincere lesson on home cooking to a dog. They're different species, and they speak different languages. But in a medium so transactional, it can be exciting to see someone throw off the straitjacket of convention. Weird Twitter users are also making themselves less valuable to the platform; it's hard to sell ads against this kind of madcap persona or to push products at people who will respond, if at all, by speaking in tongues. I don't think that the Weird Twitter ethos is an inherently anticorporate one—one self-identified Weird Twitter user told me that his participation in a *Wall Street Journal* article, which identified his coterie as a group of satirists using "insincere engagement" to push back against encroaching advertisements, was mostly a prank, a way to create a deliberately false impression of a group that doesn't really care enough to stand for anything. In this gleeful nihilism, though, is a kernel of anticorporate dissent. But they're not upset about what Twitter has become; they just don't care, and so, operating from this pose of radical indifference, they try to have the best time they can by operating on their own incomprehensible terms. That makes them, at least as advertising vectors, mostly worthless to Twitter Inc.

Praised be the hoaxers and the malcontents of social media, then, for they show us our self-seriousness, our submission to one another's expectations and those of the platform owners. I find myself turning to these types of accounts regularly, as if there's some message lying in wait. They, along with the digital artists and the outcasts, can tell us something about our behaviors that have quickly ossified into dim routine. Glitchr is one of my favorites. Run by Laimonas Zakas, an artist from Lithuania, Glitchr appears on a number of platforms—Facebook, Twitter, Google+, Tumblr—but its messages are never the same. Zakas uses "scripting tricks, Unicode characters, and diacritics" to make cascading shapes of black text, lines, and symbols that break out of the simple message boxes that are supposed to contain them. Go to twitter.com/glitchr_ or facebook.com/glitchr and you might think that your computer is having a seizure. There are squiggles and dashes running all over the page, which sometimes stutters as it scrolls. Zakas's creations seem to move of their own volition, especially on Twitter, where small black characters spill out from an @glitchr_ tweet and run down the timeline, draping other tweets in a canopy. Zakas has found loopholes and bugs—glitches—in the programming of these systems, and he's using them to spread small scraps of art that have little meaning beyond the fact that they shouldn't exist. If Facebook had done its job properly, Glitchr's page, which has more than 32,000 fans, wouldn't be crowding my screen with what look like runic etchings piled upon one another. The very existence of his work is the result of a mistake, and that's what's so great about it. It's also fated to disappear: eventually, the layouts and designs that Zakas has broken will be altered or fixed—a number of his posts have been deleted by the services themselves. Each crackdown causes him to find new ways to break the system. It seems possible that one day he might be banned entirely from these services, charged with violating a terms of service agreement in the name of subversive art. But in the meantime, he has created something strangely profound, a sly and ephemeral outsider art.

All this might sound a little Soviet—only art sanctioned by the state, in the appropriate styles, may be displayed—but that's essentially the case. Social media is the staid, white-walled showroom. You may only hang your paintings in the designated areas, and please don't touch the exhibits. Facebook's mantras of "Move Fast and Break Things" and "The Hacker Way" apply only to its engineers, not to Facebook's users. In the summer of 2013, a Palestinian hacker named Khalil Shreateh tried to report a security hole on Facebook that would allow someone with the proper knowledge to post on anyone's wall. Because Facebook, like many tech companies, offers cash bounties to white-hat hackers who alert them to security issues, Shreateh submitted a report. When Facebook replied that Shreateh hadn't found a real bug—a Facebook engineer later claimed that Shreateh's English was poor, making communication difficult, but acknowledged that the company should've pursued the issue more—Shreateh did something about it. He hacked into Mark Zuckerberg's Facebook page and posted information (along with an apology for violating his privacy) about the security issue directly on the wall of the company's CEO. Within minutes, a Facebook representative got in touch with Shreateh, but the company also temporarily blocked his account. And because Shreateh violated Facebook's terms of service, they didn't reward him for his discovery. His prize came later, through a crowdfunding campaign that raised more than $13,000.

- - - - - - - - - - - - - - - - -
DELIBERATE OBSCURITY

A comic book called *The Private Eye*, written by Brian K. Vaughan, takes place in Los Angeles, about sixty years from now. The cloud (the Internet kind) has "burst," as the narrator puts it. In this unexplained but recent digital apocalypse, private information comes gushing over the levees, viewable to all. "When the flood happened, it wasn't people's private letters or chat transcripts that destroyed their lives," explains a librarian, who, like his colleagues, has become a dedicated guardian of his patrons' privacy. "It was their search histories." The Internet, as a global network, collapsed. Popular conceptions of identity, both public and private, change dramatically. People start using fake names or "nyms"—separate, legal public identities that conceal their real selves, which are immensely valuable and vulnerable to attack. Upon reaching adulthood, young people are allowed to shed their past identities and take on new nyms. When walking in public, people wear masks, makeup, and elaborate costumes; technology—camouflage coats, holographic headpieces—is used to conceal one's self, not to reveal. Photographing someone without consent is a serious felony, a crime enforced by journalists, who act as a kind of cross between police officers and civic-minded private investigators.

In its metaphor of a cloud "bursting," Vaughan's story is a touch melodramatic. His use of fake public identities, masks, and the ability to change one's name at age eighteen, however, are deeply grounded in contemporary notions of privacy, identity, and what it means when everyone is a public figure. With companies and governments invested in improving facial-recognition technology—in mid-2012, Facebook paid about $60 million for an Israeli facial-recognition company called Face.com—one's face, and the right to control how it's seen and processed, is becoming a serious privacy battleground. By disguising themselves, Vaughan's characters are practicing something that could be called "deliberate obscurity." The methods of protection deployed in *The Private Eye* are not that much different from those being considered by activists and artists today. One might even point to the Nambikwara's use of secret names (doxed by Claude Lévi-Strauss) or Eric Schmidt's comment that young adults should be allowed to take on new legal identities, free of a data trail, when they turn eighteen as possible sources of inspiration for Vaughan. As cameras and facial-recognition software proliferate throughout our cities, designers have begun considering how people can become obscure, or opaque, to these systems.

The artist Adam Harvey created something he calls CV Dazzle, a form of "expressive interference" that uses makeup, hairstyling, and other tools to make faces unreadable by facial-recognition software. Designed to obscure speed, size, and direction, dazzle was a type of camouflage used on World War I warships. The patterns were deliberately varied and tended to feature large slashing stripes of color, almost zebra-like. Meant to confuse rather than to completely conceal, dazzle produced some extraordinary aesthetics: comparisons are often made to cubism, with its multiple, clashing perspectives and distorted representations of three-dimensional space. With CV Dazzle, Harvey's models are given slashing asymmetric haircuts, heavy applications of makeup that contrast with natural skin tones, and hair or other prosthetics that conceal the nose bridge—an important identifier for many facial-recognition algorithms. Harvey's work attempts to fool the software which, for instance, allows Facebook, Google+, and Picasa to detect faces in photos and recommend tags (or auto-tag). A wearer becomes less visible to machines but more present and vivid in the eyes of humans, while also being able to preserve a sense of fashion and self-expression.

Simone C. Niquille, a Dutch artist, also drew on the idea of dazzle, as well as novelist William Gibson's idea of a shirt that could be invisible to CCTV, to create REALFACE Glamouflage T-shirts. Looking like something that Salvador Dalí might have dashed off had he come of age online, these shirts are covered with "distorted faces of celebrity impersonators," which Facebook's facial-recognition/auto-tagging system might interpret as real faces. The obscuring effect is double or even triple: the system doesn't recognize your face, but it is also picking up on mangled photos of people who look like well-known celebrities.

ObscuraCam uses facial-recognition technology toward more productive (and protective) ends. It recognizes faces in photos and videos and allows you to blur or pixelate individuals or, alternatively, to highlight one individual and blur everything and everyone else. The app, which also allows you to strip identifying metadata from media, could be particularly useful for activists and dissidents. YouTube has instituted a similar measure, allowing people to blur faces in videos before publishing them. An artist named Leo Selvaggio has gone so far as to create prosthetic masks of his own face, which he encourages others to order and wear, allowing them, he said, "to present an alternative identity when in public."

Deliberate obscurity need not only be about protection or revealing the tracking, targeting, and sorting systems dictating digital life. It can also be about art or entertainment or an expression of one's politics. Rachel Law, a graduate student at the New School in New York, created a browser extension called Vortex, which she described as "a game that allows you to control how you are seen by the network." Vortex lets you choose from different data profiles, essentially creating your own custom, false online persona to confuse ad targeters and e-commerce sites that might charge you higher prices because of your demographic data. Players could also swap or share profiles. (By default, Vortex gives "Narnia" as a user's location, though that can be adjusted as part of the game.) Although Vortex wasn't released into the wild—it was a thesis project, it had some security holes, and browser companies would've been unlikely to allow it—it was social-media rebellion at its best. As Law wrote, "The 'Internet' does not exist. Instead, it is many overlapping filter bubbles which selectively curate us into data objects to be consumed and purchased by advertisers." Her program, even if it never was made publicly available, brilliantly illuminated these points. It's the kind of project that's deeply revealing of how the surveillance economy works: by arranging us into limited categories and subgroups that can easily be managed and monitored, with our data and attention bought and sold accordingly. Vortex was also a potential terror for advertisers: the industry publication *Advertising Age* reported on Law's invention with the headline, "This Student Project Could Kill Digital Ad Targeting."

Vortex is part of a growing crop of art projects, stunts, games, programs, and other tools that aim to wrest some control away from targeting systems. Crypstagram uses "glitchy photo filters" to disguise photos from prying eyes; only you and your intended recipient can decrypt the photo. Cryptagram accomplishes something similar by using a Chrome browser extension to password protect images. BlinkLink allows you to create a link that disappears after a set number of views. A South Korean graphic designer named Sang Mun created a typeface that fools the OCR technology used by computer systems around the world, including those of major intelligence agencies. During his time in the South Korean military, Mun had worked with the NSA, and he used his knowledge to take advantage of some of the flaws in OCR. After a year of work, the result was a typeface called ZXX (a Library of Congress classification meaning "No linguistic content; Not applicable") which, despite the noise and scratches and blobs intentionally introduced, is still highly readable by humans but, crucially, not by machines.

There are more of these projects—some only crass provocations, others potentially useful. Hell Is Other People is an app that takes Foursquare check-in data from friends and recommends destinations where you're likely not to run into them; its creator called it "an experiment in anti-social media." Facebook Demetricator, a browser plug-in, removes all numbers from Facebook—number of notifications, friends, likes, shares, and other metadata. Ben Grosser, the artist who created the app, said that he was worried that he cared more about these metrics than the content of messages or the quality of his relationships. "The site's relentless focus on quantity leads us to continually measure the value of our social connections within metric terms," he explained, "and this metricated viewpoint may have consequences on how we act within the system." With Demetricator activated, instead of seeing that "25 people like this"—and perhaps feeling compelled to respond to be part of the group bonhomie—you only see that "people like this." You also can't see how many people liked and commented on ads. Grosser's other creations include a browser extension called ScareMail, which adds to every Gmail message terms that the NSA might be monitoring. Drawing on a list of keywords from a list produced by the Department of Homeland Security, the program creates nonsensical phrases around these words. By possibly consuming the attention (and bandwidth) of the NSA's systems, he's making this surveillance more costly and, consequently, performing an act of civil disobedience. We live in an attention economy, Grosser seems to be saying, and manipulating it isn't just a crass moneymaking scheme (as with viral media); it can also be a way to prove a political point.

F.A.T., or Free Art and Technology, may be the leaders in this genre. They are culture jammers par excellence. A collection of about twenty artists and pranksters, F.A.T. distributes all of their material free of copyright. Often, they produce 3-D models, instruction kits, masks, and other materials that they encourage people to disseminate and to make at home. F.A.T. has produced a fake Google self-driving car that they drove around New York City. They built a fake Google Street View car and took it around Berlin, where concern about Google's privacy and surveillance practices runs high. (The car also appeared in New York.) They created BRICKiPhone, a functional case for the iPhone ("the most ubiquitous device of the yuppie class"), which turns the phone into a blocky, gray, late-eighties-style cell phone, complete with protruding black antenna. The uglified device makes a statement about how we use our sleek gadgets to inform our identities and how we tend to feel a put-on kind of nostalgia for antiquated technologies, like early mobile phones.

In a piece called "Clearing 4 Months of Internet Cache," F.A.T. printed months' worth of Internet browsing, compressed the paper into a large cube, and displayed it in a gallery—a physical instantiation of our insatiable informational appetite. When Twitter introduced the ability for any user to send a direct message to another, the artists of F.A.T. equipped a Twitter account with a script that caused it to automatically tweet any direct messages it received. Riffing off of Twitter's impending IPO, the piece was called "Going Public." FriendFlop, a browser extension, jumbled the names and avatars of people on your Twitter timeline and Facebook News Feed, "dissolving your biases and reminding you that everyone is saying the same shit anyway." Social Roulette was a F.A.T.-produced Facebook app that allowed you to generate a random number, and if it landed on 1, your Facebook account would be deleted. There were a couple of problems, though: Facebook makes it extremely hard to delete your account (apps certainly aren't allowed this functionality), and the company blocked the app within hours. The app was also rigged to always land on 1, but the whole thing—what one F.A.T. member called "a performance disguised as a game"—existed to prove a valuable point: "Even suggesting that we own our digital data will get you shut down." Facebook decides how you will manage your digital identity, and if you want to get rid of it, you're going to have do it their way.

In an inspired piece, F.A.T. harnessed the iconography and philosophy of TED ("Ideas Worth Spreading") to cleverly subvert that organization. Led by artist Evan Roth, the group created realistic but fake TED stages—the black curtain fronted by a large screen, the imposing red TED letters the size of a small child, a sturdy reminder of the brand bringing you these life-changing promises of technological solutionism. Their fake TED speakers wore Secret Service–style earpieces, and some adopted the look of the maverick techie: hoodie, dark jeans, I'm-about-to-blow-your-mind hand gestures. A few of the speakers appeared deliberately ridiculous; one gustily flicked off the camera, another grabbed his own crotch. So what's the scam, what's the joke? It's that, once published, images of these scenes—much in line with TED's motto—began to spread through the Internet. And now, performing a Google image search on TED talks, some of the fake images come up alongside the real. (I found one of the fakes on the Web site of a well-known magazine.) The message seems to be not that just a simulacrum can come to be mistaken for the real thing, but that something about TED's sudden ubiquity, its ease of replication, franchised all over the world like some fast-food chain, calls the whole enterprise into question. These images only appear in Google image search results because people have started linking to and using them, and if no one cares enough to check their veracity, then perhaps that's because TED itself is just an exercise in wishful thinking, preaching betterment through the viral sharing of uplifting ideas. Watch this YouTube video and you'll become a smarter, more thoughtful person—it doesn't matter if what's being dispensed to you (and the audience that paid thousands for the event) is true or plausible. TED exists to make a privileged class feel better about its own cultural supremacy, and these images are a shorthand for that. We share them to show that we're in tune with the latest techno-utopian trends. That some of these images are an art collective's contrivances only reinforces the sense that it is, at bottom, a charade, a mock-up.

We need these exercises in digital critique as much as acts of refusal or boycotts. The fact that their points are mostly political or aesthetic, when there is any point at all, only makes them more effective, particularly in a fully commoditized media space where everything is tracked, everything must scale, and everyone must be a good digital serf, always producing data. F.A.T.'s first act of rebellion was to declare that they care about none of this. Sure, you can take and share their stuff, but they want nothing in return, not money, not data. (Many of the projects discussed here, including those made by F.A.T., are open source or without copyright.) By leveraging the everyday materials of our digital lives, including viral channels, they produce work that mingles with the very things they're satirizing. At times, it's hard to tell the real from the virtual apart, which is precisely the point. And in tune with the tenor of the viral Web, their work is designed to be shareable and disposable, with most of their projects created in a few minutes or few hours. Their "open source activism" becomes stronger, and more inclusive, as it spreads.

At the Rhizome Seven on Seven conference, an annual showcase in which a technologist and an artist spend a day working together to create something new—an app, a video, a presentation—the emphasis is usually on novel deployments of existing technologies. For example, one year the late Aaron Swartz and the artist Taryn Simon created Image Atlas, a site that shows Google image results from seventeen different countries, illustrating how search results are contextually dependent. The effect is mutual, with search results also potentially shaping local attitudes. An Afghan searching for "Americans" will see photos of soldiers and George W. Bush. A Chinese search produces images of happy couples and a man standing in front of a flag. The Iranian search is nearly all U.S. soldiers, while a German search is photos of a type of cookie called an Amerikaner.

At the 2013 version of Seven on Seven, Dalton Caldwell, the founder of App.net, an ad-free social network designed to be interoperable with other networks, worked with the artist and musician Fatima Al Qadiri. They created a video, less than two minutes long, featuring a succession of various alert noises, including ones you know from AOL, the iPhone, and AIM, as well as others that might seem naggingly familiar. These are the antic sounds of the network, prodding you to return, offering you connection. As the sounds accumulate, words such as "constant," "durable," "permanent," "uninterrupted," "nonstop," and "continual" appear and dissolve onscreen. Hosted on a site called ConstantUpdate.net, the video ended with the insistent rumble of a phone vibrating, demanding to be answered.

"We wanted to do something that will never update," said Caldwell, describing his team's entry. "This is something that is finished. Don't like it on Facebook. Don't subscribe to our e-mail list."

It was just a quickly conceived art project, cheeky and deliberately nonfunctional, but the virtue of ConstantUpdate.net was that it was mostly useless. It wasn't constant, it was easily stoppable. You could watch the video on a loop and try to read something profound into the noise, but you could just as easily get the message the first time. It's finished, it's done. Here it is. It won't adopt different formats depending on your preferences; it won't try to tell you what you want. It won't refresh itself with new content or push notifications at you. It's that rare Internet phenomenon: static. It's already "an artifact," as Caldwell called it. It has no relationship to the circulatory systems of the Web. It'll never send you an e-mail out of the blue, asking why you haven't visited in so long and telling you what recent news you've missed. But the noises and message it communicates *are* constant; they are the backing track to our digital lives. We hear them everywhere.

Artifacts are increasingly rare in today's Web. The sheer amount of material produced online means that some of it will inevitably fall out of date, become orphan work, or fall prey to changing technologies. Hosting services lapse into disarray or shut down, becoming potter's fields of Web real estate, the remains of once vibrant online communities. Each blog you stumble upon that hasn't been updated in a year or three is a kind of artifact. I find abandoned personal blogs unexpectedly poignant, especially those containing only a few posts. Someone started something here, perhaps with a plan and excitement for the future, the prospect of establishing himself as a writer scrolling out in front of him. Blogs are the quintessential ever-updating commodity: they're not supposed to go fallow. Each one that's abandoned (I've let a few go) is a sign that something shifted the writer's attentions elsewhere, or that he lost faith in himself or the project. And so this personal document sits there, aging without gathering any dust, the only marker of the passing years being the lonely time stamp on the most recent post.

Unless we dig through archive.org, an essential repository of Web history, we rarely stumble upon these artifacts, in part because our filter bubbles emphasize the new and the heavily trafficked. Bob Dole's 1996 campaign site, dolekemp96.org, is still up as of this writing. Encountering it today is remarkable, an immediate encounter with the past. The site is like a museum exhibit (in fact, it remains up due to the efforts of an entity called 4President.org). It sits there, preserved in the Web's formaldehyde, its primitive design a callback to an earlier era of connectivity, even as its use of animated GIFs and the issues it emphasizes (freedom from Internet spying, the decline of education) remind us that some things spring eternal.

There is something ominous here, too: the reminder that nothing done online truly disappears. Data tends not to have a half-life; it only becomes more visible or more obscure. That is why more progressive regulatory regimes promise a right to be forgotten: the right to have one's personal data expire or, in some circumstances, deleted by request. It's a right that must be coded into law, because the architects of these systems have neglected to code it themselves. Even the services that emphasize privacy and ephemerality, such as Snapchat, have become distinguished by their failures to deliver just that. The persistent irony of all this is that lying and obfuscation have become forms of protection, insincerity a shield. Armed with these weapons, the good citizens among us find some way to rebel, but it is the wise ones who make an art of it.

- - - - - - - - - - - - - - - - - -
A DIGITAL BILL OF RIGHTS

All of this agitating and culture jamming doesn't obviate the need for changes in the law. For that, we need to look to people such as Kade Crockford, who runs the Technology and Liberty project at the ACLU.

Based out of the ACLU's Massachusetts office, Crockford is a fierce advocate, but she also has some of the rebelliousness and utter disregard for convention that makes many of the other personalities in this chapter, along with their work, so appealing. Whether on Twitter as @onekade or on PrivacySOS.org, a site she runs in connection with her ACLU work, Crockford is smart, blunt, angry, and occasionally, wonderfully profane. Or as she said to me, "I'm kind of an asshole and have a loud mouth."

Crockford gives the impression of being someone who's both extraordinarily committed to the issues of privacy and civil liberties and fearless in advocating on behalf of the kind of widespread reforms that are dearly needed. Her main goal, through lobbying, advocacy, and public education, is to "bring the Fourth Amendment into the digital age"—to require the government to get a warrant any time it seeks to seize personal data. That's no easy task, and there is plenty of work to do. In June 2014, the United States Supreme Court, in a unanimous decision, declared that law enforcement must get a warrant to access an arrestee's cell phone. That decision, in a case known as *Riley v. California*, was "hugely significant," Crockford said. "It's also about fifteen years too late."

"We can't wait that long for new laws governing powerful technologies that contain the most private details of our lives," she said. "We really need a twenty-first-century Bill of Rights. Instead, the government's moving in the opposite direction."

This sense that we are well behind the curve when it comes to protecting people's rights in the digital age animates all of Crockford's work. It also reflects a deeper cultural malaise, in which we, as ordinary citizens, have not taken the responsibility to educate ourselves about the perils of electronic surveillance and new digital technologies. Crockford's group is working to pass five privacy bills in Massachusetts. If they're successful, one hopes that their efforts could be replicated nationally and that Congress might finally begin to rein in government surveillance, provide checks on the power of law enforcement, and pass legislation that will better protect consumers and their data.

Despite her obvious ambition, Crockford called this plan a

straightforward and "not at all radical solution to a very serious problem." The ACLU, however, has had "a surprising amount of difficulty" getting these laws passed, which has been both frustrating and disconcerting for Crockford. Some colleagues compare their job to Whac-a-Mole—always trying to respond to new technologies, new legal loopholes, and other novel threats. Crockford thinks that a more dire metaphor is in order: "What it's really is like trying to catch a tsunami with a spoon. We're really struggling against just a really massive surveillance-military-prison-industrial complex. The ACLU looks really good, but we punch above our weight. We're a relatively small organization. There aren't really very many organizations doing the work we do, as far as cutting across the digital and [physical world] boundaries." Indeed, the ACLU's slate is a full one, tackling cases and policies related to anything from mass incarceration to drone strikes, the drug war to immigration, LGBT rights to CIA torture.

Despite any limitations it might have in terms of funding or number of personnel, the ACLU is well served by this expansive portfolio, particularly when its representatives, such as Crockford, acknowledge that many of these issues remain interconnected. For example, addressing government surveillance inevitably involves considering issues of corporate surveillance and data collection, the so-called war on terror, and civil liberties. All the same, Crockford lamented that they're often "stuck at the level of trying to obtain warrant protections for technologies that we've been using for twenty years." That is less a problem of lobbying or public will than of Congress's slow pace, its general technological illiteracy, and the tremendous influence that law enforcement and makers of surveillance technologies can have in pushing back against proposed civil liberties protections. "Cops are very influential in legislative bodies," she said. "It's very difficult for legislators to move statutes that take away any power from law enforcement."

And all this is just designed to catch up on the last two decades of technological change. As Crockford said, "The threats on the horizon are very serious as well. The things that legislatures on every level have really not grappled with at all yet are problems related to the crossing over of digital and physical surveillance," such as biometric surveillance.

Although she's dedicated to legal reform, Crockford isn't a lawyer—something that's a source of pride for her. "I like what I do," she said. "I feel like I can say certain things that my colleagues who are lawyers can't say." This grants her a measure of independence to say whatever she thinks—whether it's cussing out the FBI on Twitter or writing passionately and learnedly on PrivacySOS about innocent Muslim Americans who can't pass through an airport without being pulled aside and questioned for hours. Crockford doesn't need to worry about decorum, nor about ever appearing in court against one of her online sparring partners. It's this same independence, and her insouciant attitude, that make her one of the most interesting activists for the cause of surveillance reform. She also reflects the kind of intense commitment that is necessary if we are to make any headway on these issues.

Crockford started at the ACLU in 2009, spending ten hours a week going through boxes of information on fusion centers that had been obtained through public records requests. Set up in the wake of 9/11, fusion centers are Department of Homeland Security–run "spy centers," as Crockford calls them. There are about eighty of them across the country, and they're tasked with coordinating surveillance and information sharing between various law enforcement organizations from the federal level on down to state and local entities. Crockford's task was deceptively difficult, as research in the intelligence field tends to be. She was supposed to address some basic questions: What are these centers doing? What are their policies? What are their budgets? The answers, she hoped, were in those boxes.

Crockford came to the ACLU with a political background—she's steeped in the history of U.S. government surveillance, from COINTELPRO to the NSA's attempts to promote the Clipper chip in the 1990s—but she admits that she was unaware of the full extent of what the government had authorized after 9/11. Fusion centers, in particular, have become an ominous example of what's known as "mission creep." Developed to coordinate information sharing for counter-terrorism investigations, these facilities have since been used to track petty criminals, Occupy activists, and even U.S. politicians. "It was really just incredibly eye-opening to me," Crockford said. There was "a massive industry being established in the shadows."

That research experience launched Crockford on her subsequent career as a public advocate and an essential source about the scope of government and corporate surveillance. After that, she said, "I was just obsessed, basically, with surveillance."

Still, with five years on this beat, she said that along with a finely honed sense of outrage, she can't help but sometimes feel fatigue and despair. "It wears on your soul," she said, offering a self-deprecating laugh. "It's also really fun and I love my job."

Her work has affected how she uses digital technologies—but not always in the ways that one might expect. "In some ways, I've actually gotten less meticulous about my digital habits," she admitted. "I used to be really, really careful about posting things about myself online." For a while, she wouldn't even post photos of herself on the Internet. "And then something strange happened," she said, "which is that I became a quasi-public figure." She was invited to speak at conferences and to write op-eds for the *Guardian*, which meant that these organizations wanted a headshot to present alongside her latest speech or column. She also started appearing on television, from local channels to international news broadcasts. It began to seem stupid, she said, to try to control the spread of her image. Anyone with access to Google's image search could find her. That produced a peculiar irony: "The deeper I got into work on surveillance and privacy, the less relevant it was for me really to protect my face, because it was already out there."

That doesn't mean, though, that she's lost faith in the cause or hasn't made other moves to protect her privacy. On the contrary, she's an enthusiastic proponent of encryption. She likes PGP—the widely respected encryption method invented by Phil Zimmerman—and Adium, a chat app. She's also joined Riseup, which "provides online communication tools for people and groups working on liberatory social change."

But nothing is secure 100 percent of the time. The ACLU technologist Christopher Soghoian is fond of saying, "There is no technology that can protect you from a $10 billion intelligence agency." Even if you're just trying to avoid the prying eyes of data brokers, hackers, or certain people you know, maintaining privacy requires education, vigilance, and work, and it *still* might not keep all of your communications secure. That's not a reason to give up; using these tools can be helpful, not least for increasing your own knowledge and making you invested in your own security. As Edward Snowden has repeatedly said, "Encryption works." By that he doesn't mean that all forms of encryption are uncrackable, or that some entity might not find another way to obtain your personal information, but encryption works well enough that it raises the cost of doing business for those who would otherwise indiscriminately vacuum up your data. It also offers support to the idea that privacy and security are worth upholding for everyone, not just for those people with technical know-how. One reason why it's so important for big tech companies to start encrypting all of their data links—something that most of them only began doing after the Snowden revelations—is that it provides this kind of social benefit for a huge number of people. And as large actors such as Yahoo and Google increasingly get behind encryption, that boosts public awareness and creates more of a market for the kinds of privacy enhancing technologies that many people could benefit from.

Crockford recognizes that using these tools requires work and that some people might take them up with more gusto than others. "Like anything else," she said, "there's a cost-benefit analysis." She compares it to having unprotected sex—people might forgo condoms because they like how it feels, but they should know that they're taking a risk. "The same thing is true with using services like Gmail and Facebook, which is that you need to look very closely at what you're getting from those services" and at "what you're giving up" in return. For Crockford, Facebook "is not useful to me. It doesn't meet the standard of being more useful than harmful." On the other hand, Google Search is still tremendously helpful for research, but she's given up Gmail. She recommends that "people think very seriously about what they actually need and why, and what they want and why." Finally, the best alternatives tend to be those products that are open source and whose code has been studied by experts in the field.

Talking to Crockford, I often returned to the idea that privacy and security require vigilance, and that can be enervating. It's difficult to feel as if we have any agency—whether we're facing the prospect of government surveillance or the social pressures of living on public networks. Social media already tends to induce a great deal of self-consciousness, a by-product of that constant modulation of identity and selfhood that goes on when we choose to share bits and pieces of ourselves online. To take that burden and add the pressures of becoming one's own counterintelligence operation can seem exhausting, if not paranoia-inducing.

Crockford, despite her obsession with surveillance, seemed remarkably even-keeled about all this. "You want to enjoy life," she said. "You do what you can basically and try not to sweat. Part of the problem with surveillance . . . is that everybody is so freaked out constantly about it." In other words, you have to be smart and judicious about your behavior online, but you can't let these anxieties rule you.

"People need to live their lives," she reminded me. "Clearly part of living our lives, I think, needs to be resisting those systems that are dominating us and controlling us and making us afraid to speak our minds. But I would say that people should get offline sometimes. Leave your phone at home. Go for a walk with your friend or your lover. Go into the woods . . . Swim in a pond where nobody can see you. Try to actually enjoy privacy sometimes. Get away from the Internet and have a life that's independent of that kind of shit."

She was offering a mantra whose simplicity is belied only by how difficult it is to uphold: be who you want to be. In a better world, constant connection wouldn't represent a burden; it wouldn't ask us to sacrifice our autonomy or the most private details of our lives. But what would be the fun in that?

Acknowledgments

Thank you to Liz Applegate, Scott Beauchamp, Stephanie Butnick, Brendan Byrne, Adam Chandler, Emily Cunningham, Thomas de Zengotita, Will Evans, Chris Greens, David Greenwald, Rachel Gutkin, Gary Gutkin, HarperCollins, Harry, Evan Hughes, Inkwell Management, Dan Kois, Koshka, Chris Lehmann, Liel Leibovitz, Michael Libby, Lou Mathews, Jesse Moskowitz, Reyes Deli, Mike Schapira, Matthew Shaer, Alex Shephard, Loretta Siciliano, Rocco Siciliano, Scott Silverman, Elaine Silverman, David Silverman, Joel Smith, Lauren Smythe, John Summers, and Bari Weiss.

Notes

vii   Instant messages: Nicholas Carlson. "Well, These New Zuckerberg IMs Won't Help Facebook's Privacy Problems." *Business Insider*. May 13, 2010. businessinsider.com/well-these-new-zuckerberg-ims-wont-help-facebooks-privacy-problems-2010-5.

viii   "catchall term": Richard Byrne. "A Nod to Ned Ludd." *Baffler*. Summer 2013. thebaffler.com/past/a_nod_to_ned_ludd.

viii   "within an hour of being born": James Vincent. "Most Babies Make Facebook Debut Within an Hour of Being Born." *Independent*. Aug. 27, 2013. independent.co.uk/news/uk/home-news/most-babies-make-facebook-debut-within-an-hour-of-being-born-8785721.html.

ix   "you're never out of ideas": Tom Simonite. "Google's Boss Envisions a Utopian Future." *MIT Technology Review*. Sept. 28, 2010. technologyreview.com/view/420962/googles-boss-envisions-a-utopian-future.

xii   "essence of technology": Trebor Scholz. "Why Does Digital Labor Matter Now?" In *Digital Labor: The Internet as Playground and Factory*. Trebor Scholz, ed. New York: Routledge, 2013, 3.

xiv   "governmental power": David Golumbia. "Cyberlibertarians' Digital Deletion of the Left." *Jacobin*. Dec. 4, 2013. jacobinmag.com/2013/12/cyberlibertarians-digital-deletion-of-the-left.

xiv   "That job is left to others": Claire Cain Miller. "Addicted to Apps." *New York Times*. Aug. 24, 2013. nytimes.com/2013/08/25/sunday-review/addicted-to-apps.html.

xiv   "their values": Natasha Tiku. "Why Pax Dickinson Matters." *Valleywag*, a blog on *Gawker*. Sept. 11, 2013. gawker.com/why-pax-dickinson-matters-1293728062.

1   "from a bizarre fusion": Richard Barbrook and Andy Cameron. "The Californian Ideology." *Alamut*. alamut.com/subj/ideologies/pessimism/califIdeo_I.html.

2   "all the ills of society": Tim Wu. *The Master Switch*. New York: Random House, 2010, 36.

2   Marconi's prediction: Ivan Narodny. "Marconi's Plans for the World." *Technical World Magazine*, October 1912, 145–50. earlyradiohistory.us/1912mar.htm.

5   Peter Thiel's views: George Packer. "No Death, No Taxes." *New Yorker*. Nov. 28, 2011. newyorker.com/reporting/2011/11/28/111128fa_fact_packer.

6   "transforming relationships": Sheryl Sandberg. "Sharing to the Power of 2012." *Economist*. Nov. 17, 2011. economist.com/node/21537000.

6   "a triumph of humanity": Lauren Dugan. " 'Twitter Is Not to Be a Triumph

of Technology but of Humanity'—Biz Stone." *AllTwitter*, a blog on *Mediabistro*. Nov. 10, 2010. mediabistro.com/alltwitter/twitter-is-not-to-be-a-triumph-of-technology-but-of-humanity-biz-stone_b169.

6   Apple profile: "Twitter. Triumph of Humanity." Apple. apple.com/se/business/profiles/twitter.

6   connectivity is a human right: Tamar Weinberg. "SXSW: Mark Zuckerberg Keynote (the Edited Liveblogged Version)." Techipedia. March 11, 2008. techipedia.com/2008/mark-zuckerberg-sxsw-keynote.

6   "bringing your vision to the world": Katherine Losse. *The Boy Kings: A Journey into the Heart of the Social Network*. New York: Simon & Schuster, 2012, 201.

6   "we know what you like": Alexia Tsotsis. "Eric Schmidt: 'We Know Where You Are, We Know What You Like.'" *TechCrunch*. Sept. 7, 2010. techcrunch.com/2010/09/07/eric-schmidt-ifa.

7   "the phone tells you": Jacob Ward. "Innovation of the Year: Google Now." *Popular Science*. November 2012. popsci.com/bown/2012/product/google-now.

8   "even if they hadn't consented": Losse. *The Boy Kings*, 88.

8   "one voice": Sheryl Sandberg. "Sharing to the Power of 2012."

9   "everyone could be a movie star": Holman W. Jenkins Jr. "Technology = Salvation." *Wall Street Journal*. Oct. 9, 2010. wsj.com/article/SB10001424052748704696304575537882643165738.html.

12   "frictionless sharing": Adrian Short. "Why Facebook's New Open Graph Makes Us All Part of the Web Underclass." *Guardian*. Sept. 27, 2011. guardian.co.uk/technology/2011/sep/27/facebook-open-graph-web-underclass.

12   a single organism: Losse. *The Boy Kings*, 166.

12   400 billion actions: Harrison Weber. "Facebook: More Than 400 Billion Open Graph Actions Have Now Been Shared." *The Next Web*. March 10, 2013. thenextweb.com/facebook/2013/03/10/facebook-more-than-400-billion-open-graph-actions-have-now-been-shared.

12   "self-censoring": Jennifer Golbeck. "On Second Thought . . ." *Slate*. Dec. 13, 2013. slate.com/articles/technology/future_tense/2013/12/facebook_self_censorship_what_happens_to_the_posts_you_don_t_publish.html.

12   "for tech to overcome": Losse. "As I wrote, the goal is to create everyone as 'cells in a single organism.' For FB the privacy of thought is a problem for tech to overcome." August 10, 2013, 11:34 a.m. Tweet (@fake_train).

14   2013 revenue: Google Investor Relations. "2014 Financial Tables." investor.google.com/financial/tables.html.

14   2012 revenue: Tim Peterson. "Google Finally Crosses $50 Billion Annual Revenue Mark: Company Begins to Reverse CPC Declines, Mobile Pricing Stabilizing." *Adweek*. Jan. 22, 2013. adweek.com/news/technology/google-finally-crosses-50-billion-annual-revenue-mark-146710.

14   Google+ signups: Jon Brodkin. "Google Doubles Plus Membership with Brute-Force Signup Process." *ArsTechnica*. Jan. 22, 2012.

arstechnica.com/gadgets/2012/01/google-doubles-plus-membership-with-brute-force-signup-process.

14   Search results: Sean Gallagher. "Google 'Plus-ifies' Search with Social Features in Effort to Un-plus Facebook." *ArsTechnica*. Jan. 10, 2012. arstechnica.com/business/2012/01/google-plus-ifies-search-with-social-features-in-effort-to-un-plus-facebook.

14   Personalization: Harry McCracken. "Search, Plus Your World: Google's Risky Gambit." *Time*. Jan. 19, 2012. techland.time.com/2012/01/19/search-plus-your-world-googles-risky-gambit.

15   YouTube founder's comments: Nick Summers. "YouTube co-founder Jawed Karim asks: 'Why the f*** do I need a Google+ account to comment on a video?'" The Next Web. Nov. 8, 2013. thenextweb.com/google/2013/11/08/youtube-co-founder-jawed-karim-f-need-google-account-comment-video.

15   Engagement statistics: Brodkin. "Google Doubles Plus Membership."

16   Google+ actions: Amir Efrati. "There's No Avoiding Google+." *Wall Street Journal*. Jan. 2, 2013. wsj.com/article/SB1000142412788732473130457819 3781852024980.html.

16   Click-through rates: ibid.

16   Twitter's early success: Jon Russell. "Twitter Retires API v1, Finally Killing Off TweetDeck for iOS, Android, AIR and Other Apps." *The Next Web*. June 12, 2013. thenextweb.com/twitter/2013/06/12/twitter-retires-api-v1-finally-killing-off-tweetdeck-for-ios-android-air-and-other-apps.

16   Invention of hashtag: Zachary M. Seward. "The First-Ever Hashtag, @-Reply and Retweet, as Twitter Users Invented Them." *Quartz*. Oct. 15, 2013. qz.com/135149/the-first-ever-hashtag-reply-and-retweet-as-twitter-users-invented-them.

22   over-sharing: Elizabeth Bernstein. "Thank You for Not Sharing." *Wall Street Journal*. May 6, 2013. wsj.com/article/SB1000142412788732382680 4578466831263674230.html.

22   disinhibition: Russell W. Belk. "From Embarrassing Facebook Posts to Controversial Tweets, Why Are Consumers Oversharing Online?" Phys.org. July 26, 2013. phys.org/news/2013-07-facebook-controversial-tweets-consumers-oversharing.html.

23   "institutionalizes envy": Christopher Lasch. *The Culture of Narcissism: American Life in an Age of Diminishing Expectations*. New York: W. W. Norton, 1991, 73.

24   "The purpose of publicity": John Berger. *Ways of Seeing*. New York: Penguin, 1972, 142.

24   "Dissenters have to work": Robert W. Gehl. "A History of Like." *New Inquiry*. March 27, 2013. thenewinquiry.com/essays/a-history-of-like.

24   "liking studies": ibid.

25   "compulsive responses": Alexis C. Madrigal. "The Machine Zone: This Is Where You Go When You Just Can't Stop Looking at Pictures on Facebook." *Atlantic*. July 31, 2013. theatlantic.com/technology/

archive/2013/07/the-machine-zone-this-is-where-you-go-when-you-just-cant-stop-looking-at-pictures-on-facebook/278185.

25   Awesome button: Andrew Bosworth. "What's the History of the Awesome Button (That Eventually Became the Like Button) on Facebook?" Quora. Oct. 5, 2010. quora.com/Facebook-Inc-company/Whats-the-history-of-the-Awesome-Button-that-eventually-became-the-Like-button-on-Facebook.

25   "a bad review": ibid.

26   "psychological barrier": Pete Cashmore. "Should Facebook Add a Dislike Button?" CNN. July 22, 2010. cnn.com/2010/TECH/social.media/07/22/facebook.dislike.cashmore.

26   "abhor" and "hate": Jason Gilbert. "Facebook 'Dislike' Button Still Not Allowed For New Apps." *Huffington Post.* Oct. 2, 2011. huffingtonpost.com/2011/10/02/facebook-dislike-button-not-allowed-apps_n_989515.html.

26   General Mills terms of service: Stephanie Strom. "When 'Liking' a Brand Voids Your Right to Sue." *New York Times.* April 16, 2014. nytimes.com/2014/04/17/business/when-liking-a-brand-online-voids-the-right-to-sue.html.

27   "happier, healthier, and more productive": "The Science behind Happier." Happier. happier.com/science.

27   "your feelings hurt": Rob Walker. "Can Tumblr's David Karp Embrace Ads without Selling Out?" *New York Times Magazine.* July 12, 2012. nytimes.com/2012/07/15/magazine/can-tumblrs-david-karp-embrace-ads-without-selling-out.html.

27   "Everybody loves everybody": ibid.

27   Firing of Tumblr journalists: Adam Weinstein. "If Tumblr 'Couldn't Be Happier' with Its Journalists, Why'd It Just Fire Them All?" *Gawker.* April 10, 2013. gawker.com/5994232/if-tumblr-couldnt-be-happier-with-its-journalists-whyd-it-just-fire-them-all.

28   "marketing as journalism": Joe Pompeo. "Is It Marketing or Is It Journalism? The Case of Tumblr's 'Storyboard.'" *Capital New York.* Dec. 10, 2012. capitalnewyork.com/article/media/2012/12/6816545/it-marketing-or-it-journalism-case-tumblrs-storyboard.

28   "11 percent": Sarah Perez. "Tumblr's Adult Fare Accounts for 11.4% of Site's Top 200K Domains, Adult Sites Are Leading Category of Referrals." *TechCrunch.* May 20, 2013. techcrunch.com/2013/05/20/tumblrs-adult-fare-accounts-for-11-4-of-sites-top-200k-domains-tumblrs-adult-fare-accounts-for-11-4-of-sites-top-200k-domains-adults-sites-are-leading-category-of-referrals.

28   "22 percent": Joshua Brunstein. "If Yahoo Buys Tumblr, What Will It Do with All That Porn?" *Bloomberg Businessweek.* May 17, 2013. businessweek.com/articles/2013-05-17/if-yahoo-buys-tumblr-what-will-it-do-with-all-that-porn.

28   "to police" porn: Lauren O'Neil. "Tumblr Users Rile Against Porn Crackdown." CBC News. July 23, 2013. cbc.ca/newsblogs/

yourcommunity/2013/07/tumblr-porn-crackdown-halted-by-users.html.

29   "opposed to advertising": Mark Milian. "Tumblr: 'We're Pretty Opposed to Advertising.'" *Los Angeles Times*. April 17, 2010. latimesblogs.latimes.com/technology/2010/04/tumblr-ads.html.

29   "heroes": John Reynolds. "Tumblr Founder Criticises Rivals' Publishing of Performance Stats." *Guardian*. June 20, 2013. theguardian.com/media/2013/jun/20/tumblr-david-karp-attacks-twitter.

29   "Palo Alto or Sunnyvale": Matthew Creamer. "Tumblr's David Karp Makes His Ad Pitch at Cannes." *Advertising Age*. June 20, 2013. adage.com/article/special-report-cannes-2013/tumblr-s-david-karp-makes-ad-pitch-cannes/242740.

30   Tagging emotions: Roddy Lindsay and Raylene Yung. "Adding What You're Doing to Status Updates." Facebook. April 10, 2013. newsroom.fb.com/News/600/Adding-What-Youre-Doing-to-Status-Updates.

30   Expanded data collection: Will Oremus. "Facebook's Cute New Emoticons Are a Fiendish Plot. Don't Fall For It." *Slate*. April 10, 2013. slate.com/blogs/future_tense/2013/04/10/facebook_emoji_status_update_emoticons_are_bad_for_privacy_good_for_advertisers.html.

32   "drum of lube": Nick Bergus. "How I Became Amazon's Pitchman for a 55-Gallon Drum of Personal Lubricant on Facebook." *Nick Bergus* (blog). Feb. 23, 2012. nbergus.com/2012/02/how-i-became-amazons-pitchman-for-a-55-gallon-drum-of-personal-lubricant-on-facebook.

32   "boosting that engagement": Somini Segupta. "On Facebook, 'Likes' Become Ads." *New York Times*. May 31, 2012. nytimes.com/2012/06/01/technology/so-much-for-sharing-his-like.html.

33   "$20 million settlement": Joe Miller. "Facebook to Compensate Users for Sharing Details on Ads." BBC News. Aug. 27, 2013. bbc.co.uk/news/technology-23848323.

33   Dating site ad: Helen A.S. Popkin. "Bullied Dead Girl's Image Used in Dating Ad on Facebook." NBC News. Sept. 18, 2013. nbcnews.com/technology/bullied-dead-girls-image-used-dating-ad-facebook-4B11187466.

34   Comic Con controversy: Brian Crecente. "New York Comic Con Using Attendee Twitter Accounts to Send Promo Tweets (Update)." *Polygon*. Oct. 10, 2013. polygon.com/2013/10/10/4826150/new-york-comic-con-promotional-tweets-twitter.

34   BookVibe complaint: Nick Bilton. "Warning: Do not sign up for @BookVibe they just Tweeted from my account without my permission. That start-up is dead to me." May 13, 2014, 1:22 p.m. Tweet (nickbilton). twitter.com/nickbilton/status/444161625226899456.

36   Dataminr: Elizabeth Dwoskin. "Twitter's Data Business Proves Lucrative." *Wall Street Journal*. Oct. 6, 2013. online.wsj.com/article/SB10001424052702304441404579118531954483974.html.

36   Thomson Reuters statistic: Tom Groenfeldt. "Thomson Reuters Gets More Sentimental with News and Social." *Forbes*. March 8, 2012. forbes.com/

sites/tomgroenfeldt/2012/03/08/thomson-reuters-gets-more-sentimental-with-news-and-social.

37   Derwent Capital tweets: Amy Or. "Now Trending: Turning Tweets into Trades." *MarketBeat*, a blog on WSJ.com*Stretal*. D 2011. blogs.wsj.com/marketbeat/2011/12/12/now-trending-turning-tweets-into-trades.

37   Derwent Capital returns: Groenfeldt. "Thomson Reuters."

37   Investors' reservations: Rachael King. "Trading on a World of Sentiment." *Bloomberg Businessweek*. March 1, 2011. businessweek.com/stories/2011-03-01/trading-on-a-world-of-sentimentbusinessweek-business-news-stock-market-and-financial-advice.

37   Gnip's customers: Matthew Philips. "How Many HFT Firms Actually Use Twitter to Trade?" *Bloomberg Businessweek*. April 24, 2013. businessweek.com/articles/2013-04-24/how-many-hft-firms-actually-use-twitter-to-trade.

38   Bots statistics: Kate Kaye. "Hill Holliday's Ilya Vedrashko on the Trouble with Social Listening." *Advertising Age*. July 30, 2013. adage.com/article/datadriven-marketing/q-a-social-listening-cracked/243363.

38   Twitter's bot statistics: Jeff Elder. "Inside a Twitter Robot Factory." *Wall Street Journal*. Nov. 24, 2013. online.wsj.com/news/articles/SB10001424052702304607104579212122084821400.

39   "sarcasm and sincerity": IBM. "IBM Social Sentiment Index." ibm.com/analytics/us/en/conversations/social-sentiment.html.

39   Ad targeting: Katherine Rosman and Elizabeth Dwoskin. "Marketers Want to Know What You Really Mean Online." *Wall Street Journal*. March 23, 2014. wsj.com/news/article_email/SB10001424052702303369904579423402132106512-lMyQjAxMTA0MDEwMjExNDIyWj.

40   PhoneID Score: Kashmir Hill. "Your Phone Number Is Going to Get a Reputation Score." *Forbes*. Nov. 11, 2013. forbes.com/sites/kashmirhill/2013/11/13/your-phone-number-is-going-to-be-scored.

40      $214 million industry: Natasha Singer. "In a Mood? Call Center Agents Can Tell." *New York Times*. Oct. 12, 2013. nytimes.com/2013/10/13/business/in-a-mood-call-center-agents-can-tell.html.

40   Beyond Verbal background: Beyond Verbal. "App Developers." beyondverbal.com/join-us/app-developers.

41   "measure psychological traits": Michal Kosinski, David Stilwell, and Thor Graepel. "Private Traits and Attributes Are Predictable from Digital Records of Human Behavior." *Proceedings of the National Academy of Sciences*. March 11, 2013, 110(15), 5802-5. pnas.org/content/early/2013/03/06/1218772110.

42   "a win-win-win": Nicholas Carr. "Automating the Feels." *Rough Type*. Aug. 20, 2013. roughtype.com/?p=3693.

42   "The more proactive": Betsy Morais. "Can Humans Fall in Love with Bots?" *New Yorker*. Nov. 19, 2013. newyorker.com/online/blogs/elements/2013/11/her-film-spike-jonze-can-humans-fall-in-love-with-bots.html.

42   Suggest responses: BBC News. "Google Patents Robot Help for Social Media Burnout." Nov. 22, 2013. bbc.co.uk/news/technology-25033172.

43   Robot sales pitch: Zeke Miller and Denver Nicks. "Meet the Robot Telemarketer Who Denies She's a Robot." *Time*. Dec. 10, 2013. newsfeed.time.com/2013/12/10/meet-the-robot-telemarketer-who-denies-shes-a-robot.

47   "The point of being": Rob Horning. "Affective Privacy and Surveillance." *New Inquiry*. April 30, 2013. thenewinquiry.com/blogs/marginal-utility/affective-privacy-and-surveillance.

47   "There's no such thing": Author's notes. Rhizome Seven on Seven conference. April 20, 2013.

48   "a giant scoreboard": Rob Horning. "Fragments on Microcelebrity." *New Inquiry*. Oct. 1, 2012. thenewinquiry.com/blogs/marginal-utility/fragments-on-microcelebrity.

48   "it becomes interesting": Losse. *The Boy Kings*, 114.

50   "data dread": Author's notes. Rhizome Seven on Seven conference. April 20, 2013.

51   "continuous partial attention": Linda Stone. "Continuous Partial Attention." *Linda Stone* (blog). lindastone.net/qa/continuous-partial-attention.

51   Frequent interruptions: Bob Sullivan and Hugh Thompson. "Brain, Interrupted." *New York Times*. May 3, 2013. nytimes.com/2013/05/05/opinion/sunday/a-focus-on-distraction.html.

51   Interruption statistics: Rachel Emma Silverman. "Workplace Distractions: Here's Why You Won't Finish This Article." *Wall Street Journal*. Dec. 11, 2012. wsj.com/article/SB10001424127887324339204578173252223022388.html.

51   when we expect interruptions: Bob Sullivan and Hugh Thompson. "Brain, Interrupted." *New York Times*. May 3, 2013. nytimes.com/2013/05/05/opinion/sunday/a-focus-on-distraction.html.

52   Sysomos study: Frederic Lardinois. "The Short Lifespan of a Tweet: Retweets Only Happen within the First Hour." *ReadWriteWeb*. Sept. 29, 2010. readwrite.com/2010/09/29/the_short_lifespan_of_a_tweet_retweets_only_happen.

52   "Ninety-two percent of retweets": ibid.

53   "more than ten faves": Katherine Rosman. "On Twitter, More 'Favoriting.'" *Wall Street Journal*. July 24, 2013. wsj.com/news/articles/SB10001424127887324564704578626070775502176.

54   "fretting about reciprocity": Henry Alford. "Twitter Shows Its Rude Side." *New York Times*. April 26, 2013. nytimes.com/2013/04/28/fashion/nasty-comments-on-twitter.html.

54   "no compunction to re-tweet": ibid.

54   "feeling very guilty": ibid.

55   "furnish evidence": Susan Sontag. *On Photography*. New York: MacMillan, 2011, 5.

55 "record of your living": ibid., 183.

56 "acquisitive": ibid., 4.

57 "travel becomes a strategy": ibid., 9.

61 "a 'Facebook Eye'": Nathan Jurgenson. "The Facebook Eye." *Atlantic*. Jan. 13, 2012. theatlantic.com/technology/archive/2012/01/the-facebook-eye/251377.

65 "But telling the truth kills virality": Fahrad Manjoo. "Why Everyone Will Totally Read This Column." *Wall Street Journal*. December 1, 2013. wsj.com/news/articles/SB10001424052702304579404579231772007379090.

65 "boorish, arrogant": *Smoking Gun*. "Meet the Horrible Florida Woman Who Filmed Herself Berating Dunkin' Donuts Workers." June 10, 2013. thesmokinggun.com/buster/taylor-chapman-dunkin-donuts-viral-video-856341.

65 *Gawker* quote: Neetzan Zimmerman. "Worst Person Ever Gives Dunkin' Donuts Worker Hell over Missing Receipt." June 10, 2013. gawker.com/worst-person-ever-gives-dunkin-donuts-worker-hell-over-512298174.

66 Reddit decision: Reddit. "Crazy girl threatens to fly to mars and blow up a Dunkin' Doughnuts because she did not receive a receipt the night before." reddit.com/r/videos/comments/1fy4zm/crazy_girl_threatens_to_fly_to_mars_and_blow_up_a.

66 "the new American dream": Jeff Jarvis. *Public Parts: How Sharing in the Digital Age Improves the Way We Work and Live*. New York: Simon & Schuster, 2011, 134.

67 "a valuable asset": Tim Wu. *The Master Switch: The Rise and Fall of Information Empires*. New York: Atlantic Books, 2010, 230.

67 definition of celebrity: Daniel J. Boorstin. *The Image: A Guide to Pseudo-Events in America*. New York: Vintage, 2012, 217.

67 Chapman mental illness: Terrence McCoy. "Taylor Chapman: Dunkin' Donuts Ranter Has Mental Issues." *New Times* (Broward/Palm Beach, Florida). June 27, 2013. browardpalmbeach.com/2013-06-27/news/taylor-chapman-dunkin-donuts-ranter-has-mental-issues.

68 "local racist": ibid.

68 "digital scarlet letter": ibid.

68 "definitely get in": Terrence McCoy. "Taylor Chapman, Dunkin' Donuts Hater, Says She's a Stripper." *New Times* (Broward/Palm Beach, Florida). June 28, 2013. blogs.browardpalmbeach.com/pulp/2013/06/taylor_chapman_stripper_interivew.php.

70 Sweet Brown: Aisha Harris. "The Troubling Viral Trend of the 'Hilarious' Black Neighbor." *Slate*. May 7, 2013. slate.com/blogs/browbeat/2013/05/07/charles_ramsey_amanda_berry_rescuer_becomes_internet_meme_video.html.

70 Michelle Clark: Bercy Blaire. "#KABOOYAW Lady Is the New Sweet Brown." HelloGiggles. March 26, 2013. hellogiggles.com/kabooyaw-lady-is-the-new-sweet-brown.

70 Viewer comments: "KABOOYOW! Michelle Clark Original Hail

Story KPRC." YouTube. March 21, 2013. youtube.com/watch?v=_
KbHKB5IiSU.

71   "borrowing money": Lynna Lai. "Charles Ramsey: *Daily Mail*
      Report Is 'Full of Lies.'" WKYC. July 16, 2013. wkyc.com/story/
      local/2013/07/15/3266887.

71   "the aesthetics of black poverty": Zoe Chace. "'Bed Intruder
      Song' Climbs the Charts.'" NPR. Aug. 13, 2010. npr.org/blogs/
      therecord/2010/08/13/129178815/bed-intruder-song-climbs-the-charts.

73   "I couldn't help but feel": Jonathan Trudel. "10 Years Later, 'Star Wars Kid'
      Speaks Out." *Maclean's*. May 9, 2013. macleans.ca/2013/05/09/10-years-
      later-the-star-wars-kid-speaks-out.

74   "negative feedback": Alex Pasternack. "Impossible Is Something: An
      Interview with Aleksey Vayner." *Motherboard*, a blog on Vice.com. May
      9, 2012. motherboard.vice.com/read/motherboard-tv-impossible-is-
      something-an-interview-with-aleksey-vayner—2.

74   "They don't know the story": ibid.

74   Death of Alex Vayner: Alex Pasternack. "Aleksey Vayner, Whose Tale the
      Internet Mocked, Has Died at 29." *Motherboard*, a blog on Vice.com. Jan.
      24, 2013. motherboard.vice.com/blog/aleksey-vayner-death-video.

75   "My intended audience": Author interview with Lena Chen. Oct. 15, 2013.

75   "compulsive over-sharers": Sheila. "The Compulsive Oversharers of the
      Internet: A Field Guide." *Gawker*. April 1, 2008. gawker.com/378892/the-
      compulsive-oversharers-of-the-internet-a-field-guide.

75   "get off the Internet": Hamilton Nolan. "It's Official: Oversharing Makes
      You Crazy." *Gawker*. Sept. 11, 2008. gawker.com/5048536/its-official-
      oversharing-makes-you-crazy.

75   "small Asian woman": Randall Patterson. "Students of Virginity." *New
      York Times Magazine*. March 30, 2008. nytimes.com/2008/03/30/
      magazine/30Chastity-t.html.

76   "The whole system": Amanda Hess. "Lena Chen on Assault by Photograph."
      Washington City Paper. April 1, 2010. washingtoncitypaper.com/blogs/
      sexist/2010/04/01/lena-chen-on-assault-by-photograph.

76   "why did I get anxious": Author interview with Lena Chen. Oct. 15, 2013.

76   "unmitigated glee": Lena Chen. "I Was the Harvard Harlot." *Salon*. May
      23, 2011. salon.com/2011/05/24/harvard_harlot_sexual_shame.

77   "When slut-shaming works": Hess. "Lena Chen."

79   "Most of the postings": Author interview with Miriam Lazewatsky. Dec. 6, 2013.

80   "totally untrue": ibid.

80   "The one I remember": ibid.

80   "I share less information": ibid.

80   "my indignation": ibid.

81   "a walking case study": Lena Chen. Lenachen.com.

84   "a really great mood ring": Author interview with Jared Keller. March 25,
      2013.

84   "most trends in Sina Weibo": Louis Yu, Sitaram Asur, and Bernardo A.

Huberman. "Dynamics of Trends and Attention in Chinese Social Media." Social Science Research Network. Dec. 2, 2013. papers.ssrn.com/sol3/papers.cfm?abstract_id=2362561.

84   "when one greedy industry": Evgeny Morozov. *To Save Everything, Click Here: The Folly of Technological Solutionism.* New York: Public Affairs, 2013, 157.

85   YTView menu: Chase Hoffberger. "I Bought Myself 60,000 YouTube Views for Christmas." *Daily Dot.* Jan. 3, 2013. dailydot.com/entertainment/how-to-buy-youtube-views.

85   Revenue: ibid.

87   2013 Twitter statistics: Karen Wickre. "Celebrating #Twitter7." Twitter blog. March 21, 2013. blog.twitter.com/2013/celebrating-twitter7.

87   2014 Twitter statistics: Yoree Koh. "Report: 44% of Twitter Accounts Have Never Sent a Tweet." *Digits*, a blog on WSJ.com. April 11, 2014. http://blogs.wsj.cos/2014/04/11/new-data-quantifies-dearth-of-tweeters-on-twitter.

87   Gingrich buying followers: John Cook. "Most of Newt Gingrich's Twitter Followers Are Fake." *Gawker.* Aug. 1, 2011. gawker.com/5826645/most-of-newt-gingrichs-twitter-followers-are-fake.

89   Paris Hilton's ad rate: Mat Honan. "How to Use Social Media to Juice Your Story's Popularity." *Wired.* July 16, 2013. wired.com/gadgetlab/2013/07/cheat-page.

89   "human-computation engine": Edwin Chen and Alpa Jain. "Improving Twitter Search with Real-Time Human Computation." Twitter blog. January 8, 2013. Archived at https://web.archive.org/web/20131021002654/https://blog.twitter.com/2013/improving-twitter-search-real-time-human-computation.

90   "difficult to work with": ibid.

90   "full-time job": ibid.

91   Ruckus's promises: Simon Owens. "Facebook User Starts a Group in Order to Get His Girlfriend to Allow a Threesome." *Bloggasm.* Sept. 9, 2006. bloggasm.com/facebook-user-starts-a-group-in-order-to-get-his-girlfriend-to-allow-a-threesome.

92   Ruckus e-mail scam: Justin Appel. "Ruckus Upsets College Music Scene." *eSchool News.* Oct. 27, 2006. Archived at web.archive.org/web/20070312161005/http://www.eschoolnews.com/news/showStory.cfm?ArticleID=6662.

94   "50 percent of ads": Russell Brandon. "Who Owns the Hashtag? (It Isn't Twitter)." *The Verge.* Feb. 7, 2013. theverge.com/2013/2/7/3960580/hashtags-are-bigger-than-twitter-vine-tumblr-instagram.

94   Ann Coulter shenanigans: "#BreakingBad Hijacked by Companies Desperately Trying to Be Cool." *Huffington Post.* Sept. 30, 2013. huffingtonpost.com/2013/09/29/breakingbad-hijacked-twitter-companies_n_4014318.html.

95   Tumblr advice for brands: "Tumblr + Brands." Tumblr. brands.tumblr.com/start.

95   "How are we supposed to": Author's notes. Rhizome Seven on Seven conference. April 2013.

96 *Gawker* traffic stats: Tom Scocca. "More People Than Ever Are Reading This Post, Maybe." *Gawker.* Aug. 26, 2013. gawker.com/more-people-than-ever-are-reading-this-post-maybe-1202309434.

96 Google and Yahoo stats: Adrianne Jeffries. "Yahoo Is Bigger Than Google, But Google Is Bigger Than Yahoo." *The Verge.* Aug. 27, 2013. theverge.com/2013/8/27/4660994/googles-global-traffic-still-beats-the-pants-off-yahoos-report-says.

96 Burberry stats: Jeff Elder. "Twitter Posts Inaccurately High Metrics about Its Ads, Changes Them After Questions." blog.sfgate.com/techchron/2013/09/18/twitter-posts-inaccurately-high-metrics-about-its-ads-changes-them-after-questions.

97 50 percent of all video ads: David Segal. "The Great Unwatched." *New York Times.* May 3, 2014. nytimes.com/2014/05/04/business/the-great-unwatched.html.

97 Fraud stats: Mike Shields. "Online Ad Fraudsters Are Stealing $6 Billion from Brands." *Adweek.* Oct. 1, 2013. adweek.com/news/advertising-branding/online-ad-fraudsters-are-stealing-6-billion-brands-152823.

98 "unforeseen manipulations": Brian Abelson. "Whither the Pageview Apocalypse?" *Brian Abelson* (blog). Oct. 9, 2013. brianabelson.com/open-news/2013/10/09/Whither-the-pageview_apocalypse.html.

102 engagement or engaged minutes: Lauryn Bennett. "Metrics that Matter & the Death of the Page View." *Chartbeat* (blog). Jan 10, 2012. blog.chartbeat.com/2012/01/10/metrics-that-matter-and-death-of-the-page-view.

103 "55 percent of Web users": Tony Haile. "What You Think You Know About the Web Is Wrong." *Time.* March 9, 2014. time.com/12933/what-you-think-you-know-about-the-web-is-wrong.

104 "did not return a call": Ben Jacobs. "Cory Booker Visiting Iowa In August (CORRECTED)." *Daily Beast.* July 29, 2013. thedailybeast.com/articles/2013/07/29/cory-booker-visiting-iowa-in-august.html

106 "Obama Is Wrong": Danny Hayes. "Obama is wrong. Traditional journalism isn't dead." *Wonkblog,* a blog on Washingtonpost.com. Aug. 4, 2013. washingtonpost.com/blogs/wonkblog/wp/2013/08/04/obama-is-wrong-traditional-journalism-isnt-dead.

106 *Washington Post* correction: ibid.

106 *Wonkblog* retraction: Will Oremus. "Why Bloggers Fell for a Fake *TechCrunch* Story about Self-Driving Cars." *Slate.* Aug. 27, 2013. slate.com/blogs/future_tense/2013/08/27/uber_google_and_driverless_cars_fake_techcrunch_story_widely_reported_as.html.

107 "the trouble of figuring out": ibid.

107 "it all depends": Dan Alexander. "Can Houston Astros Really Be Losing Money Despite Rock-Bottom Payroll?" *Forbes.* Aug. 29, 2013. forbes.com/sites/danalexander/2013/08/29/can-houston-astros-really-be-losing-money-despite-rock-bottom-payroll.

109   "Some of this is on you": Alex Goldman. "The Breaking News Consumer's Handbook." *On the Media*. Sept. 20, 2013. onthemedia.org/story/breaking-news-consumers-handbook-pdf.

110   "When everyone is monitoring": James Gleick. "'Total Noise,' Only Louder." *New York*. April 20, 2013. nymag.com/news/intelligencer/boston-manhunt-2013-4.

110   "practically anything": ibid.

111   "But now we should assume": John Herrman and Ben Smith. "The Media Doesn't Own the Story Anymore." *BuzzFeed*. April 18, 2013. buzzfeed.com/jwherrman/the-media-doesnt-own-the-story-anymore.

114   "content is more viral": Jonah Peretti. "Mormons, Mullets, and Maniacs." New York Viral Media Meetup. Aug. 12, 2010. socialmediagovernance. com/blog/team-building/the-bored-at-work-network-by-jonah-peretti.

116   Fort Hood post page views: Benny Johnson. "Inside Fort Hood, the Site of Tragedy and Everyday American Life." *BuzzFeed*. April 6, 2014. buzzfeed. com/bennyjohnson/inside-fort-hood-the-site-of-tragedy-and-everyday-american-l.

117   "Axe's owner Unilever": Mark Duffy. "Top 10 Best Ever WTF OMG Reasons *BuzzFeed* Fired Me, LOL!" *Gawker*. Nov. 25, 2013. gawker.com/top-10-best-ever-wtf-omg-reasons-buzzfeed-fired-me-lol-1471409834.

118   they micro-target: Will Oremus. "The Rise of the Demolisticle." *Slate*. July 18, 2013. slate.com/articles/technology/future_tense/2013/07/demolisticles_buzzfeed_lists_crafted_for_specific_demographics_are_social.html.

118   "thisness": Sasha Weiss. "What We're Reading: *BuzzFeed*, "Pulphead," Chekhov, and More." NewYorker.com. April 2012. newyorker.com/online/blogs/books/2012/04/what-were-reading-buzzfeed-pulphead-chekhov-and-more.html.

119   A common affirmative response: Chadwick Matlin. "The 1 Way to Get Readers to Share Your Web Content." "A Tumblr." May 2, 2013. chadwickmatlin.tumblr.com/post/49453256731/the-1-way-to-get-readers-to-share-your-web-content.

119   "We adore being targeted": Alice Gregory. "Nicole Holofcener Nails It." *New Yorker*. Sept. 20, 2013. newyorker.com/culture/culture-desk/nicole-holofcener-nails-it.

119   "it's intellectually disingenuous": ibid.

120   "the flattery intrinsic": Thomas de Zengotita. *Mediated: How the Media Shapes Your World and the Way You Live in It*. New York: Bloomsbury, 2005, 14.

122   "the fastest growing": Anya Kamenetz. "How Upworthy Used Emotional Data to Become the Fastest Growing Media Site of All Time." *Fast Company*. June 7, 2013. fastcompany.com/3012649/how-upworthy-used-emotional-data-to-become-the-fastest-growing-media-site-of-all-time.

123   Headline testing: Nitsuh Abebe. Watching Team Upworthy Work Is Enough to Make You a Cynic. Or Lose Your Cynicism. Or

Both. Or Neither." *New York*. March 23, 2014. nymag.com/daily/intelligencer/2014/03/upworthy-team-explains-its-success.html.

126 An opinion . . . he doesn't hold: Joe Eskenazi. "Top 5 Ways Bleacher Report Rules the World!" *SF Weekly*. Oct. 3, 2012. sfweekly.com/2012-10-03/news/bleacher-report-sports-journalism-internet-espn-news-technology/full.

126 "don't expect to find any original reporting": ibid.

126 aggregate . . . search engine optimization: ibid.

127 "hyperbolic headlines": ibid.

127 "Here's a labor force": Ryan Chittum. "Bleacher Report and the Race to the Bottom." *Columbia Journalism Review*. Oct. 5, 2012. cjr.org/the_audit/sf_weekly_on_what_bleacher_rep.php.

128 $200 million: Eskenazi. "Top 5 Ways Bleacher Report Rules the World!"

129 "emerging culture of surveillance": David Lyon. "The Culture of Surveillance: Who's Watching Whom, Now?" Lecture, University of Sydney. March 1, 2012. sydney.edu.au/sydney_ideas/lectures/2012/professor_david_lyon.shtml.

129 "They quite literally": Barton Gellman and Laura Poitras. "U.S., British Intelligence Mining Data from Nine U.S. Internet Companies in Broad Secret Program." *Washington Post*. June 7, 2013. washingtonpost.com/investigations/us-intelligence-mining-data-from-nine-us-internet-companies-in-broad-secret-program/2013/06/06/3a0c0da8-cebf-11e2-8845-d970ccb04497_story.html.

130 Utah data center: James Bamford. "The NSA Is Building the Country's Biggest Spy Center (Watch What You Say)." *Wired*. March 15, 2012. wired.com/threatlevel/2012/03/ff_nsadatacenter.

131 "Any transactional information": Author's notes. MoMA PS1 event. June 30, 2013.

132 "We look for things": Pete Thamel. "Tracking Twitter, Raising Red Flags." *New York Times*. March 30, 2012. nytimes.com/2012/03/31/sports/universities-track-athletes-online-raising-legal-concerns.html.

133 Glendale school district contract: Kelly Corrigan. "Glendale Is Paying Service to Monitor Students Online." *Glendale News-Press* (Los Angeles). Aug. 24, 2013. glendalenewspress.com/news/tn-gnp-me-monitoring-20130824,0,4640365.story.

134 "You can't have 100 percent security": Matt Spetalnick and Steve Holland. "Obama Defends Surveillance as 'Trade-Off' for Security." Reuters. June 7, 2013. reuters.com/article/2013/06/08/us-usa-security-records-idUSBRE9560VA20130608.

138 "searchable and shareable": Nick Bilton. "Disruptions: At Odds over Privacy Challenges of Wearable Computing." *Bits*, a blog on NYTimes.com. May 26, 2013. bits.blogs.nytimes.com/2013/05/26/disruptions-at-odds-over-privacy-challenges-of-wearable-computing.

138 "point-of-view lifestyle" device: Panasonic. A100: Point-of-View Lifestyle Wearable Full HD Camcorder. shop.panasonic.com/shop/model/HX-A100D.

138 "a vibrant social life": *Economist*. "Very Personal Finance." June 2, 2012.
     economist.com/node/21556263.

138 "information underrepresents reality": Jaron Lanier. *You Are Not a Gadget*.
     New York: Random House, 2010, 69.

139 data-mining is ineffective: Ryan Singel. "Data-Mining for Terrorists Not
     'Feasible,' DHS-Funded Study Finds." *Wired*. Oct. 7, 2008.
     wired.com/2008/10/data-mining-for.

142 "hid from the telescreens": Walter Kirn. "Little Brother Is Watching."
     New York Times Magazine. Oct. 15, 2010. nytimes.com/2010/10/17/
     magazine/17FOB-WWLN-t.html.

144 British CCTV stat: "We're watching you: 'Britons Caught on CCTV 70 Times
     a Day.'" *London Evening Standard*. March 3, 2011. standard.co.uk/news/were-
     watching-you-britons-caught-on-cctv-70-times-a-day-6573202.html.

144 Hayden Twitter eavesdropping: Brian Fung. "Inside Former NSA Chief
     Michael Hayden's 'Interview' with an Amtrak Live-Tweeter." *The Switch*, a
     blog on Washingtonpost.com Oct. 24, 2013. washingtonpost.com/blogs/
     the-switch/wp/2013/10/24/inside-former-nsa-chief-michael-haydens-
     interview-with-an-amtrak-live-tweeter.

147 Justin Bieber server stat: Samuel Axon. "3% of Twitter's Servers Dedicated
     to Justin Bieber." *Mashable*. Sept. 7, 2010. mashable.com/2010/09/07/
     justin-bieber-twitter/

149 "Merely being on social media": Rob Horning. "Affective privacy and
     surveillance." *New Inquiry*. April 30, 2013. thenewinquiry.com/blogs/
     marginal-utility/affective-privacy-and-surveillance.

152 Robert O'Donnell: Jim Hopkins and Charisse Jones. "Disturbing Legacy
     of Rescues: Suicide." *USA Today*. Sept. 23, 2003. usatoday.com/news/
     nation/2003-09-22-legacy-usat_x.htm.

152 Adopting micro-fame: Rob Horning. "Hi Haters!" *New Inquiry*. Nov. 27,
     2012. thenewinquiry.com/essays/hi-haters.

155 "Anonymity is a shield": *McIntyre v. Ohio Elections Commission*, 514
     U.S. 334. Supreme Court of the United States. 1995. Legal Information
     Institute. 27 Aug. 2013.

156 "a burden": Sherry Turkle. "In Constant Digital Contact, We Feel 'Alone
     Together.'" *Fresh Air*, National Public Radio. Oct. 17, 2012.
     npr.org/2012/10/18/163098594/in-constant-digital-contact-we-feel-alone-
     together.

156 "it's the Facebook identity": ibid.

158 "account for their past": Charles Arthur. "Google's Eric Schmidt: Drone
     Wars, Virtual Kidnaps and Privacy for Kids." *Guardian*. Jan. 29, 2013.
     theguardian.com/technology/2013/jan/29/google-eric-schmidt-drone-wars-
     privacy.

158 Changing names: Holman W. Jenkins Jr. "Google and the Search for the
     Future." *Wall Street Journal*. Aug. 14, 2010. wsj.com/article/SB1000142405
     2748704901104575423294099527212.html.

159 "the social web can't exist": Julianne Pepitone. "Facebook Is Now Too Big to

Buy." CNNMoney. Nov. 8, 2011. money.cnn.com/2011/11/08/technology/zuckerberg_charlie_rose/index.htm.

159 "Having two identities": Miguel Helft. "Facebook, Foe of Anonymity, Is Forced to Explain a Secret." *New York Times*. May 13, 2011. nytimes.com/2011/05/14/technology/14facebook.html

159 "I think anonymity": Bianca Bosker. "Facebook's Randi Zuckerberg: Anonymity Online 'Has to Go Away.'" *Huffington Post*. July 27, 2011. huffingtonpost.com/2011/07/27/randi-zuckerberg-anonymity-online_n_910892.html.

159 "people behave a lot better": ibid.

159 malicious comments: Gregory Fernstein. "Surprisingly Good Evidence That Real Name Policies Fail to Improve Comments." *TechCrunch*. July 29, 2012. techcrunch.com/2012/07/29/surprisingly-good-evidence-that-real-name-policies-fail-to-improve-comments.

160 Facebook's PR firm: Dan Lyons. "Facebook Busted in Clumsy Smear on Google." *Daily Beast*. May 11, 2011. thedailybeast.com/articles/2011/05/12/facebook-busted-in-clumsy-smear-attempt-on-google.html.

162 Christopher Poole on identity: Aleks Krotoski. "Online Identity: Is Authenticity or Anonymity More Important?" *Guardian*. April 19, 2012. guardian.co.uk/technology/2012/apr/19/online-identity-authenticity-anonymity.

163 "We should no longer see": Anders Colding-Jørgenson. "Goodbye to the Global Information Society." *Scenario*. Copenhagen Institute for Futures Studies. 2012. scenariomagazine.com/goodbye-to-the-global-information-society.

167 "Out of revenge": Claude Lévi-Strauss. *Triste Tropiques*. New York: Penguin, 2012, 279.

168 "SEO-shaming": Laura Hudson. "Why You Should Think Twice Before Shaming Anyone on Social Media." *Wired*. July 24, 2013. wired.com/underwire/2013/07/ap_argshaming.

168 "the same tactic that trolls use": ibid.

170 @PublicShaming background: Author interview. Oct. 23, 2013.

171 "needs to be entertaining": ibid.

171 "Social justice blogs": ibid.

171 "The point of the blog": ibid.

171 "If you saw a kid": ibid.

171 Binder's criteria: ibid.

172 "I really don't want": Author interview. Oct. 18, 2013.

172 "It's really opened my eyes": ibid.

173 "I'm not at the forefront": ibid.

174 The democratic promise of online speech: Michael Eisen. "The Court of Public Opinion Is About Mob Justice and Reputation as Revenge." *Wired*. Feb. 26, 2013. wired.com/opinion/2013/02/court-of-public-opinion.

176 Selling personal fitness data: WXII12. "Privacy Concerns with Health Apps." Aug. 28, 2013. wxii12.com/news/local-news/north-carolina/Privacy-concerns-with-Health-Apps/21679798.

177   Messenger permissions: Sam Fiorella. "The Insidiousness of Facebook Messenger's Mobile App Terms of Service." *Huffington Post.* Dec. 1, 2013. huffingtonpost.com/sam-fiorella/the-insidiousness-of-face_b_4365645.html.

181   Criticism of LinkedIn: David Veldt. "LinkedIn: The Creepiest Social Network." *Interactually.* May 9, 2013. interactually.com/linkedin-creepiest-social-network.

182   "He responded by saying": Author interview with Franklin Leonard. Oct. 27, 2013.

183   Racist price discrimination: Benjamin Edelman and Michael Luca. "Digital Discrimination: The Case of Airbnb.com." Harvard Business School working paper. Jan. 10, 2014. www.hbs.edu/faculty/Publication%20 Files/14-054_e3c04a43-c0cf-4ed8-91bf-cb0ea4ba59c6.pdf.

184   "We have to have our dark corners": Chris Heath. "Mad German Auteur, Now in 3-D!" *GQ.* May 2011. gq.com/entertainment/movies-and-tv/201105/werner-herzog-profile-cave-of-forgotten-dreams.

184   "I welcome it": ibid.

187   "They're becoming increasingly wary": Somini Sengupta. "No U.S. Action, So States Move on Privacy Law. *New York Times*, October 30, 2013. nytimes.com/2013/10/31/technology/no-us-action-so-states-move-on-privacy-law.html.

187   Finding an Uber score: Aaron Landy. "How to Find Your Uber Passenger Rating." Medium. June 27, 2014. medium.com/@aaln/how-to-find-your-uber-passenger-rating-4aa1d9cc927f.

188   Uber passenger rating: Kevin Roose. "Uber Anxiety: When Your Car Service Is Judging You Back." *New York.* June 4, 2014. nymag.com/daily/intelligencer/2014/06/uber-anxiety.html.

189   Percent of fake reviews: Dave Streitfeld. "Give Yourself 5 Stars? Online, It Might Cost You." *New York Times.* Sept. 22, 2013. nytimes.com/2013/09/23/technology/give-yourself-4-stars-online-it-might-cost-you.html.

189   Average Yelp review: Seth Graham-Felsen. "Starstruck." *New York.* March 18, 2012. nymag.com/news/intelligencer/yelp-2012-3.

189   Herding effect: Daniel Akst. "When 'Likes' Bias Ratings." *Wall Street Journal.* Sept. 6, 2013. wsj.com/news/articles/SB100014241278873241230 04579055081221503924.

189   Value of an extra star: Ray Fisman. "Should You Trust Online Reviews?" *Slate.* Aug. 14, 2012. slate.com/articles/business/the_dismal_science/2012/08/tripadvisor_expedia_yelp_amazon_are_online_reviews_trustworthy_economists_weigh_in_.2.html.

190   Honestly's deception: Phil Freo. "Honestly.com—Not Acting so Honestly." PhilFreo.com. May 14, 2011. philfreo.com/blog/honestly-com-not-acting-so-honestly.

190   Searching for passives: Anthony Ha. "Honestly.com Becomes a Talent Search Engine, Renames Itself TalentBin." *TechCrunch.* May 15, 2012. techcrunch.com/2012/05/15/honestly-becomes-talentbin.

192 Origin of Rateocracy: Joe McKendrick. "What Technology Has Wrought: The Rise of the 'Rateocracy.'" *SmartPlanet*. July 31, 2013. smartplanet.com/blog/bulletin/what-technology-has-wrought-the-rise-of-the-rateocracy.

192 As one critic put it, *Stop the Cyborgs*. "Rateocracy: When Everyone and Everything Has a Rating." Oct. 18, 2013. stopthecyborgs.org/2013/10/18/rateocracy-when-everyone-and-everything-has-a-rating.

192 Size of reputation-management field: Graeme Wood. "Scrubbed." *New York*. June 16, 2013. nymag.com/news/features/online-reputation-management-2013-6.

193 "Scarlet SEO": Kashmir Hill. "Scarlet SEO: Zumba Prostitution List Published to Shame Johns." *Forbes*. Oct. 16, 2012. forbes.com/sites/kashmirhill/2012/10/16/scarlet-seo-zumba-prostitution-list-published-to-shame-johns.

194 "The Internet has democratized": Ben Popper. "Your Klout Score Must Be Greater Than 35 to Read This." *The Verge*. Oct. 8, 2012. theverge.com/2012/10/8/3461226/your-klout-score-must-be-greater-than-35-to-read-this-post.

195 Klout usage: Casey Newton. "Under the Influence: Bing Lifts Klout to the Top of Search Results." *The Verge*. May 8, 2013. theverge.com/2013/5/8/4311366/under-the-influence-bing-lifts-klout-to-the-top-of-search-results.

195 American Airlines perks: Emily Price. "How Klout Can Score You the VIP Airport Treatment." *Mashable*. May 7, 2013. mashable.com/2013/05/07/klout-american-airlines.

195 More Klout perks: John Scalzi. "Why Klout Scores Are Possibly Evil." CNNMoney. Nov. 15, 2011. money.cnn.com/2011/11/15/technology/klout_scores.

195 Klout in hiring decisions: Seth Stevenson. "What Your Klout Score Really Means." *Wired*. April 24, 2012. wired.com/business/2012/04/ff_klout.

196 Klout sale price: JP Mangalindan. "Klout Acquired for $200 Million by Lithium Technologies." *Fortune*. March 26, 2014. tech.fortune.cnn.com/2014/03/26/klout-acquired-for-200-million-by-lithium-technologies.

196 Salesforce using Klout: Stevenson. "What Your Klout Score Really Means."

196 Genesys using Klout: Alex Knapp. "A High Klout Score Can Lead to Better Customer Service." *Forbes*. June 12, 2012. forbes.com/sites/alexknapp/2012/06/12/a-high-klout-score-can-lead-to-better-customer-service.

196 Klout uses Klout scores for hiring: Jeanne Meister. "Will Your Klout Score Get You Hired? The Role of Social Media in Recruiting." *Forbes*. May 7, 2012. forbes.com/sites/jeannemeister/2012/05/07/will-your-klout-score-get-you-hired-the-role-of-social-media-in-recruiting.

197 "Brands couldn't get in touch": Stevenson. "What Your Klout Score Really Means."

197 "Individuals tend to believe": Chad Edwards, Patric R. Spence, Christina J. Gentile, America Edwards and Autumn Edwards. "How Much Klout Do You Have . . . A Test of System Generated Cues on Source Credibility."

Computers in Human Behavior 29, no. 5 (September 2013):A12–A16. sciencedirect.com/science/article/pii/S0747563212003767.

199   "real answers from people": Quora. "Quora's Mission Is to Share and Grow the World's Knowledge." quora.com/about.

199   the logic of advertising: Michael Thomsen. "How to Get a Job with Twitter and Why That's a Horrible Thing." *Complex*. Aug. 7, 2013. complex.com/tech/2013/08/twitter-job.

200   Klout encourages "cadence": Stevenson. "What Your Klout Score Really Means."

200   "visibility" and "empowerment": Taina Bucher. "Want to Be on the Top? Algorithmic Power and the Threat of Invisibility on Facebook." *New Media & Society* 14, no. 7 (April 8, 2012):1164–80. nms.sagepub.com/content/14/7/1164.

201   "a certain platform logic": ibid.

202   Facebook's acknowledgement: Jennifer Slegg. "Facebook Admits: Expect Organic Reach for Pages to Continue Declining." *Search Engine Watch*. Dec. 6, 2013. searchenginewatch.com/article/2317757/Facebook-Admits-Expect-Organic-Reach-for-Pages-to-Continue-Declining.

202   Decline of organic reach: Sam Biddle. "Facebook Is Ending the Free Ride." *Valleywag*, a blog on *Gawker*. March 19, 2014. valleywag.gawker.com/facebook-is-about-to-make-everyone-pay-1547309811.

203   "100,000 individual weights": Matt McGee. "EdgeRank Is Dead: Facebook's News Feed Algorithm Now Has Close to 100K Weight Factors." *Marketing Land*. Aug. 16, 2013. marketingland.com/edgerank-is-dead-facebooks-news-feed-algorithm-now-has-close-to-100k-weight-factors-55908.

204   "When we *tell* people": Christian Rudder. "We Experiment on Human Beings!" OkCupid. July 28, 2014. blog.okcupid.com/index.php/we-experiment-on-human-beings.

204   "That's how Web sites work": ibid.

204   Variety of credit data: Evgeny Morozov. "Your Social Networking Credit Score." *Slate*. Jan. 30, 2013. slate.com/articles/technology/future_tense/2013/01/wonga_lenddo_lendup_big_data_and_social_networking_banking.single.html.

205   "all data is credit data": Quentin Hardy. "Just the Facts. Yes, All of Them." *New York Times*. March 24, 2012. nytimes.com/2012/03/25/business/factuals-gil-elbaz-wants-to-gather-the-data-universe.html.

205   "Given how much they know": Morozov, "Your Social Networking Credit Score."

205   "most receptive to brand messages": Elizabeth Dwoskin. "Web Giants Threaten End to Cookie Tracking." *Wall Street Journal*. Oct. 28, 2013. wsj.com/news/articles/SB10001424052702304682504579157780178992984.

205   they may already know: Adrianne Jeffries. "As Banks Start Nosing Around Facebook and Twitter, the Wrong Friends Might Just Sink Your

Credit." *BetaBeat*. Dec. 13, 2011. betabeat.com/2011/12/as-banks-start-nosing-around-facebook-and-twitter-the-wrong-friends-might-just-sink-your-credit.

208 "Google Removal Requests": Just Mugshots.

208 "extortion or defamation business": "Disclaimer (Please Read Before Making Threats!)." Just Mugshots. support.justmugshots.com/entries/22059837-Legal-Disclaimer-Please-read-before-making-threats.

209 doing a public service: David Segal. "Mugged by a Mugshot Online." *New York Times*. Oct. 5, 2013. nytimes.com/2013/10/06/business/mugged-by-a-mug-shot-online.html.

209 Lawyer death threats: David Segal. "Mug-Shot Websites, Retreating or Adapting." *New York Times*. Nov. 9, 2013. nytimes.com/2013/11/10/your-money/mug-shot-websites-retreating-or-adapting.html.

209 "end your humiliation": ImageMax PR. "How to Remove Mugshot from Google." YouTube. Jan. 24, 2012. youtube.com/watch?v=knDzXWT1IEw.

209 "much like a document service": RemoveArrest. removearrest.com.

209 Mugshot site owner's criminal record: David Kravets. "Mug-Shot Industry Will Dig Up Your Past, Charge You to Bury It Again." *Wired*. Aug. 2, 2011. wired.com/threatlevel/2011/08/mugshots.

211 Corona's prison stint: Eric J. Lyman. "Italy's Reviled Paparazzo Fabrizio Corona Facing 10 Years in Prison." *Hollywood Reporter*. April 11, 2013. hollywoodreporter.com/news/italys-reviled-paparazzo-fabrizio-corona-437761.

213 Reputation.com's clients: Tim Dowling. "Search Me: Online Reputation Management." *Guardian*. May 24, 2013. theguardian.com/technology/2013/may/24/search-me-online-reputation-management.

213 "personal brand": Hamza Shaban. "LinkedIn's New Network for Teens Is a Wasted Opportunity." *New Republic*. Oct. 30, 2013. newrepublic.com/article/115406/linkedins-new-network-teens-wasted-opportunity.

213 Metal Rabbit Media rates: Wood. "Scrubbed."

213 body doubles: ibid.

213 "data has a life": Matt Waite. "Public Info Doesn't Always Want to Be Free." *Source*. March 11, 2013. source.opennews.org/en-US/learning/public-info-doesnt-always-want-be-free.

214 "Because you can": ibid.

215 "identity verification": Lois Beckett. "Everything We Know About What Data Brokers Know About You." *ProPublica*. Sept. 13, 2013. propublica.org/article/everything-we-know-about-what-data-brokers-know-about-you.

215 TSA sells to debt collectors: Susan Stellin. "Security Check Now Starts Long Before You Fly." *New York Times*. Oct. 21, 2013. nytimes.com/2013/10/22/business/security-check-now-starts-long-before-you-fly.html.

215 Experian investigation: Brian Krebs. "Experian Sold Consumer Data

to ID Theft Service." KrebsonSecurity. Oct. 20, 2013. krebsonsecurity.com/2013/10/experian-sold-consumer-data-to-id-theft-service.

215   Identity thieves in Vietnam: Sean Vitka. "Experian-Acquired Data Broker Sold Social Security Numbers to Identity Thieves." *Slate*. Oct. 25, 2013. slate.com/blogs/moneybox/2013/10/25/experian_data_broker_social_security_numbers_sold_to_identity_thieves.html.

215   Experian revenue: "Investor Centre." Experian. Oct. 31, 2013. experianplc.com/investor-centre.aspx.

216   "detail about behaviors and proclivities": Woodrow Hartzog and Evan Selinger. "Big Data in Small Hands." 66 *Stanford Law Review Online* 81. Sept. 3, 2013. stanfordlawreview.org/online/privacy-and-big-data/big-data-small-hands.

219   Nandini Balial background and Rabbit experience: Author interviews with Nandini Balial. July and August 2014.

226   Mechanical Turk earnings: Jeremy Wilson. "My Grueling Day as an Amazon Mechanical Turk." *Kernel*. Aug. 28, 2013. kernelmag.com/features/report/4732/my-gruelling-day-as-an-amazon-mechanical-turk.

228   "a cloud-computing cross": Quentin Hardy. "Elance Pairs Hunt for Temp Work with Cloud Computing." *Bits*, a blog on NYTimes.com. Sept. 24, 2013. bits.blogs.nytimes.com/2013/09/24/elance-pairs-hunt-for-temp-work-with-cloud-computing.

228   Mechanical Turk survey: Panos Ipeirotis. "Demographics of Mechanical Turk." Stern School of Business. March 2010. archive.nyu.edu/bitstream/2451/29585/2/CeDER-10-01.pdf.

228   "hyperdata": Quentin Hardy. "Big Data's Little Brother." *New York Times*. Nov. 10, 2013. nytimes.com/2013/11/11/technology/gathering-more-data-faster-to-produce-more-up-to-date-information.html.

229   "the performance of the workers": Ayhan Aytes. "Return of the Crowds." In *Digital Labor: The Internet as Playground and Factory*. Trebor Scholz, ed. New York: Routledge, 2013, 81.

229   "artificial artificial intelligence": ibid.

230   "there is a sewer channel": Adrian Chen. "Inside Facebook's Outsourced Anti-Porn and Gore Brigade, Where 'Camel Toes' Are More Offensive Than 'Crushed Heads.'" *Gawker*. Feb. 16, 2012. gawker.com/5885714/inside-facebooks-outsourced-anti-porn-and-gore-brigade-where-camel-toes-are-more-offensive-than-crushed-heads.

231   trauma for content moderators: Brad Stone. "Policing the Web's Lurid Precincts." *New York Times*. July 19, 2010. nytimes.com/2010/07/19/technology/19screen.html.

232   "If you are taking more than thirty minutes": Andrew Leonard. "The Scary New Labormetrics of Work." *Salon*. Aug. 16, 2013. salon.com/2013/08/16/the_scary_new_labormetrics_of_work.

232   chance of Gigwalk success: Caroline Winter. "Gigwalk Does Temp-Worker Hiring without Job Interviews." *Bloomberg Businessweek*. June 24, 2013. businessweek.com/articles/2013-06-24/gigwalk-does-temp-worker-hiring-without-job-interviews.

234 "people as businesses": Caleb Garling. "Taking Care with the Internet Axiom." *San Francisco Chronicle*. Nov. 12, 2013. blog.sfgate.com/techchron/2013/11/12/internet-axiom-github-airbnb.

235 "Prime Time amount": Salvador Rodriguez. "Lyft Also Will Instate Fares in California, Ditching Donation System." *Los Angeles Times*. Nov. 15, 2013. latimes.com/business/technology/la-fi-tn-lyft-minimum-fares-california-20131115,0,1699156.story.

235 Rating drivers: "A Sense of Place." *Economist*. Oct. 25, 2012. economist.com/news/special-report/21565007-geography-matters-much-ever-despite-digital-revolution-says-patrick-lane.

236 "That's part of the strategy": Alyson Shontell. "My Nightmare Experience as a TaskRabbit Drone." *Business Insider*. Dec. 7, 2011. businessinsider.com/confessions-of-a-task-rabbit-2011-12.

236 deactivating drivers' accounts: Rachel Swan. "Chopped Livery: Start-Ups Revolutionize the Cab Industry." *SF Weekly*. March 27, 2013. sfweekly.com/2013-03-27/news/uber-lyft-sidecar-cabs-sfmta.

238 San Francisco evictions: Steven T. Jones and Parker Yesko. "Into Thin Air." San Francisco Bay Guardian. Aug. 6, 2013. sfbg.com/2013/08/06/thin-air.

238 "grassroots organization": "About Peers." Peers. peers.org/about.

238 "sharing economy start-ups": Andrew Leonard. "Who Owns the Sharing Economy?" *Salon*. Aug. 2, 2013. salon.com/2013/08/02/who_owns_the_sharing_economy.

239 "It's like the United Nations": Jones and Yesko. "Into Thin Air."

239 "a shared identity": Tom Slee. "Why the Sharing Economy Isn't." *Whimsley*. Aug. 30, 2013. tomslee.net/2013/08/why-the-sharing-economy-isnt.html.

239 "unreasonable obstacles": ibid.

239 Atkin's connection to Peers: Nitasha Tiku. "Airbnb's Industry Mouthpiece Astroturfs for Donations." *Valleywag*, a blog on *Gawker*. Dec. 11, 2013. valleywag.gawker.com/airbnbs-industry-mouthpiece-astroturfs-for-donations-1481305550.

239 Pierre Omidyar and Peers: Ryan Chittum. "*Fortune* Flacks for the 'Sharing Economy.'" *Columbia Journalism Review*. Dec. 10, 2013. cjr.org/the_audit/fortune_flacks_for_the_sharing.php.

239 "We'll provide everything": Peers. "Help Fix the Law in New York." action.peers.org/page/signup/office-drop-by-in-new-york-.

240 "You are responsible": Airbnb. "Terms of Service." May 22, 2012. airbnb.com/home/terms.

241 Uber financing program: Mark Milian. "Uber Drivers to Get GM and Toyota Financing Deals." Bloomberg News. Nov. 25, 2013. bloomberg.com/news/2013-11-25/uber-drivers-to-get-gm-and-toyota-financing-deals.html.

242 Uber fixes the market: Matthew Stoller. "Observations on Credit and Surveillance: "Uber's Algorithmic Monopoly." April 9, 2014. mattstoller.tumblr.com/post/82233202309/ubers-algorithmic-monopoly-we-are-not-setting.

242   TaskRabbit worker almost fired: Shontell. "My Nightmare Experience as a
       TaskRabbit Drone."

243   Airbnb taxes: Nitasha Tiku. "Airbnb Is Suddenly Begging New York City to
       Tax Its Hosts $21 Million." *Valleywag*, a blog on *Gawker*. March 28, 2014.
       valleywag.gawker.com/airbnb-is-suddenly-begging-new-york-city-to-tax-its-
       hos-1553889167.

244   "we literally stand on the brink": Tom Slee. "Why the Sharing Economy Isn't."

245   Nandini Balial background and TaskRabbit experience: Author interviews
       with Nandini Balial. July and August 2014.

249   "a human right": Queena Kim. "Mark Zuckerberg: Internet Connectivity Is
       a Human right." *Marketplace*. Aug. 21, 2013. marketplace.org/topics/tech/
       mark-zuckerberg-internet-connectivity-human-right.

249   "Companies are transcending power": Kevin Roose. "The Government
       Shutdown Has Revealed Silicon Valley's Dysfunction Fetish." *New York*.
       Oct 16, 2013. nymag.com/daily/intelligencer/2013/10/silicon-valleys-
       dysfunction-fetish.html.

250   "the paper belt": Nick Statt. "A Radical Dream for Making Techno Utopias
       a Reality." *CNET*. Oct. 19, 2013. cnet.com/8301-1023_3-57608320-93/a-
       radical-dream-for-making-techno-utopias-a-reality.

250   "without having to deploy them": Claire Cain Miller. "Larry Page Gets
       Personal at Google's Conference." *Bits*, a blog on NYTimes.com. May 15,
       2013. bits.blogs.nytimes.com/2013/05/15/larry-page-gets-personal-at-
       googles-conference.

251   "a public construction project": John Perry Barlow. "A Declaration of the
       Independence of Cyberspace." Electronic Frontier Foundation. Feb. 8,
       1996. projects.eff.org/~barlow/Declaration-Final.html.

254   "at no cost": ibid.

255   Amazon pulls *1984*: Brad Stone. "Amazon Erases Orwell Books from
       Kindle." *New York Times*. July 17, 2009. nytimes.com/2009/07/18/
       technology/companies/18amazon.html.

256   Zuckerberg's color-blindness: Jose Antonio Vargas. "The Face
       of Facebook." *New Yorker*. Sept. 20, 2010. newyorker.com/
       reporting/2010/09/20/100920fa_fact_vargas.

257   transcend our conditions: Jeremy Antley. "From Data Self to Data Serf."
       Peasant Muse. June 19, 2012. peasantmuse.com/2012/06/from-data-self-to-
       data-serf.html.

260   "This digital labor": Scholz, "Why Does Digital Labor Matter Now?,"2.

260   "the micro-division of labor": Andrew Ross. "In Search of the Lost
       Paycheck." In *Digital Labor: The Internet as Playground and Factory*. Trebor
       Scholz, ed. New York: Routledge, 2013, 20.

260   Apps similar to Twitch: Rachel Metz. "The Next Frontier in
       Crowdsourcing: Your Smartphone." *MIT Technology Review*. March
       12, 2014. technologyreview.com/news/525481/the-next-frontier-in-
       crowdsourcing-your-smartphone.

262   *Wall Street Journal* praise for *BuzzFeed*: Farhad Manjoo. "*BuzzFeed*'s

Brazen, Nutty, Growth Plan." *Wall Street Journal*. Oct. 14, 2013. wsj.com/article/SB10001424052702304500404579129590411867328.html.

262 "digital volunteers": Curt Hopkins. "The Smithsonian Is Outsourcing Transcription . . . to You." *Daily Dot*. Nov. 8, 2013. dailydot.com/lifestyle/smithsonian-transcription-volunteer-project.

262 Google promises exposure: Andrew Adam Newman. "Use Their Work Free? Some Artists Say No to Google." *New York Times*. June 14, 2009. nytimes.com/2009/06/15/business/media/15illo.html.

263 AOL Community Leaders: Johanna Francisca Theodora Maria van Dijck ("José"). "Users Like You? Theorizing Agency in User-Generated Content." *Media, Culture & Society* 31, no 1. (Jan. 2009): 41–58. mcs.sagepub.com/content/31/1/41.refs.

263 savings from community leaders: Tiziana Terranova. "Free Labor." In *Digital Labor: The Internet as Playground and Factory*. Trebor Scholz, ed. New York: Routledge, 2013, 48.

263 AOL settlement: Lauren Kirchner. "AOL Settled with Unpaid 'Volunteers' for $15 Million." *Columbia Journalism Review*. Feb. 10, 2011. cjr.org/the_news_frontier/aol_settled_with_unpaid_volunt.php.

263 "a gift economy *and* an advanced capitalist economy": Terranova. "Free Labor," 50.

263 "mistake this coexistence": ibid.

263 "variable scale": van Dijck. "Users Like You?"

264 Craigslist revenue per employee: Peter M. Zollman. "Craigslist Revenue, Profits Soar." Aim Group. April 30, 2010. aimgroup.com/2010/04/30/craigslist-revenue-profits-soar.

264 "social factory": Mark Andrejevic. "Estranged Free Labor." In *Digital Labor: The Internet as Playground and Factory*. Trebor Scholz, ed. New York: Routledge, 2013, 159.

264 "We do tiny bits of work": Ian Bogost. "Hyperemployment, or the Exhausting Work of the Technology User." *Atlantic*. Nov. 8, 2013. theatlantic.com/technology/archive/2013/11/hyperemployment-or-the-exhausting-work-of-the-technology-user/281149.

265 "We do the work": Will Oremus. "Facebookers of the World, Unite!" *Slate*. May 3, 2013. slate.com/articles/technology/books/2013/05/jaron_lanier_s_who_owns_the_future_review_facebookers_of_the_world_unite.single.html.

268 Resurgence of flânerie: Evgeny Morozov. "The Death of the Cyberflâneur." *New York Times*. Feb. 4, 2012. nytimes.com/2012/02/05/opinion/sunday/the-death-of-the-cyberflaneur.html.

270 Consumer's self-service: Ross. "In Search of the Lost Paycheck," 25.

270 "McDonaldization of society": George Ritzer and Nathan Jurgenson. "Production, Consumption, Prosumption: The Nature of Capitalism in the Age of the Digital 'Prosumer." *Journal of Consumer Culture* 10, no 1 (March 2010):13–36. joc.sagepub.com/content/10/1/13.abstract.

272 Changes in support staff: Craig Lambert. "Our Unpaid, Extra Shadow

Work." *New York Times*. Oct. 29, 2011. nytimes.com/2011/10/30/opinion/sunday/our-unpaid-extra-shadow-work.html.

272   Time spent e-mailing: Jena McGregor. "How Much Time You Really Spend Emailing on Work." *PostLeadership*, a blog on Washingtonpost.com. July 31, 2012. washingtonpost.com/blogs/post-leadership/post/how-much-time-you-really-spend-emailing-at-work/2012/07/31/gJQAI50sMX_blog.html.

275   EU data protection law: David Jolly. "European Union Takes Steps Toward Protecting Data." *New York Times*. March 12, 2014. nytimes.com/2014/03/13/business/international/european-union-takes-steps-toward-protecting-data.html.

279   "Why would they need that?": Nesita Kwan. "OfficeMax Apologizes for 'Daughter Killed in Car Crash' Letter." NBC 5 Chicago. Jan. 20, 2014. nbcchicago.com/news/local/OfficeMax-Apologizes-Illinois-Family-Letter-241147581.html.

281   "Privacy is mostly an illusion": Linton Weeks. "Privacy 2.0: The Garbo Economy." NPR. April 27, 2011. npr.org/2011/04/27/135623137/privacy-2-0-the-garbo-economy.

281   a "new asset class": "Personal Data: The Emergence of a New Asset Class." World Economic Forum. January 2011. www3.weforum.org/docs/WEF_ITTC_PersonalDataNewAsset_Report_2011.pdf.

282   "trust economy": Carlos Dominguez. "The Monetization of Privacy—Birth of a 'Trust Economy.'" Cisco Blogs. May 25, 2012. blogs.cisco.com/news/the-monitization-of-privacy-birth-of-a-trust-economy.

282   AT&T rates: Dan Gillmor. "AT&T Wants to Know: How Much Would You Pay for a Little Online Privacy?" *Guardian*. Dec. 13, 2013. theguardian.com/commentisfree/2013/dec/13/at-t-austin-uverse-experiment-user-data.

282   "your individual web browsing information": ibid.

284   "conceptual murkiness": Helen Nissenbaum. *Privacy in Context: Technology, Policy, and the Integrity of Social Life*. Palo Alto: Stanford University Press, 2010, 2.

284   "the opposite of being in public": Author's notes. Cyber-bullying panel at SXSW 2013.

284   "the ability to control": ibid.

284   "What people care about most": Nissenbaum. *Privacy in Context*, 2.

286   "Privacy has a social value": Daniel J. Solove. "'I've Got Nothing to Hide' and Other Misunderstandings of Privacy." *San Diego Law Review* 44 (November 2007):745–72. GWU Law School Public Law Research Paper No. 289. Available at Social Science Research Network: ssrn.com/abstract=998565.

286   Queer Chorus outing: Geoffrey A. Fowler. "When the Most Personal Secrets Get Outed on Facebook." *Wall Street Journal*. Oct. 13, 2012. wsj.com/article/SB10000872396390444165804578008740578200224.html.

287   "off the table": Tracie Hunte, Annmarie Fertoli, and Colby Hamilton.

"NYPD Commissioner Calls for More Surveillance Cameras." WNYC. April 22, 2013. wnyc.org/articles/wnyc-news/2013/apr/22/kelly-cameras.

288 "the right to be alone": Samuel Warren and Louis Brandeis. "The Right to Privacy." *Harvard Law Review* 4, no. 5 (Dec. 15, 1890). groups.csail.mit .edu/mac/classes/6.805/articles/privacy/Privacy_brand_warr2.html.

288 "instantaneous photographs": ibid.

289 "any other modern device": ibid.

291 "networked privacy": Danah Boyd. "Networked Privacy." Lecture, Personal Democracy Forum, New York City, June 6, 2011. danah.org/papers/ talks/2011/PDF2011.html.

291 "collective privacy": Lior Jacob Strahilevitz. "Collective Privacy." In *The Offensive Internet*, Saul Levmore and Martha C. Nussbaum, eds. Cambridge, Mass.: Harvard University Press, 2010, 217–36.

292 "social norm": Bobbie Johnson. "Privacy No Longer a Social Norm, Says Facebook Founder." *Guardian*. Jan 10, 2010. guardian.co.uk/ technology/2010/jan/11/facebook-privacy.

292 "We decided": Marshall Kirckpatrick. "Facebook's Zuckerberg Says the Age of Privacy Is Over." *ReadWriteWeb*. Jan. 9, 2010. readwrite .com/2010/01/09/facebooks_zuckerberg_says_the_age_of_privacy_is_ov.

292 "simpler": Declan McCullagh. "Facebook event outlines 'simpler' privacy controls." CNet. May 26, 2010. http://news.cnet.com/8301-13578_3- 20005976-38.html

293 Facebook claims the right: Turow. *The Daily You*, 145.

293 "Facebook trains you": Ciara Byrne. "Cory Doctorow: Tech Companies Exploit the Way We Undervalue Privacy." *VentureBeat*. Sept. 23, 2011. venturebeat.com/2011/09/23/cory-doctorow-tech-companies-exploit-the- way-we-undervalue-privacy.

293 Schrems's data file: Kevin J. O'Brien. "Austrian Law Student Faces Down Facebook." *New York Times*. Feb. 5, 2012. nytimes.com/2012/02/06/ technology/06iht-rawdata06.html.

294 "We record some of this": Facebook Help Center. "What Information Does Facebook Get When I Visit a Site with the Like Button or Another Social Plugin?" facebook.com/help/186325668085084.

294 "a pretty radical social change": Anil Dash. "The Facebook Reckoning." *Anil Dash* (blog). Sept. 13, 2010. dashes.com/anil/2010/09/the-facebook- reckoning-1.html.

294 "I have the privilege to do so": ibid.

295 Street View fine: Alyssa Newcomb. "Google to Pay $7 Million Fine for Street View Privacy Breach." ABC News. March 13, 2013. abcnews .go.com/Technology/google-hit-million-fine-street-view-privacy-breach/ story?id=18717950.

295 FCC fine: David Goldman. "Google Fined $25,000 for 'Willfully' Stonewalling FCC." CNNMoney. April 16, 2012. money.cnn.com/2012/04/16/technology/google-fcc/index.htm.

296 Do Not Track stats: Natasha Singer. "Do-Not-Track Talks Could Be

Running Off the Rails." *Bits*, a blog on NYTimes.com. May 3, 2013.
bits.blogs.nytimes.com/2013/05/03/do-not-track-talks-could-be-running-
off-the-rails.

296   Web sites ignoring Do Not Track: David Goldman. "Turning on Do
Not Track in Chrome." "CNNMoney Tech Tumblr." Nov. 29, 2012.
cnnmoneytech.tumblr.com/post/36831929685/turning-on-do-not-track-in-
google-chrome.

296   "on life support": David Goldman. "Do Not Track Is Dying." CNNMoney.
Nov. 30, 2012. money.cnn.com/2012/11/30/technology/do-not-track/
index.html.

296   "permit data collection": Natasha Singer. "Do Not Track? Advertisers Say
'Don't Tread on Us." *New York Times*. Oct. 13, 2012.
nytimes.com/2012/10/14/technology/do-not-track-movement-is-drawing-
advertisers-fire.html.

296   Internet Explorer criticism: ibid.

296   "Consumers have a right": Office of the Press Secretary, White House.
"Fact Sheet: Plan to Protect Privacy in the Internet Age by Adopting a
Consumer Privacy Bill of Rights." Feb. 23, 2012. whitehouse.gov/the-press-
office/2012/02/23/fact-sheet-plan-protect-privacy-internet-age-adopting-
consumer-privacy-b.

296   Confusion over Do Not Track: Ryan Singel. "White House Privacy Bill of
Rights Brought to You by Years of Online Debacles." *Wired*. Feb. 23, 2012.
wired.com/threatlevel/2012/02/privacy-bill-of-rights.

297   "retargeters": Somini Sengupta. "Start-Up Lets Users Track Who Tracks
Them." *Bits*, a blog on NYTimes.com. April 16, 2013. bits.blogs.nytimes
.com/2013/04/16/palo-alto-start-up-lets-users-track-who-tracks-them.

297   Blocking third-party cookies: John M. Simpson. "Microsoft Should
Act Now to Protect Online Privacy." *Consumer Watchdog*. May 1, 2013.
consumerwatchdog.org/blog/microsoft-should-act-now-protect-online-
privacy.

297   Ghostery selling data: Tom Simonite. "A Popular Ad Blocker Also Helps the
Ad Industry." *MIT Technology Review*. June 17, 2013. technologyreview
.com/news/516156/a-popular-ad-blocker-also-helps-the-ad-industry.

297   "This faith in information": Nissenbaum. *Privacy in Context*, 44.

298   Plan UK billboard: Yi Chen. "Facial Recognition Billboard Only Lets
Women See the Full Ad." PSFK. Feb. 21, 2012. psfk.com/2012/02/facial-
recognition-billboard.html.

299   NEC billboards: Michael Fitzpatrick. "Advertising Billboards Use Facial
Recognition to Target Shoppers." *Guardian*. Sept. 27, 2010.
guardian.co.uk/media/pda/2010/sep/27/advertising-billboards-facial-
recognition-japan.

299   "an intelligent sensor": Immersive. imrsv.com/faq/cara-basics.

300   ANAR ad: *DIYPhotography*. "Lenticular Photo Used to Secretly Convey
Hot Line Number to Abused Kids." May 3, 2013. diyphotography.net/
lenticular-photo-used-secretly-convey-hot-line-number-abused-kids.

301 Google location information: John McDermott. "Google Takes Its Tracking into the Real World." *Digiday*. Nov. 6, 2013. digiday.com/platforms/google-tracking.

301 Stores tracking shoppers: Molly Mulshine. "Department Stores Now Using Your Smartphone to Track Your Every Move." *Betabeat*. July 15, 2013. betabeat.com/2013/07/department-stores-now-using-your-smartphone-to-track-your-every-move.

301 Tracking over repeated visits: Stephanie Clifford and Quentin Hardy. "Attention, Shoppers: Store Is Tracking Your Cell." *New York Times*. July 14, 2013. nytimes.com/2013/07/15/business/attention-shopper-stores-are-tracking-your-cell.html.

301 Information sold to brands: Elizabeth Dwoskin. "Tracking Companies Agree to Notify Shoppers, but Retailers Demur." *Digits*, a blog on WSJ.com. Oct. 22, 2013. http://blogs.wsj.cos/2013/10/22/tracking-companies-agree-to-notify-shoppers-but-retailers-demur.

301 EyeSee mannequins: Jennifer O'Mahony. "Store Mannequins 'Spy on Customers' with Hidden Cameras." *Telegraph*. Nov. 26, 2012. telegraph.co.uk/technology/news/9702683/Store-mannequins-spy-on-customers-with-hidden-cameras.html.

301 recognize VIP customers: Brenda Salinas. "High-End Stores Use Facial Recognition Tools to Spot VIPs." *All Tech Considered*, a blog on NPR.com. July 21, 2013. npr.org/blogs/alltechconsidered/2013/07/21/203273764/high-end-stores-use-facial-recognition-tools-to-spot-vips.

301 "personalized deals": Red Pepper. "Facedeals." redpepperland.com/lab/details/check-in-with-your-face.

302 "These apps are bridgeheads": Cyrus Farivar. "Facial Detection Cameras Ready to Creep Out San Francisco Bar Patrons." *Ars Technica*. May 18, 2012. arstechnica.com/business/2012/05/scenetap-poised-to-creep-out-san-francisco-bar-patrons.

302 "pay per gaze": Sophie Curtis. "Google Patents 'Pay-per-Gaze' Advertising Technology." *Telegraph*. Aug. 19, 2013. telegraph.co.uk/technology/google/10252111/Google-patents-pay-per-gaze-advertising-technology.html.

302 "the more context it has": Matthew Ingram. "Google Wants to Build Maps that Customize Themselves Based on What They Know about You." *GigaOm*. Nov. 5, 2013. gigaom.com/2013/11/05/google-wants-to-build-maps-that-customize-themselves-based-on-what-they-know-about-you.

304 Self-esteem study: Rebecca J. Rosen. "Is This the Grossest Advertising Strategy of All Time?." *Atlantic*. Oct. 3, 2013. theatlantic.com/technology/archive/2013/10/is-this-the-grossest-advertising-strategy-of-all-time/280242.

304 MediaBrix campaign: Michael Carney. "MediaBrix Enables Targeted Advertising During Moments of Positive and Negative Emotion in Social Games." *Pando Daily*. April 18, 2013. pando.com/2013/04/18/mediabrix-

enables-targeted-advertising-during-moments-of-positive-and-negative-
emotion-in-social-games.

304 "Advertisers provide targeting data": Wook Jin Chung. "Patent Application
Title: Targeting Advertisements Based on Emotion." Patentdocs, Internet
FAQ Archives. June 7, 2012. faqs.org/patents/app/20120143693.

305 "It appears that Samsung": Steve Brachmann. "Samsung Seeks Patents
on Sharing User Emotion on a Social Network, Fragrant Mobile Phone."
IPWatchdog. June 26, 2013. ipwatchdog.com/2013/06/26/samsung-seeks-
patents-on-sharing-user-emotion-on-a-social-network-fragrant-mobile-
phone/id=41625.

305 FTC study: Kate Kaye. "FTC: Fitness Apps Can Help You Shred
Calories—and Privacy." *Advertising Age*. May 7, 2014. adage.com/
article/privacy-and-regulation/ftc-signals-focus-health-fitness-data-
privacy/293080.

305 Moves data sharing: Reed Albergotti. "After Facebook Deal, Moves App
Changes Privacy Policy." *Digits*, a blog on WSJ.com. May 5, 2014. http://blogs
.wsj.cos/2014/05/05/after-facebook-deal-moves-app-changes-privacy-policy./

306 Recycling bin stats: Siraj Datoo. "This Recycling Bin Is Following You."
Quartz. Aug. 8, 2013. qz.com/112873/this-recycling-bin-is-following-you.

306 "cookie for the real world": Zachary M. Seward and Siraj Datoo. "City of
London Halts Recycling Bins Tracking Phones of Passers-By." *Quartz*. Aug.
12, 2013. qz.com/114174/city-of-london-halts-recycling-bins-tracking-
phones-of-passers-by.

307 "the role of feedback mechanism": Andrejevic. "Estranged Free Labor," 158.

307 steering their behavior: ibid.

308 Acxiom data portal: Natasha Singer. "A Data Broker Offers a Peek Behind
the Curtain." *New York Times*. Aug. 31, 2013. nytimes.com/2013/09/01/
business/a-data-broker-offers-a-peek-behind-the-curtain.html.

308 "it's simply the idea": Author interview with Joseph Turow. Oct. 14, 2013.

308 "companies are defining us": ibid.

308 Nathalie Blanchard insurance story: CBC News. "Depressed Woman
Loses Benefits over Facebook Photos." Nov. 21, 2009. cbc.ca/news/canada/
montreal/depressed-woman-loses-benefits-over-facebook-photos-1.861843.

309 "People will worry": Author interview with Joseph Turow. Oct. 14, 2013.

310 "Facebook's privacy tools": *Consumer Reports*. "Facebook & Your Privacy."
June 2012. consumerreports.org/cro/magazine/2012/06/facebook-your-
privacy/index.htm.

311 EFF praise for Twitter: Electronic Frontier Foundation. "Who Has Your
Back?" April 2013. eff.org/who-has-your-back-2013.

311 Scroogled ads: Tom Warren. "Microsoft's Latest 'Scroogled' Ad Attacks
Android with Privacy Fears." *The Verge*. April 9, 2013. theverge
.com/2013/4/9/4206486/scroogled-android-google-play-campaign.

313 Data production stat: Erez Aiden and Jean-Baptiste Michel. *Uncharted: Big
Data as a Lens on Human Culture*. New York: Penguin, 2013, 11.

313 "the surveillance possibilities": Steve Lohr. "Big Data Is Opening Doors,

but Maybe Too Many." *New York Times*. March 23, 2013. nytimes
.com/2013/03/24/technology/big-data-and-a-renewed-debate-over-privacy
.html.

314   "You need the haystack": Barton Gellman and Ashkan Soltani. "NSA
Collects Millions of E-Mail Address Books Globally." *Washington Post*.
Oct. 14, 2013. washingtonpost.com/world/national-security/nsa-collects-
millions-of-e-mail-address-books-globally/2013/10/14/8e58b5be-34f9-
11e3-80c6-7e6dd8d22d8f_story.html.

314   ODNI mission: Kashmir Hill. "U.S. Spy Rocket Has Octopus-Themed
'Nothing Is Beyond Our Reach' Logo. Seriously." *Forbes*. Dec. 5, 2013.
forbes.com/sites/kashmirhill/2013/12/05/u-s-spy-rocket-launching-today-
has-octopus-themed-nothing-is-beyond-our-reach-logo-seriously.

315   "the practice of apophenia": Danah Boyd and Kate Crawford. "Six
Provocations for Big Data." Social Science Research Network. Sept. 21,
2011. papers.ssrn.com/sol3/papers.cfm?abstract_id=1926431.

316   growth of risk management: David Lyon. "Surveillance, Security and
Social Sorting: Emerging Research Priorities." *International Criminal
Justice Review* 17, no. 3 (September 2007):161–70. ijc.sagepub.com/
content/17/3/161.abstract.

316   "Everything is interesting": Eliza Kern. "Facebook Is Collecting Your
Data—500 Terabytes a Day." *GigaOm*. gigaom.com/2012/08/22/facebook-
is-collecting-your-data-500-terabytes-a-day.

316   Facebook data collection stats: ibid.

316   Soliciting advertisers: Kashmir Hill. "Facebook Joins Forces with Data
Brokers to Gather More Intel about Users for Ads." *Forbes*. Feb. 27, 2013.
forbes.com/sites/kashmirhill/2013/02/27/facebook-joins-forces-with-data-
brokers-to-gather-more-intel-about-users-for-ads.

316   Third-party retailers: Somini Sengupta. "What You Didn't Post, Facebook
May Still Know." *New York Times*. March 25, 2013. nytimes
.com/2013/03/26/technology/facebook-expands-targeted-advertising-
through-outside-data-sources.html.

316   Facebook data broker partnerships: ibid.

317   "It could mean seeing": ibid.

317   improved click-through rates: Hill. "Facebook Joins Forces with Data
Brokers to Gather More Intel about Users for Ads."

317   Data brokers' anonymization practices: Turow. *The Daily You*, 7.

317   De-anonymizing location data: Larry Hardesty. "How Hard Is It to 'De-
anonymize' Cellphone Data?" *MIT News*. March 27, 2013. newsoffice.mit
.edu/2013/how-hard-it-de-anonymize-cellphone-data.

318   "barred from selling data": Chris Jay Hoofnagle and Jennifer King.
"Consumer Information Sharing: Where the Sun Still Don't Shine." Social
Science Research Network. Dec. 17, 2007. papers.ssrn.com/sol3/papers
.cfm?abstract_id=1137990.

318   Data brokers can learn your habits: Charles Duhigg. "How Companies
Learn Your Secrets." *New York Times Magazine*. Feb. 16, 2012. nytimes

.com/2012/02/19/magazine/shopping-habits.html.

318 Differential pricing: Jennifer Valentino-Devries, Jeremy Singer-Vine, and Ashkan Soltani. "Websites Vary Prices, Deals Based on Users' Information." *Wall Street Journal.* Dec. 24, 2012. wsj.com/article/SB10001424127887323 777204578189391813881534.html.

319 Medbase controversy: Elizabeth Dwoskin. "Data Broker Removes Rape-Victims List after Journal Inquiry." *Digits*, a blog on WSJ.com. Dec. 19, 2013. http://blogs.wsj.cs/2013/12/19/data-broker-removes-rape-victims-list-after-journal-inquiry./

319 "A widespread revolt": Lasch. *Culture of Narcissism*, 245.

320 "what we 'like'": Hartzog and Selinger. "Big Data in Small Hands."

322 "information consumerism": Evgeny Morozov. "The Price of Hypocrisy." *Frankfurter Allgemeine.* July 24, 2013. faz.net/aktuell/feuilleton/debatten/ueberwachung/information-consumerism-the-price-of-hypocrisy-12292374.html.

323 "more data is always good": Losse. *The Boy Kings*, 43.

323 "collect everything": Michael Kelley. "CIA Chief Tech Officer: Big Data Is the Future and We Own It." *Business Insider.* March 21, 2013. businessinsider.com/cia-presentation-on-big-data-2013-3.

323 "all human generated information": ibid.

323 Google's mission: Google. "About Google." google.com/about/company

323 "Imagine a platform": Hamza Shaban. "LinkedOut." *New Inquiry.* July 19, 2013. thenewinquiry.com/essays/linkedout.

323 Factual's mission: Hardy. "Just the Facts." March 24, 2012.

324 "every business in the world": Chris Welch. "Facebook's Check-In Based Free Wi-Fi Rolling Out Nationally with Cisco's Help." *The Verge.* October 2, 2013. theverge.com/2013/10/2/4795732/facebook-cisco-partner-free-wi-fi-check-ins.

324 "augment the human condition": Sam Biddle. "If Your Startup Isn't Improving the Entire Human Condition, Fuck It." *Valleywag*, a blog on *Gawker.* Aug. 19, 2013. valleywag.gawker.com/if-your-startup-isnt-improving-the-entire-human-condit-1167263834.

324 "all data is credit data": Hardy. "Just the Facts." March 24, 2012.

325 "The life of your data": Matt Waite. "Public Info Doesn't Always Want to Be Free." *Source.* March 11, 2013. source.opennews.org/en-US/learning/public-info-doesnt-always-want-be-free.

325 "What could happen": ibid.

326 "If you try to keep up": Author interview with Joseph Turow. Oct. 14, 2013.

327 "The engineering ideology": Losse. *The Boy Kings*, 137.

327 Factual's informational appetite: Quentin Hardy. "Just the Facts. March 24, 2012

328 "haven't even reached the level": Author's notes. SXSW, April 2013.

328 "human genomics marketplace": Daniela Hernandez. "Selling Your Most Personal Item: You." *Wired.* March 27, 2013. wired.com/business/2013/03/miinome-genetic-marketplace.

329 "a person's data": "Personal Data: The Emergence of a New Asset Class."

World Economic Forum.

329   "Think of personal data": ibid.

329   Datacoup offer: Tom Simonite. "Sell Your Personal Data for $8 a Month." *Technology Review*. Feb. 12, 2014. technologyreview.com/news/524621/sell-your-personal-data-for-8-a-month.

329   ScreenWise program: Matt McGee. "Google Screenwise: New Program Pays You to Give Up Privacy & Surf the Web with Chrome." *Search Engine Land*. Feb. 8, 2012. http://searchengineland.com/google-screenwise-panel-open-110716.

330   "very important new research study": Wolf Richter. "How Much Is My Private Data Worth? (Google Just Offered Me $$)." *Testosterone Pit*. Nov. 28, 2013. testosteronepit.com/home/2013/11/28/how-much-is-my-private-data-worth-google-just-offered-me.html.

330   How much Google pays for data: ibid.

330   Data managers: World Economic Forum. "Unlocking the Value of Personal Data: From Collection to Usage." Feb. 2013. www3.weforum.org/docs/WEF_IT_UnlockingValuePersonalData_CollectionUsage_Report_2013.pdf.

331   "the functional end of the app": Liz Gannes. "Lyft and Uber Price Wars Leave Some Drivers Feeling Crunched." *Recode*. April 30, 2014. recode.net/2014/04/30/lyft-and-uber-price-wars-leave-some-drivers-feeling-crunched.

339   Bradbury's comments on the Internet: Jennifer Steinhauer. "A Literary Legend Fights for a Local Library." *New York Times*. June 19, 2009. nytimes.com/2009/06/20/us/20ventura.html.

341   "I think in tweets": David Roberts. "Goodbye for Now." *Grist*. Aug. 19, 2013. grist.org/article/goodbye-for-now.

342   "an escalating cycle": Lasch. *Culture of Narcissism*, 90.

343   "the performing self": ibid.

345   "Awareness commenting on awareness": ibid.

346   "like television": Steven Levy. "Inside the Science That Delivers Your Scary-Smart Facebook and Twitter Feeds." *Wired*. April 22, 2014. wired.com/2014/04/perfect-facebook-feed.

346   "method actors": de Zengotita. *Mediated*, 11.

347   *"conflictual ways"*: "Fetishism of Digital Commodities and Hidden Exploitation: The Cases of Amazon and Apple." Wu Ming Foundation. Oct. 10, 2011. www.wumingfoundation.com/english/wumingblog/?p=1895.

348   "a form of labor dispute": Author interview with Liel Leibovitz. Dec. 5, 2013.

349   "acts of obfuscation": ibid.

353   "insincere engagement": Katherine Rosman. "Some Twitter Users Push Back on Ads." *Wall Street Journal*. Aug. 4, 2013. wsj.com/news/articles/SB10001424127887324328904578623811676679672.

353   "scripting tricks": Secret Thirteen. "Interview—Laimonas Zakas—Glitchr." June 10, 2012. secretthirteen.org/secret-thirteen-interview-laimonas-zakas-glitchr.

354   Shreateh bug dispute: Tom Warren. "Facebook Ignored Security Bug, Researcher Used It to Post Details on Zuckerberg's Wall." *The Verge*. Aug. 18, 2013. theverge.com/2013/8/18/4633046/facebook-security-bug-let-anyone-post-on-walls.

355   Shreateh fund-raising: Marc Maiffret. "Khalil Shreateh—Facebook Bounty." Go Fund Me. Aug. 19, 2013.gofundme.com/3znhjs.

355   Face.com purchase price: Alexia Tsotsis. "Facebook Scoops Up Face .com for $55–60M to Bolster Its Facial Recognition Tech (Updated)." *TechCrunch*. June 18, 2012. techcrunch.com/2012/06/18/facebook-scoops-up-face-com-for-100m-to-bolster-its-facial-recognition-tech.

356   "expressive interference": Adam Harvey. "CV Dazzle." ahprojects.com/projects/cv-dazzle.

357   "distorted faces": Kyle Vanhemert. "Weird T-Shirts Designed to Confuse Facebook's Auto-Tagging." *Wired*. Oct. 2, 2013. wired.com/design/2013/10/thwart-facebooks-creepy-auto-tagging-with-these-bizarre-t-shirts.

357   ObscuraCam background: The Guardian Project. "ObscuraCam: Secure Smart Camera." guardianproject.info/apps/obscuracam.

357   "an alternative identity": Leslie Katz. "Anti-surveillance Mask Lets You Pass as Someone Else." *CNET*. May 8, 2014. cnet.com/news/urme-anti-surveillance-mask-lets-you-pass-as-someone-else.

357   "control how you are seen": Rachel Law. "Vortex." Parsons the New School for Design. mfadt.parsons.edu/2013/projects/vortex.

357   Vortex data profiles: Kate Kaye. "This Student Project Could Kill Digital Ad Targeting." *Advertising Age*. July 3, 2013. adage.com/article/privacy-and-regulation/student-project-kill-digital-ad-targeting/242955.

358   "This Student Project": ibid.

358   "glitchy photo filters": Crypstagram. cryptstagram.com.

358   Cryptagram: cryptogram.prglab.org.

358   Disappearing link: Blinklink.me.

358   ZXX typeface: Kyle Vanhemert. "An NSA Whiz Designs 4 Fonts to Foil Google's All-Seeing Eye." *Wired*. Sept. 24, 2013. wired.com/design/2013/09/you-can-read-these-4-fonts-but-your-computer-cant.

358   "anti-social media": Chris Welch. "Anti-social: 'Hell Is Other People' Keeps You as Far Away from Your 'Friends' as Possible." *The Verge*. June 16, 2013. theverge.com/2013/6/16/4435718/hell-is-other-people-helps-you-avoid-Foursquare-friends.

359   "metricated viewpoint": Ben Grosser. "Facebook Demetricator." bengrosser.com/projects/facebook-demetricator.

359   ScareMail NSA terms: Ben Grosser. "ScareMail." bengrosser.com/projects/scaremail.

359   civil disobedience: Ryan Gallagher. "ScareMail : Benjamin Grosser's Gmail Extension Aims to Drown NSA in Nonsense." *Slate*. Oct. 9 , 2013. slate.com/blogs/future_tense/2013/10/09/scaremail_benjamin_grosser_s_gmail_extension_aims_to_drown_nsa_in_nonsense.html.

359   "the most ubiquitous device": Addie Wagenknecht. "BRICKiPhone." April 24, 2012. fffff.at/brickiphone.

360   "dissolving your biases": Kyle McDonald. "FriendFlop." F.A.T. Oct. 16, 2013. fffff.at/friendflop.

360   "we own our digital data": Kyle McDonald. "When You Don't Own Yourself." F.A.T. May 22, 2013. fffff.at/when-you-dont-own-yourself.

362   "open source activism": Off Book, PBS Arts. "Hacking Art & Culture with F.A.T. Lab." Sept. 14, 2011. youtube.com/watch?v=-b0rlJvO1BQ.

362   ConstantUpdate video: ConstantUpdate. "Constant Update." April 19, 2013. youtube.com/watch?v=fxxzZrSpc5Q.

363   "never update": Author's notes. Rhizome Seven on Seven Conference. April 20, 2013.

363   "an artifact": ibid.

365   "I'm kind of an asshole" (and subsequent remarks): Author interview with Kade Crockford. July 2014.

369   "liberatory social change": Riseup. help.riseup.net.

369   "There is no technology": Christopher Soghoian. "There is no technology that can protect you from a $10 billion intelligence agency. Doesn't require a conspiracy, just math." Aug. 6, 2014, 2:49 a.m. Tweet (csoghoian). twitter.com/csoghoian/status/496910953230114816.

Index

About the Author

Jacob Silverman's work has been published in the *New York Times*, the *Los Angeles Times*, *Slate*, the *Atlantic*, the *New Republic*, and many other publications. He is on the board of Deep Vellum, a publisher of international literature. In 2008, the *Virginia Quarterly Review* recognized him as one of the top literary critics under thirty. He lives in Brooklyn, New York.